AGROBIODIVERSITY AND THE LAW

A wide range of crop genetic resources is vital for future food security. Loss of agricultural biodiversity increases the risk of relying on a limited number of staple food crops. However, many laws, such as seed laws, plant varieties protection and access and benefit-sharing laws, have direct impacts on agrobiodiversity, and their effects have been severely underestimated by policy-makers. This is of concern not only to lawyers, but also to agronomists, biologists, and social scientists, who need clear guidance as to the relevance of the law to their work. This book analyzes the impact of the legal system on agrobiodiversity (or agricultural biodiversity) – the diversity of agricultural species, varieties, and ecosystems. Using an interdisciplinary approach, it takes up the emerging concept of agrobiodiversity and its relationship with food security, nutrition, health, environmental sustainability, and climate change. It assesses the impacts on agrobiodiversity of key legal instruments, including seed laws, the International Convention for the Protection of New Varieties of Plants, plant breeders' rights, the Convention on Biological Diversity (regarding specifically its impact on agrobiodiversity), and the International Treaty on Plant Genetic Resources for Food and Agriculture. It also reviews the options for the implementation of these instruments at the national level in several countries. It discusses the interfaces between the free software movement, the 'commons' movement, and seeds, as well as the legal instruments to protect cultural heritage and their application to safeguard agrobiodiversity-rich systems. Finally, it analyzes the role of protected areas and the possibility of using geographical indications to enhance the value of agrobiodiversity products and processes.

Juliana Santilli is a lawyer and public prosecutor in the Federal District of Brazil, specialized in Environmental and Cultural Heritage Law and Public Policies. She has a PhD in Environmental Law, and is an associate researcher in Environmental Law at the University of Brasília Center for Sustainable Development. She is a co-founding member of the Brazilian civil society organization Instituto Socioambiental.

AGROBIODIVERSITY AND THE LAW

Regulating genetic resources, food security and cultural diversity

Juliana Santilli

earthscan
from Routledge

Routledge
Taylor & Francis Group
LONDON AND NEW YORK

First edition published 2012
by Earthscan
2 Park Square, Milton Park, Abingdon, Oxon OX14 4RN

Simultaneously published in the USA and Canada
by Earthscan
711 Third Avenue, New York, NY 10017

Routledge is an imprint of the Taylor & Francis Group, an informa business

First issued in paperback 2016

British Library Cataloguing in Publication Data
A catalogue record for this book is available from the British Library

Library of Congress Cataloging in Publication Data
A catalog record for this book has been applied for

ISBN13: 978-1-84971-372-6 (hbk)
ISBN13: 978-1-138-68033-3 (pbk)

Typeset in Sabon by Prepress Projects Ltd, Perth, UK

CONTENTS

TABLES

ABOUT THE AUTHOR

Juliana Santilli is a lawyer and a public prosecutor in the Federal District of Brazil, specialized in Environmental and Cultural Heritage Law and Public Policies. She has a PhD in Environmental Law, and has participated in research programs in both Brazil and internationally. She is also an associate researcher in Environmental Law at the University of Brasília, Center for Sustainable Development, and is a co-founding member of the Brazilian civil society organization Instituto Socioambiental (ISA; www.socioambiental.org).

Juliana Santilli is the author of two books: *Socio-environmentalism and New Rights: Legal Protection of Biological and Cultural Diversity* (São Paulo: Peirópolis/ISA/IEB, 2005) and *Agrobiodiversity and Farmers' Rights* (São Paulo: Peirópolis, 2009). She has also organized and co-authored the book, *Indigenous Peoples and the Law* (Brasília: Núcleo de Direitos Indígenas). She has also published several articles on environmental and cultural heritage law and public policies, in Portuguese, English, French, and Spanish.

Juliana is associated with the multidisciplinary research program "Local communities, agrobiodiversity and traditional knowledge in the Brazilian Amazon," developed by Institut de Recherche pour le Développement (IRD) and University of Campinas (Unicamp). She graduated from Law School (Federal University of Rio de Janeiro) in 1988 and became a public prosecutor, in the Federal District of Brazil, in 2000. In 2004, she obtained a master's degree in Public Law (University of Brasília, Brazil), and in 2009 she obtained a PhD in Environmental Law (Catholic University of Paraná, Brazil).

She was the rapporteur of the Working Group on Traditional Knowledge and Biodiversity, created by the Brazilian Genetic Resources Management Council, which is responsible for the implementation of the Brazilian access and benefit-sharing national legislation (in accordance with the Convention on Biological Diversity). She was also a legal advisor for GTZ (now GIZ, the German Technical Cooperation) on legal issues relating to Indigenous peoples and environmental resources, in the context of the elaboration of the Indigenous component of the Pilot Programme for the Protection of Tropical Forests in Brazil (PP-G7), and a legal advisor for the World Wildlife Fund (WWF) in Brazil, advising on Environmental Law and Legal Protection to Brazilian Wetlands (Pantanal). She was also a legal advisor for the Green Protocol Program, developed by the Secretariat of Sustainable Development Policies/Ministry of Environment – Brazil.

Juliana is also a member of the council of the Brazilian civil society organization Instituto Internacional de Educação do Brasil (IEB; www.iieb.org.br).

PREFACE

Loss of agricultural diversity, at varying levels, is associated with the industrialization of agriculture, particularly after the Green Revolution, and cannot be attributed to legal systems only. However, many laws and regulations have direct impacts on agrobiodiversity, and these have been underestimated and sometimes not even considered by policy-makers and legislators. Agrobiodiversity and associated cultural diversity are major values that must be encompassed in all regulations that affect biological, genetic, and agroecosystem diversity.

The main objective of this book is to analyze the impacts that international and national legal instruments have on agrobiodiverse farming systems and on the local small-scale farmers who conserve and manage them. However, before analyzing legal instruments in themselves, we discuss what agrobiodiversity is, and show its important interfaces with food security, nutrition, health, social equity, environmental sustainability, and climate change. Understanding the multiple connections and implications of agrobiodiversity is essential to comprehend why it is so important that agricultural and rural development laws prioritize its conservation and sustainable use.

The book analyzes seed production, utilization, and sales laws, IP rights over plant varieties (and the so-called plant breeders' rights), farmers' rights, livestock keepers' rights, and access and benefit-sharing regulations and their impacts on plant and animal genetic resources for food and agriculture, among others. On the other hand, it shows that some legal tools that were not conceived primarily for agrobiodiversity conservation can also be used for this purpose, such as biological open source and creative commons, protected areas, and geographical indications. Positive and negative effects in different countries are also discussed.

We comprehend crops and agroecosystems as cultural artifacts, made and molded by man as much as ceramics, buildings, monuments, and other works of art. Therefore, the book discusses how legal instruments conceived to safeguard cultural heritage can also be used to recognize and promote agrobiodiverse farming systems and all of their components, both tangible (such as landscapes and cultivated plants) and intangible (agricultural techniques, practices, and knowledge), as well as the social and cultural processes that maintain and enrich them. Intangible cultural heritage registries and cultural landscapes recognition and protection, as well as the safeguarding of globally important ingenious agricultural heritage systems, are analyzed from the perspective of agrobiodiversity conservation and sustainable use.

Historically, public policies have focused mainly on wild biodiversity, and cultivated biodiversity has received very little attention not only from public officials but also from environmentalists, civil society organizations, and the general public. However, protecting species and varieties of maize, rice, beans, wheat, etc., as well as the diversity of agroecosystems, is no less important than saving the Amazon rainforest and threatened wild species such as panda bears, birds, turtles, etc. It is my sincere hope that this book will contribute to the development of more effective policies, legislation, and regulations governing *in situ* and on-farm management of agrobiodiversity, as well as for the protection of farmers' and livestock keepers' rights.

Juliana Santilli

ACKNOWLEDGEMENTS

This book is an edited and updated version of my PhD thesis, on Environmental Law, approved by the Center for Social and Legal Studies of the Catholic University of Paraná (Brazil) in 2009. It is a result of my participation in the research project "Local Communities, Agrobiodiversity and Traditional Knowledge in the Brazilian Amazon" (PACTA), which was developed through a cooperation program between CNPQ (the Brazilian National Council for Scientific and Technological Development), Institut de Recherche pour le Développement (IRD, French Institute of Research for Development, UMR 208 "Local Heritage"/Muséum National d'Histoire Naturelle), and the University of Campinas (state of São Paulo, Brazil), with the participation of Instituto Socioambiental (ISA). This research received the generous support of CNPQ, IRD, Bureau des Ressources Génétiques (BRG), and Agence Nationale de Recherche/Institut Français de la Biodiversité. I am immensely grateful for the support of these organizations.

During my doctoral studies, I benefited from a fellowship granted by the Netherlands Fellowship Programme, of the Netherlands Organization for International Cooperation in Higher Education (Nuffic), to participate in the training program "Contemporary Approaches to Plant Genetic Resources, Conservation and Use," organized by Wageningen International, the Netherlands, in 2008. I also received financial support from the Fondation Agropolis to participate in the training program "École Thématique Internationale Agrobiodiversité: des hommes et des plantes, outils et méthodes d'analyse," organized by Centre de Coopération Internationale en Recherche Agronomique pour le Développement (CIRAD) and IRD, in 2008. I would like to express my sincere gratitude for all the support received from these institutions.

I am also deeply grateful to the Ministério Público do Distrito Federal e Territórios (the Brazilian Federal District Prosecutors' Office), for granting me the paid educational leave that was essential for the completion of my doctoral studies. My deep appreciation also goes to all graduate students and professors at the Center for Social and Legal Studies of the Catholic

University of Paraná (Brazil), for their companionship and friendship throughout the whole research project.

Let me thank all my friends at the Instituto Internacional de Educação do Brasil (IEB) and Instituto Socioambiental (ISA). I am very honored to be a co-founding member of ISA and a member of the council of the IEB, two Brazilian civil society organizations whose valuable work has been essential for the protection of biological and cultural diversity in Brazil, and I am immensely grateful for all their support for my work.

Finally, I would like to express my sincere gratitude to several people who contributed, with their insights and experiences, to this book: Laure Emperaire, Carlos Marés, Altair Toledo Machado, Patricia Bustamante, Nivaldo Peroni, Maria Rita Reis, and Luiz Cláudio Bona. Heartfelt gratitude also goes to Antônia, Ana, Andréa Chaves, Denise Nicolaidis, Branca, Xoxô, Lilinha, Duda, Aninha, and Esmeralda. I am also very grateful to Ariadna Bô Ferraz, who did the beautiful illustration for the book cover. I would also like to thank Eric Sawyer and Betsey Neal for their precious help with the translation of the original manuscript to English.

My deepest appreciation goes to my husband, Marcio Santilli, and to our son, Lucas, and to my two sisters, Ariadna and Amaryllis, who have supported me so lovingly.

ACRONYMS AND ABBREVIATIONS

ABRASEM	Associação Brasileira de Sementes e Mudas (Brazilian Seeds and Seedlings Association)
ABS	access and benefit-sharing
AIAB	Associazione Italiana per l'Agricoltura Biologica
AS-PTA	Assessoria e Serviços a Projetos em Agricultura Alternativa (Advisory Services for Alternative Agriculture Projects, Brazilian NGO)
ASSINSEL	Association Internationale des Sélectionneurs pour la Protection des Obtentions Végétales (International Association of Plant Breeders for the Protection of Plant Varieties)
BIOS	Biological Open Source
BRG	Bureau des Ressources Génétiques (Genetic Resources Bureau, France, now extinct and merged into Fondation pour la Recherche sur la Biodiversité)
CBD	Convention on Biological Diversity
CGEN	Conselho de Gestão do Patrimônio Genético (Brazilian Council on Genetic Resources Management)
CGIAR	Consultative Group on International Agricultural Research
CGRFA	Commission on Genetic Resources for Food and Agriculture
CIAT	International Center for Tropical Agriculture
CIMMYT	Centro Internacional de Mejoramiento de Maiz y Trigo (International Maize and Wheat Improvement Center)
CIP	Centro Internacional de la Papa (International Potato Center)
CIRAD	Centre de Coopération Internationale en Recherche Agronomique pour le Développement (Center of International Cooperation in Agricultural Research for Development, France)

CNPQ	Conselho Nacional de Desenvolvimento Científico e Tecnológico (Brazilian National Council for Scientific and Technological Development)
CNRS	Centre National de la Recherche Scientifique (National Center for Scientific Research, France)
CONAB	Companhia Nacional de Abastecimento (National Corporation for Food Supply, Brazil)
COP	Conference of the Parties (CBD)
EMBRAPA	Empresa Brasileira de Pesquisa Agropecuária (Brazilian Corporation for Agricultural Research)
EPAGRI	Empresa de Pesquisa Agropecuária e Extensão Rural de Santa Catarina (Company for Agricultural Research and Rural Extension of Santa Catarina, Brazil)
EPO	European Patent Office
ESA	European Seed Association
ETC Group	Action Group on Erosion, Technology and Concentration (formerly RAFI, Rural Advancement Foundation International)
EU	European Union
FAO	Food and Agriculture Organization of the United Nations
FUNAI	Fundação Nacional do Índio (National Agency for Indigenous Affairs, Brazil)
GATT	General Agreement on Tariffs and Trade
GEF	Global Environment Facility
GIAHS	Globally Important Agricultural Heritage Systems
GMO	genetically modified organism
GNP	gross national product
GPA	Global Plan of Action for the Conservation and Sustainable Utilization of Plant Genetic Resources for Food and Agriculture
GPL	General Public License
GPLPG	General Public License for Plant Germplasm
GRAIN	Genetic Resources Action International (NGO)
GTZ (now GIZ)	Deutsche Gesellschaft für Technische Zusammenarbeit (German Technical Cooperation)
IARC	International Agricultural Research Center
IBAMA	Instituto Brasileiro do Meio Ambiente e dos Recursos Naturais Renováveis (Brazilian Institute for the Environment and Renewable Natural Resources)
IBPGR	International Board for Plant Genetic Resources (later IPGRI, see below)
ICRISAT	International Crops Research Institute for the Semi-Arid Tropics

IEB	Instituto Internacional de Educação do Brasil (International Institute for Education, Brazilian NGO)
IFPRI	International Food Policy Research Institute
IICA	Instituto Interamericano de Cooperação para a Agricultura (Interamerican Institute for Cooperation in Agriculture)
IITA	International Institute of Tropical Agriculture
ILO	International Labor Organization
ILRI	International Livestock Research Center
IRRI	International Rice Research Institute
INIA	Instituto Nacional de Innovacion Agraria (National Institute for Agrarian Innovation)
INPI	Instituto Nacional de Propriedade Industrial (National Institute for Industrial Property, Brazil)
INRA	Institut National de la Recherche Agronomique (National Institute for Agricultural Research, France)
IOM	International Organization for Migration
IPAM	Instituto de Pesquisa Ambiental da Amazônia (Institute for Environmental Research of the Amazon, Brazil)
IPCC	Intergovernmental Panel on Climate Change
IPGRI	International Plant Genetic Resources Institute (now Bioversity International)
IPHAN	Instituto do Patrimônio Histórico e Artístico Nacional (Institute for National Historical and Artistic Heritage, Brazil)
IP	intellectual property
IRD	Institut de Recherche pour le Développement (Institute of Research for Development, France)
ITPGRFA	International Treaty on Plant Genetic Resources for Food and Agriculture
ISA	Instituto Socioambiental (Socioenvironmental Institute, Brazilian NGO)
IUCN	International Union for Conservation of Nature
MAPA	Ministério da Agricultura, Pecuária e Abastecimento (Brazilian Ministry of Agriculture, Livestock and Food Supply)
MDA	Ministério do Desenvolvimento Agrário (Brazilian Ministry of Agrarian Development)
MDG	Millennium Development Goal
MNHN	Muséum National d'Histoire Naturelle (National Museum of Natural History)
MPEG	Museu Paraense Emílio Goeldi (Emilio Goeldi Museum of Pará, Brazil)
NGO	nongovernmental organization

NordGen	Nordic Genetic Resource Center
PAA	Programa de Aquisição de Alimentos (Food Acquisition Program, Brazil)
PGRFA	plant genetic resources for food and agriculture
PVPA	Plant Variety Protection Act
PRONAF	Programa Nacional de Fortalecimento da Agricultura Familiar (National Program to Strengthen Family Agriculture, Brazil)
RENASEM	Registro Nacional de Sementes e Mudas (National Seeds and Seedlings Registry, Brazil)
SMTA	standard material transfer agreement
TRIPS	Trade-Related Aspects of Intellectual Property Rights
UNCED	United Nations Conference on Environment and Development
UNDP	United Nations Development Program
UNDRIP	UN Declaration on the Rights of Indigenous Peoples
UNEP	United Nations Environment Program
UNESCO	United Nations Educational, Scientific and Cultural Organization
UNFCCC	UN Framework Convention on Climate Change
UNICAMP	Universidade Estadual de Campinas (State University of Campinas, Brazil)
UPOV	Union pour la Protection des Obtentions Végétales (Union for the Protection of New Varieties of Plants)
USDA	United States Department of Agriculture
WHO	World Health Organization
WIPO	World Intellectual Property Organization
WTO	World Trade Organization
WWF	Worldwide Fund for Nature

1

AGROBIODIVERSITY

A concept under construction

The concept of "agrobiodiversity" emerged in the past 10–15 years, in an interdisciplinary context that involves various areas of knowledge (agronomy, anthropology, ecology, botany, genetics, conservation biology, etc.). It reflects the dynamic and complex relations among human societies, cultivated plants, and domestic animals and the ecosystems in which they interact. Agrobiodiversity is directly associated with food security, health, social equity, hunger alleviation, environmental sustainability, and rural sustainable development.

Biodiversity or biological diversity – the diversity of life forms – covers three degrees of variability: diversity of species, genetic diversity (variability within the set of individuals of a given species), and ecological diversity, referring to different ecosystems and landscapes. The same is true of agrobiodiversity, which comprises the diversity of species (different species of cultivated plants, such as maize, rice, pumpkins, tomatoes, etc., called interspecific diversity), genetic diversity within a given species (different varieties of maize or beans, etc., called intraspecific diversity), and the diversity of cultivated ecosystems or agroecosystems[1] (agroforestry systems, shift cultivation, home gardens, rice paddy fields, etc.). Local knowledge and culture are also integral parts of agricultural biodiversity, since it is agriculture, a human activity, that conserves biodiversity. Most crops have lost their original seed dispersal mechanisms as a result of domestication and so can no longer thrive without human input (Cromwell *et al.*, 2003). Agrobiodiversity, or agricultural diversity, is an important part of biodiversity and encompasses all the elements which interact in agricultural production, including all crops and livestock and their wild relatives, and all interacting species of pollinators, symbionts, parasites, pests, predators, and competitors, and the genetic diversity within them (Qualset *et al.*, 1995). According to Cromwell *et al.* (2003), agricultural biodiversity includes (1) higher plants: crops, wild plants harvested and managed for food, trees on farms, and pasture and rangeland species; (2) higher animals: domestic animals, wild animals hunted for food, etc., wild and farmed fish; (3) arthropods: mostly insects including pollinators (e.g., bees, butterflies), pests (e.g., wasps, beetles), and insects involved in the soil cycle (notably termites); (4) other macro-organisms (e.g., earthworms); (5) micro-organisms (e.g., rhizobia, fungi, disease-producing pathogens). Agrobiodiversity's functions are related to sustainable production of food and other agricultural products, including providing the building blocks for the evolution or deliberate breeding of useful new crop varieties; biological support to production by means of, for example, soil biota, pollinators, and predators; and wider ecological services provided by agroecosystems, such as landscape protection, soil protection and health, water cycling and quality, and air quality (Cromwell *et al.*, 2003).

According to Harold Brookfield (2001, p. 46), agrobiodiversity is the "dynamic variation in cropping systems, output and management practice that occurs within and between agroecosystems. It arises from biophysical differences, and from the many and changing ways in which farmers manage diverse genetic resources and natural variability, and organize their management in dynamic social and economic contexts." Brookfield (2001, p. 41) defines management diversity as all methods of managing the land, water, and biota for crop production and maintaining soil fertility and structure. Local knowledge, constantly modified by new information, is the foundation of this management diversity, as with agrobiodiversity. For Brookfield (2001, p. 41), biophysical diversity includes soil characteristics and their qualities and the biodiversity of natural (or spontaneous) plant life and the faunal and microbial biota. Organizational diversity is often called the socioeconomic aspect of agriculture. It includes diversity in the manner in which farms are owned and operated and in the use of resource endowments and the farm workforce. Elements include labor, household size, the differing resource endowments of households, and reliance on off-farm employment. Also included are age group and gender relations in farm work, dependence on the farm as compared with external sources of support, the spatial distribution of the farm, and differences between farmers in terms of access to land. Organizational diversity underpins and helps explain management diversity and its variation between particular farms, communities, and societies. According to Brookfield (2001, p. 44), no agricultural system can be understood independently from the manner in which farms are organized and the forces that interact to shape this organization.

Agrobiodiversity is generally associated with crops (cultivated plants). However, Cromwell *et al.* (2003) point out that wild species[2] are important nutritionally and culturally. Foods from wild species form an integral part of the daily diets of many rural households. Livestock diversity is also an important component of agrobiodiversity. Domesticated animals provide people not only with food but also with clothing, fertilizer and fuel (from manure), and draft power. Social and cultural forces are often the most important factors in diversifying livestock (and livestock production systems) and in developing distinctive breeds. Most local livestock breeds in rural environments are products of a community of breeders, and the effective management of animal genetic diversity is essential to global food security and sustainable development.

The Food and Agriculture Organization's (FAO's) Global Databank for Animal Genetic Resources for Food and Agriculture identified a total of 7,616 livestock breeds. Such diversity has enabled farmers and pastoralists to adapt to local environmental conditions and to meet specific social and cultural needs, as well as to inhabit a wide range of production environments from hot humid tropics to arid deserts and cold mountainous regions. Genetic diversity also allows livestock to adapt to diseases, parasites, and wide variations in the type and availability of food and water. Yet livestock diversity is at risk. According to the *Report on the State of the World's Animal Genetic Resources for Food and Agriculture*,[3] approximately 20 percent of the cow, goat, pig, horse, and bird breeds in the world are threatened with extinction, and in the past six years 62 breeds of domestic animals have become extinct, which represents a loss of nearly one race per month. In 2000, FAO (FAO, 2000) estimated that, of the 3,831 breeds of cows, buffalo, goats, pigs, sheep, horses and donkeys believed to have existed in the twentieth century, 16 percent had already become extinct and a further 15 percent were at risk of extinction. The most significant threat to livestock diversity is the marginalization of traditional production systems and the associated local breeds, driven mainly by the rapid spread of intensive livestock production, often large-scale and utilizing a narrow range of breeds (*Report on the State of the World's Animal Genetic Resources for Food and Agriculture*).

Cromwell *et al.* (2003) also highlight the importance of aquatic diversity as a component of agrobiodiversity. Fish and other aquatic species are integral parts of several important farming systems. For example, in the tropical rice–fish systems of Asia, fish from rice paddies may provide as much as 70 percent of dietary protein. Another important component of agrobiodiversity is the underground biodiversity: roots bring nutrients and water to plants and stabilize the soil against erosion and soil movement on steep slopes and, in tropical systems, the contribution of roots to soil organic matter is proportionately larger than the contribution from above-ground inputs. Microbial diversity is also relevant to agricultural biodiversity, as microbes contribute a wealth of gene pools that can be sources of material for transfer to plants to achieve traits such as stress tolerance and pest resistance, and large-scale production of plant metabolites. The diversity of insects (such as bees and other pollinators), spiders, and other arthropods (grasshoppers, etc.) is another important component of agrobiodiversity, since they often act as natural enemies of crop pests. Finally, there is increasing realization of the importance of agricultural biodiversity at the ecosystems level. An ecosystem consists of a dynamic complex of plant, animal and micro-organism communities and their living environment interacting as a functional unit. That is why agrobiodiversity at the ecosystems level is sometimes referred to as "functional agricultural biodiversity": that which is necessary to sustain the ecological function of the agroecosystem, its structures, and processes in support of food production (Cromwell *et al.*, 2003). In this book, we shall focus mainly on the diversity of cultivated plants and agroecosystems, and on livestock diversity, more than on other components of agricultural biodiversity, since they are the mostly regulated components of agricultural biodiversity, and the purpose of the book is to analyze impacts of legal instruments on agrobiodiversity.

The Convention on Biological Diversity (CBD)[4] does not contain a definition of agrobiodiversity, but, according to Decision V/5, adopted during the Fifth Conference of the Parties of CBD (COP-5), "agricultural biodiversity is a broad term that includes all components of biological diversity of relevance to food and agriculture, and all components of biological diversity that constitute the agroecosystem: the variety and variability of animals, plants and microorganisms, at the genetic, species and ecosystem levels, which are necessary to sustain key functions of the agroecosystem, its structure and processes." Agricultural biodiversity has the following dimensions (according to Decision V/5):

1 Genetic resources for food and agriculture, including (a) plant genetic resources, including pasture and rangeland species, and genetic resources of trees that are an integral part of farming systems; (b) animal genetic resources, including fishery genetic resources, in cases where fish production is part of the farming system, and insect genetic resources; and (c) microbial and fungal genetic resources. These constitute the main units of production in agriculture, including cultivated species, domesticated species, and managed wild plants and animals, as well as wild relatives of cultivated and domesticated species.
2 Components of agricultural biodiversity that provide ecological services. These include a diverse range of organisms in agricultural production systems that contribute, at various scales, to *inter alia*: (a) nutrient cycling, decomposition of organic matter and maintenance of soil fertility; (b) pest and disease regulation; (c) pollination; (d) maintenance and enhancement of local wildlife and habitats in their landscape; (e) maintenance of the hydrological cycle; (f) erosion control; and (g) climate regulation and carbon sequestration.
3 Abiotic factors which have a determining effect on these aspects of agricultural biodiversity.
4 Socioeconomic and cultural dimensions, since agricultural biodiversity is largely shaped by human activities and management practices. These

include (a) traditional and local knowledge of agricultural biodiversity, cultural factors, and participatory processes; (b) tourism associated with agricultural landscapes; and (c) other socioeconomic factors.

It is important to understand that a domesticated plant is not the same as a cultivated plant. Ethnobotanist Laure Emperaire (2004) explains that domestication is an evolutionary process through which a plant moves from its wild state – in which there is no human intervention – to a closer relation with humans and their agricultural activities. Domestication requires modification in the genetic heritage (composition) of the plant. As a result of selective cultivation by man, domesticated plants lose certain characteristics (which are not interesting for agricultural activities) and develop others. Some characteristics of domesticated plants include gigantism (especially of parts of the plant of interest to humans), suppression of natural mechanisms of seed dispersion, rapid and uniform seed germination, etc. Therefore, cultivated plants are not necessarily domesticated, but domesticated plants are necessarily cultivated (Emperaire, 2004). In other words, wild plants can be cultivated without being domesticated, but the inverse is not true: domesticated plants cannot be abandoned in unmanipulated landscapes because they have lost their ecological adaptations to natural environments.

Jack Harlan (1975) explains that "to domesticate means to bring into the household." According to Harlan (1975), domesticated plants and animals have been altered genetically from their wild state and have come to be at home with man. He explains that to cultivate means to conduct those activities involved in caring for a plant, such as tilling the soil, preparing a seedbed, weeding, pruning, protecting, watering, and manuring. Cultivation is concerned with human activities, whereas domestication deals with the genetic response of the plants or animals being tended or cultivated. It is therefore possible to cultivate wild plants, and cultivated plants are not necessarily domesticated. Harvested plant materials may be classified as wild, tolerated, encouraged and domesticated (Harlan, 1975).

According to Charles Clement (1999), plant domestication is a "co-evolutionary process by which human selection on the phenotypes of promoted, managed or cultivated plant populations results in changes in the population's genotypes that make them more useful to humans and better adapted to human intervention in the landscape." Human selection may be either unconscious or directed, but for plant domestication to take place there must be selection and management to cause differential reproduction and survival. According to Clement (1999), the degree of change in the targeted population can vary:

1 *Wild.* A naturally evolved population whose genotypes and phenotypes have not been modified by human intervention.

2 *Incidentally co-evolved.* A population that volunteers and adapts in a human disturbed environment, possibly undergoing genetic change, but without human selection. Many weeds are examples of incidentally co-evolved species, which can also enter the domestication process if humans start to select them for their useful traits and start to manage or cultivate them.
3 *Incipiently domesticated.* A population that has been modified by human selection and intervention (at the very least being promoted), but whose average phenotype is still within the range of variation found in the wild population for the trait(s) subject to selection. The variance of this average is probably smaller than that of the original wild population, however, as selection has started to reduce genetic variability.
4 *Semidomesticated.* A population that is significantly modified by human selection and intervention (at the very least being managed) so that the average phenotype may diverge from the range of variation found in the wild population for the trait(s) subject to selection. The variance of this phenotypic average may be larger than that of the wild population, because the phenotypic variation now includes both types that are common in the wild population and types that are novel. Underlying genetic variability, however, continues to decrease because fewer individuals meet the selection criteria and are therefore included in the next generation. The plants retain sufficient ecological adaptability to survive in the wild if human intervention ceases, but the phenotypic variation selected by humans will gradually disappear in the natural environment.
5 *Domesticated.* A plant population similar to those semidomesticated, but whose ecological adaptability has been reduced to the point that it can only survive in human-created environments, specifically in cultivated landscapes. Genetic variability is generally less than in semidomesticated plant populations, because of increased selection pressure and loss of ecological adaptation. If human intervention ceases, the population dies out in short order, depending upon its life history, stature, and the type of vegetation that invades the abandoned area. In clonally propagated crops, a single genotype may be the domesticate, but also is lost soon after it is abandoned.

Clement (1999) also describes “landscape domestication” as “a conscious process by which human manipulation of the landscape results in changes in landscape ecology and in the demographics of its plant and animal populations, resulting in a landscape more productive and congenial for humans.” According to him, the intensity of manipulation may also vary widely:

1 *Pristine.* A landscape in which humans have not manipulated plant or animal populations.

2 *Promoted.* In this category desirable plant populations and individuals are encouraged through minimal forest clearance and expansion of the forest fringes. Even though there may have been a low level of human intervention, the biotic components of this landscape may remain modified long after humans have abandoned the area.
3 *Managed.* In this category the abundance and diversity of food and other useful plant populations may be further encouraged through partial forest clearance, expansion of the forest fringes, transplanting of desirable individual plants or planting of individual seeds, addition of amendments to enhance plant growth, and reduction of competition from nonuseful plants. Again, the biotic components of this landscape may remain long after humans have abandoned the area.
4 *Cultivated.* This category involves the complete transformation of the biotic landscape to favor the growth of one or a few selected food plants and other useful populations, through forest clearance and burning, localized or extensive tillage, seedbed preparation, weeding, pruning, manuring, mulching, and watering in any combination. The biotic components of this very artificial landscape do not survive long after human abandonment because the changes that favor the growth of the human selected populations also favor the growth of weeds and the invasion of other secondary forest species; however, it takes a long time to return to a natural state. The abiotic transformations practiced in this landscape often survive for long periods.

According to Decision V/5 of CBD, the special nature of agricultural biodiversity includes the following features:

a Agricultural biodiversity is essential to satisfy basic human needs for food and livelihood security.
b Agricultural biodiversity is managed by farmers. Many components of agricultural biodiversity depend on this human influence. Indigenous knowledge and culture are integral parts of the management of agricultural biodiversity.
c There is a great interdependence among countries on the genetic resources for food and agriculture.
d For crops and domestic animals, diversity within species is at least as important as diversity between species and has been greatly expanded through agriculture.
e Because of the degree of human management of agricultural biodiversity, its conservation in production systems is inherently linked to sustainable use.
f Nonetheless, much biological diversity is now conserved *ex situ* in genebanks or breeders' materials.

g The interaction between the environment, genetic resources and management practices that occurs within agroecosystems contributes to maintaining dynamic agricultural biodiversity.

Although the terms "agrobiodiversity" and "agrodiversity" are often employed as synonyms, Brookfield and Stocking (1999) hold that agrobiodiversity and agrodiversity have different meanings. "Agrobiodiversity," an older and more common term, is used to define the biological diversity existing in cultivated ecosystems. "Agrodiversity" is a broader expression, employed to refer to the many forms with which farmers use the natural diversity of the environment for agricultural production, including not only the selection of plant species and varieties for cultivation, but also management of the soil, water and biota as a whole (Brookfield and Paddoch, 1994). Another definition of "agrodiversity" could be "the variety resulting from the interaction among the factors which determine agroecosystems: genetic resources of plants, biotic and abiotic environments and management practices" (Almekinders *et al.*, 1995). We prefer to use "agrobiodiversity."

Agrobiodiversity is essentially a product of human intervention on ecosystems, of man's inventiveness and creativity in the interaction with the natural environment. Cultural processes, knowledge, and innovations, developed and shared by farmers, are a key component of agrobiodiversity. Species management, cultivation, and selection practices developed by farmers throughout the past 10,000 to 12,000 years are responsible, in large part, for the enormous diversity of cultivated plants and agroecosystems. All over the world, farmers innovate, experiment, and learn to adapt and diversify agricultural practices, and knowledge about biodiversity is closely tied to diverse and dynamic cultural beliefs, customs, and practices. Therefore, agrobiodiversity cannot be treated separately from the cultural and socioeconomic processes and contexts which determine it.

According to Brookfield (2001, pp. 21, 286) the adaptive dynamism of agrobiodiversity is its most essential property for survival, and for restoration of what has been lost. "Centrally important (to agrobiodiversity) is the internal dynamism of so many small-farming systems, yielding a constantly changing patchwork of relationships between people, plants and the environment," according to him.

Farmers adapt to difficulties and to opportunities, and learning and experimenting processes are constantly renewed. There are societies that adapt rice varieties to aquatic cultivation, submerged in water, in humid regions, while others adapt rice varieties for cultivation in dry regions. Different varieties of maize can be used for eating directly from the cob, for feeding animals, for making popcorn and flour, or to brew beer. They can also be used for ornamental (especially those with colorful pigmentation), medicinal, or religious purposes. Agronomist Jack Harlan (1975, p. 164) says that he once noticed that an Ethiopian farmer selected crook-necked sorghum varieties from his field. On asking the farmer the reason for this, he replied that such varieties were easier to hang from the tukel roof. (In Ethiopia, it is common to store seed inside the house [tukel] above the hearth, where the smoke from the kitchen fires provides some protection from weevils.) Other farmers selected sorghum varieties with a sweet taste for chewing. Other varieties of sorghum were set apart for making bread and beer, and the varieties with more resistant fibers were used to make baskets and as construction materials. The same species can be used for food or medication, and different parts of the same plant can also serve distinct purposes. Plants are also used in religious rituals and ceremonies, and many different names may be given to the same species by different local communities. Agricultural diversity can also be expressed both in characteristics visible to the human eye, such as variations in color, shape, or height or in the size and shape of the leaves, and in genetic variations, such as resistance to droughts, pests, and diseases, nutritional value, etc., and losses are hard to assess with precision. Extinction of agricultural knowledge is even harder to assess.

Even if the losses cannot be precisely estimated, agricultural diversity is threatened, and it constitutes the basis for survival of rural populations, particularly those of lower income. The *First Report on the State of the World's Plant Genetic Resources for Food and Agriculture*, presented during the Fourth International Technical Conference on Plant Genetic Resources, held in Leipzig, Germany, from June 17 to 23, 1996, was an important warning for the genetic and cultural erosion provoked by modern agricultural systems. The report, published by FAO in 1998, was the first global and systematic assessment of the state of conservation and use of global plant genetic resources. According to the report, in the past hundred years, farmers lost between 90 percent and 95 percent of their agricultural varieties. The report goes further:

1 In South Korea, only one-fourth of the 14 native plant varieties grown in gardens in 1985 were still in existence in 1993. Only 20 percent of maize varieties which existed in Mexico in the 1930s exist today.
2 In the United States, 95 percent of cabbage varieties, 94 percent of pea varieties, and 81 percent of tomato varieties ceased to exist in the past century. Of the 7,098 varieties of apples in existence between 1804 and 1904, 86 percent no longer exist.
3 In China, of the 10,000 varieties of wheat used in 1949, only 1,000 were still used in the 1970s. Up until the 1970s, approximately 5,000 rice varieties were grown in India, of which only 500 still exist, and between 10 and 20 varieties take up most of the Indian farmland.

Loss of agricultural biodiversity is caused mainly by the replacement of local and traditional varieties, which are characterized by large genetic variability, by "modern" varieties, with high yields and narrow genetic bases. According to the report, this is the main cause of genetic erosion (mentioned in 81 percent of the national reports presented by participating countries). Both species and varieties derived from these species have disappeared, and this is not limited to species domesticated by humans, as wild relatives continue to disappear as a result of rapid devastation of natural ecosystems. In some cases, disappearance of a variety may not necessarily lead to loss of genetic diversity, as its genes may exist in other varieties as well, but varieties themselves represent a unique combination of genes, with equally unique value and utility. It is also estimated that loss of a plant can cause disappearance of 40 types of animals and insects that depend on it for survival, in addition to singular genetic combinations and molecules (Kloppenburg and Kleinman, 1987).

The *Second Report on the State of the World's Plant Genetic Resources for Food and Agriculture* was launched by FAO in 2009, and published in 2010.[5] Its main objective was to update the first report, to assess the status and trends of plant genetic resources and to identify the most significant gaps and needs. It was prepared with information provided by 113 countries, and was the basis for updating the Global Plan of Action for the Conservation and Sustainable Utilization of Plant Genetic Resources for Food and Agriculture. It describes the most significant changes that have occurred since the publication of the first report in 1998.[6]

According to the second report, much more consensus exists on the occurrence of genetic erosion as a result of the total shift from traditional production systems (depending on farmers' varieties) to modern production systems (depending on released varieties), and major causes of genetic erosion were the same as identified in the first report: replacement of local varieties, land clearing, overexploitation, population pressures, environmental degradation, changing agricultural systems, overgrazing, and inappropriate legislation and policy, as well as pests, diseases, and weeds. However, the report points out that more attention is now being paid in several countries to increasing genetic diversity within production systems as a way to reduce risk, particularly in light of changes in climate, pests, and diseases. According the the second report, significant progress has also been made in the development of tools and techniques to assess and monitor plant genetic resources for food and agriculture (PGRFA) within agricultural production systems, and countries now report a greater understanding of the amount and distribution of genetic diversity in the field, as well as the value of local seed systems in maintaining such diversity.

According to the report, 60 countries state that genetic vulnerability is significant, and many mention the need for a greater deployment of genetic diversity in order to counter the potential threat to agricultural production. However, the report points out that the true extent of genetic erosion is very complex to assess. Some examples of genetic erosion mentioned in the report are:

- In Benin, there is concern that the current agricultural system is dominated by monocultures, in particular of yam and commercial crops.
- China has reported cases in which rice and maize varieties have become more uniform and thus more genetically vulnerable.
- Ecuador reports that endemic plants are particularly vulnerable owing to their restricted distribution. In the Galapagos Islands, at least 144 species of native vascular plants are considered rare; 69 of these are endemic to the Archipelago, including 38 species which are restricted to a single island.
- In Lebanon, the decrease in national production of almonds has been attributed to the genetic vulnerability of the few varieties grown.
- Madagascar has reported that the rice variety Rojomena, appreciated for its taste, is now rare, whereas the Botojingo and Java varieties of the northeastern coastal area have disappeared. The cassava variety Pelamainty of Taolagnaro (a city in Madagascar) and certain varieties of bean have disappeared from most producing areas and, in the case of coffee, 100 clones out of 256, as well as five species (*Coffea campaniensis*, *C. arnoldiana*, *C. rostandii*, *C. tricalysioides*, and *C. humbertii*) have disappeared from collections in the last 20 years. Wild yam species are also considered likely to disappear soon.
- Costa Rica reports that *Phaseolus* spp. (beans), including *P. vulgaris*, are threatened by serious genetic erosion; the same is true of the Indigenous crop *Sechium tacaco* and four related species: *S. pittieri*, *S. talamancense*, *S. venosum*, and *S. vellosum*.
- In India, a large number of rice varieties in Orissa, some rice varieties with medicinal properties in Kerala, and a range of millet species in Tamil Nadu, are no longer cultivated in their native habitats.
- Yemen reports that varieties of finger millet (*Eleusine coracana*) and *Eragrostis tef* as well as oil rape (*Brassica napus*), which used to be among the most important traditional crop varieties grown in the country, are no longer grown or are grown only in very specific areas and that the cultivation of wheat, including *Triticum dicoccum*, has drastically decreased.
- In Albania, all primitive wheat cultivars and many maize cultivars, have reportedly been lost.

Other major conclusions of the second report are:

- The total number of accessions conserved *ex situ* worldwide has increased by approximately 20 percent since 1996, reaching 7.4 million. While new collecting accounted for at least 240,000 accessions, and possibly considerably more, much of the overall increase is the result of exchange and unplanned duplication. It is estimated that less than 30 percent of the total number of accessions are distinct. Although the number of accessions of minor crops and crop wild relatives has increased, these categories are still generally under-represented. There is still a need for greater rationalization among collections globally.

- Scientific understanding of the on-farm management of genetic diversity has increased. The number of on-farm management projects carried out with the participation of local stakeholders has also increased somewhat. Although this approach to the conservation and use of PGRFA is becoming increasingly mainstreamed within national programs, further efforts are needed in this regard.
- In many countries public sector plant breeding has continued to contract, with the private sector increasingly taking over. Agriculture in many developing countries that reduced their support to public sector crop development, leaving, instead, the sustainable use of PGRFA to the private sector, is more vulnerable than in the past as private sector breeding and seed enterprise is largely restricted to a few crops for which farmers buy fresh seed each season. Considerably more attention and capacity-building is urgently needed to strengthen plant breeding capacity and the associated seed systems in most developing countries, where most of the important crops are not, and will not be, the focus of private enterprise.

Finally, it is important to stress that agrobiodiversity is valuable not only for small-scale traditional farming, but also for large-scale commercial production systems. Genetic diversity (including the genes of wild relatives) continues to be vital for modern agriculture, plant and livestock breeding, and for new methods of bioengineering and biotechnology. Increasing numbers of large commercial producers are also beginning to recognize and profit from the benefits of agroecosystem diversity, such as using intercropping and crop rotation and enhancing soil and insect diversity, and examples can be found among major producers of grapes, vegetables, and rice in Europe, the United States, and parts of Asia. These producers generally are motivated to try such changes after experiencing significant losses with monocultural, uniform systems (Thrupp, 1998, pp. 19–20). Biodiversity is essential to any agricultural system, and to ecological stability.

Notes

1 According to Conway (1987), an agroecosystem is an ecological and socioeconomic system, comprising domesticated plants and/or animals and the people who husband them, intended for the purpose of production of food, fiber, or other agricultural products.

2 Wood and Lenné (1999) exclude wild plants and animals outside the agroecosystem. According to them, although such wild food is often of critical importance to farm families, it is not part of the farm and is not agrobiodiversity.

3 The *State of the World's Animal Genetic Resources for Food and Agriculture* is the first global assessment of livestock biodiversity. It was announced during the First International Technical Conference on Animal Genetic Resources for

Food and Agriculture, held in Interlaken, Switzerland, between December 3 and 7, 2007, (http://www.fao.org/docrep/010/a1260e/a1260e00.HTM, accessed December 15, 2010).

4 The Convention on Biological Diversity (CBD) was approved during the Second Conference of the United Nations on Environment and Development, held in Rio de Janeiro in 1992. It entered into force on December 29, 1993, and has three main objectives: conservation of biological diversity, sustainable use of the components of biological diversity, and the fair and equitable sharing of the benefits arising out of the utilization of genetic resources. The Fifth Conference of the Parties of the CBD was held in Nairobi between May 15 and 26, 2000 (http://www.cbd.int/decision/cop/?id=7147; accessed December 10, 2010). Other decisions relevant to agricultural biodiversity are COP 10 Decision X/34, COP 8 Decision VIII/23, COP 7 Decision VII/3, COP 6 Decision VI/5, COP 5 Decision V/5, COP 4 Decision IV/6, and COP 3 Decision III/11 (http://www.cbd.int/agro/decisions.shtml; accessed December 10, 2010).

5 The *Second Report on the State of the World's Plant Genetic Resources for Food and Agriculture* is divided into eight chapters, which address the state of diversity; the state of *in situ* management; the state of *ex situ* conservation; the state of use; the state of national programs, training needs, and legislation; the state of regional and international collaboration; access to plant genetic resources, the sharing of benefits arising out of their utilization and the realization of farmers' rights; and the contribution of plant genetic resources to food security and sustainable agricultural development (http://www.fao.org/docrep/013/i1500e/i1500e00.htm; accessed December 15, 2010).

6 As a strategic framework, the Global Plan of Action needs to be periodically reviewed and updated. An updated version of the Global Plan of Action was approved by the Commission on Genetic Resources for Food and Agriculture in July 2011, based primarily on the gaps and needs identified in the Second Report, and in the light of new challenges such as climate change.

Bibliography

Almekinders, C., Fresco, L. and Struik, P. (1995) 'The need to study and manage variation in agro-ecosystems', *Netherlands Journal of Agricultural Science*, vol. 43, no. 2, pp. 127–142.

Brookfield, H. (2001) *Exploring Agrodiversity*, Columbia University Press, New York.

Brookfield, H. and Padoch, C. (1994) 'Appreciating agrodiversity: a look at the dynamism and diversity of Indigenous farming practices', *Environment*, vol. 36, pp. 8–11.

Brookfield, H. and Stocking, M. (1999) 'Agrodiversity: definition, description and design', *Global Environmental Change*, vol. 9, pp. 77–80.

Conway, G. (1987) 'The properties of agroecosystems', *Agricultural Systems*, vol. 24, no. 2, pp. 95–117.

Clement , C. (1999) '1492 and the loss of Amazonian crop genetic resources. I. The relation between domestication and human population decline', *Economic Botany*, vol. 53, pp. 188–202.

Cromwell, E., Cooper, D. and Mulvany, P. (2003) 'Defining agricultural biodiversity', in *Conservation and Sustainable Use of Agricultural Biodiversity: A Sourcebook*, Centro Internacional de la Papa (CIP) and Users' Perspective with Agricultural Research and Development (UPWARD), Manila, vol. 1, chapter 1, pp. 1–12.

Dulloo, M.E., Hunter, D. and Borelli, T. (2011) '*Ex situ* and *in situ* conservation of agricultural biodiversity: major advances and research needs', *Notulae Botanicae Horti Agrobotanici Cluj – Napoca*, vol. 38, no. 2, Special Issue, pp. 123–135

Emperaire, L. (2004) 'O que é domesticação?' in *Almanaque Brasil Socioambiental: uma nova perspectiva para entender o país e melhorar nossa qualidade de vida*, Instituto Socioambiental, São Paulo, p. 339.

Enjalbert, J., Dawson, J.C., Paillard, S., Rhoné, B., Rousselle, Y., Thomas, M. and Goldringer, I. (2011) 'Dynamic management of crop diversity: from an experimental approach to on-farm conservation', *Comptes Rendus Biologies* (doi: 10.1016/j.crvi.2011.03.005).

Food and Agriculture Organization (FAO) (2000) *World Watch List for Domestic Animal Diversity*, 3rd edn, FAO, Rome.

Eyzaguirre, P.B. and Linares, O. (eds.) (2004) *Home Gardens and Agrobiodiversity*, Smithsonian Books, Washington, DC.

Harlan, J. (1975) *Crops and Man*, American Society of Agronomy, Crop Science Society of America, Madison, WI.

Kloppenburg, J. and Kleinman, D. (1987) 'Plant germplasm controversy: analyzing empirically the distribution of the world's plant genetic resources', *BioScience*, vol. 37, no. 3, pp. 190–198.

Lenné, J.M. and Wood, D. (2011) *Agrobiodiversity Management for Food Security: A Critical Review*, Cabi International, Wallingford, UK.

Qualset, C.O., McGuire, P.E. and Warburton, M.L. (1995) 'Agrobiodiversity: key to agricultural productivity', *California Agriculture*, vol. 49, pp. 45–49.

Thrupp, L.A. (1998) *Cultivating Diversity: Agrobiodiversity and Food Security*, World Resources Institute, Washington, DC.

Thrupp, L.A. (2003) 'The central role of agricultural biodiversity', in *Conservation and Sustainable Use of Agricultural Biodiversity: A Sourcebook*, Centro Internacional de la Papa (CIP) and Users' Perspective with Agricultural Research and Development (UPWARD), Manila, vol. 1, chapter 3, pp. 20–32.

Wood, D. and Lenné, J. M. (1999) 'Why agrodiversity?', in Wood, D. and Lenné, J.M. (eds.) *Agrobiodiversity: Characterization, Utilization and Management,* Cabi Publishing, Wallingford, UK, pp. 1–13.

2

AGROBIODIVERSITY AND FOOD SECURITY, NUTRITION, HEALTH, AND ENVIRONMENTAL SUSTAINABILITY

It is the diversity of cultivated plants and domestic animals, and their capacity to adapt to adverse environmental conditions (climate, soil, vegetation, etc.) and to specific human needs, that enables farmers to survive in the harshest regions of the world. It is the cultivation of various species and varieties that protects farmers, in many circumstances, from total loss of the harvest, in cases of pests, diseases, droughts, floods, etc. With monocultures, which have an extremely narrow genetic base, the opposite occurs: pests and diseases affect the only grown species and destroy the harvest.

Genetic uniformity generates enormous risks and uncertainties to farming systems, which become particularly vulnerable. Even when a modern variety is developed to be resistant against a given pathogen,[1] any mutation in this pathogen, no matter how small, can be enough to wipe out this resistance, making the entire harvest vulnerable. According to the *First Report on the State of the World's Plant Genetic Resources for Food and Agriculture* (FAO), adopted in 1996, genetic vulnerability is the "condition that results when a widely planted crop is uniformly susceptible to a pest, pathogen or environmental hazard as a result of its genetic constitution, thereby creating a potential for widespread crop losses." Genetic erosion, on the other hand, is defined (by the first report) as

> the loss of individual genes and the loss of particular combinations of genes (i.e. of gene complexes) such as those maintained in locally adapted landraces. The term "genetic erosion" is sometimes used in a narrow sense, i.e. the loss of genes or alleles, as well as more broadly, referring to the loss of varieties.

Thus, although genetic erosion does not necessarily entail the extinction of a species or subpopulation, it does signify a loss of variability and thus a loss of flexibility.[2]

Monocultures tend to promote both genetic vulnerability and genetic erosion. One of the most famous examples of the dangers brought about by genetic uniformity was the "potato famine" in Ireland, between 1845 and 1851, caused by the devastation of potato fields by a pathogen (*Phytophthora infestans*). Ninety percent of the Irish population depended on potatoes as their main staple food, and the pathogen destroyed potato plantations. The famine killed 2 million Irish (25 percent of the population). During this period, 1.5 million Irish migrated to the United States, Australia, or New Zealand. Many died during the voyage or upon arrival, weakened by undernourishment (Woodham-Smith, 1991; Bartoletti, 2002).

There are, however, more recent examples. In the 1970s, a plant disease caused by a fungus (*Bipolaris maydis*), known as "Southern corn leaf blight," attacked corn plantations in the United States (initially in the south, later reaching the north, hitting the states of Minnesota, Michigan, and Maine). Some states lost half of their corn harvest. This also occurred in 1971, in a plantation of a single variety of wheat, known as Besostaja, in an area of 40 million hectares, extending from Kuban to the Ukraine. This variety had high yields when grown in Kuban, where temperatures were milder. In that year, Ukraine had an extremely harsh winter, which devastated its plantations and led to the loss of 20 million tons of wheat, which corresponded to 30–40 percent of the harvest. As Cary Fowler and Pat Mooney (1990) point out, in both cases neither the blight which infested the corn plantations nor the harsh winter in the Ukraine are to be blamed for the loss of corn and wheat harvests, but rather the genetic uniformity of the crops. Harvests would not have been affected so dramatically had diverse varieties been grown (Fowler and Mooney, 1990, p. IX–XI).

Another example of harvest losses caused by genetic uniformity is given by Lori Ann Thrupp (1988, 1998). According to her, in Central America, the extensive banana industry has been seriously jeopardized because of a reliance on uniform varieties grown in massive monocultural plantations. Panama disease, caused by a virulent fungus (*Fusarium oxysporum*), attacked and destroyed thousands of hectares of monocultural bananas in the Atlantic region in Costa Rica during the 1930s, and this led the nearly bankrupt industry to abandon the land, move across the country, and plant a new variety. Later on, growers again planted vulnerable monocultures, and the plantations again suffered from serious diseases – first yellow Sigatoka during the 1950s, and then black Sigatoka from the 1970s through the 1990s – costing the industry millions of dollars in losses, as well as exorbitant costs for control efforts. Banana companies continue to use a chemical-based approach to struggle against these diseases, but as long as the uniform expansive plantation systems are maintained, they are unlikely to escape the problem (Thrupp, 1988, 1998). Likewise, coffee rust has destroyed uniform coffee farms in Central America, Sri Lanka, India, Malaysia, the Philippines, and half a dozen African countries – largely due to the dependence of these countries on single varieties (Mooney, 1979).

Agrobiodiversity is an essential part of making agriculture more sustainable. Sustainable agriculture is delivered when farming (and other productive systems) produces food and other agricultural products to satisfy human needs indefinitely without unsustainable impacts on the broader environment. This requires agriculture to avoid severe or irreversible damage to the ecosystem services (such as soil fertility, water quantity and quality, genetic variability, pollinators, etc.) upon which it depends and to have acceptable impacts on the broader environment (environmental stewardship). Sustainable agriculture will rely more and more on agricultural biodiversity, making the best use of the widest range of species and varieties to build greater resilience[3] to external shocks and increase harvests, rather than relying upon external inputs (Bioversity International, 2010).

One of the principles of sustainable agriculture is crop diversification. A larger number of species in a given ecosystem, associated with other ecological factors, ensures greater stability and lesser need for external inputs, such as pesticides and nitrogen-based fertilizers. Agrobiodiversity also contributes to pest and disease management and conservation of soil fertility. Diversified agricultural systems result in harvests of different crops at alternate times of the year. A decrease in the price of a certain species or variety does not cause losses as great as in monocultural systems. Agrobiodiversity diversifies products and farmers' sources of income.

Diversification of an agroecosystem can take many forms, which include multiple cropping systems (intercropping, crop rotation, crop/livestock systems, etc.) and agroforestry systems, which integrate woody species into farming systems, conciliating agricultural production and conservation of tree species. Trees provide sources for fodder, fuel, fibers, and manures, as well as food. These systems promote an increase in organic matter in soils, reduce erosion, and preserve species diversity.

Each ecosystem, however, has distinct characteristics, and requires specific solutions. Sustainable agriculture requires deep understanding of the complex interactions among different components of agricultural systems. Every ecosystem has to find the appropriate solutions for its environmental, economic, and social conditions. Specialized production systems and the genetic homogeneity which characterizes them not only cause reduction in the diversity of species and varieties, but also the reduction of species that are important for the balance of agroecosystems, such as nitrogen-fixing bacteria, fungi which facilitate nutrient absorption, pollinators, seed dispersers, etc. They also compromise the resistance and resilience of agroecosystems, making them more vulnerable to pests, droughts, climate change, and other risk factors.

Agrobiodiversity is not only associated with sustainable food production, but also plays an essential role in promotion of food and nutrition security and human health. A diversified diet – balanced in protein, vitamins, minerals, and other nutrients – is recommended by nutritionists and is a fundamental condition for good health. An unbalanced diet may lead to what is called the "hidden hunger," that is, the lack of essential micronutrients (vitamins and minerals) in the everyday diet. The effects of "hidden hunger" are wide-ranging and may include delayed mental development, weakening of the immune system, and loss of strength and energy. It is estimated that some 2 billion people are affected by a lack of iron, and around 800 million are deficient in vitamin A, the lack of which can in severe cases lead to blindness (GTZ, 2006). Vitamin A deficiency results in the early deaths of about 1 million young children each year, according to the World Health Organization (2010). In addition, iron deficiency impairs the mental development of 40–60 percent of the developing world's children aged 6–24 months and leads to the deaths of about 50,000 women a year during pregnancy and childbirth (WHO, 2010).

The best insurance against nutrient deficiencies is eating a varied diet, thereby ensuring an adequate intake of all the macro and micronutrients needed for good health. Leafy vegetables, fruits, legumes, roots, tubers, spices, and herbs are essential for human nutrition and complement staple crops such as rice or maize. Many leguminous crops, such as cowpea and winged bean, are excellent sources of protein and micronutrients. Tropical fruits also have high vitamin and mineral content, and home gardens often accommodate a particularly rich diversity of crops. As home gardens are usually run by women, most of the production is directly used for cooking, benefiting all the family (GTZ, 2006).

Only diverse agricultural systems can offer more nutritive and balanced diets, and the fight against hunger and poverty must necessarily include more sustainable agricultural practices. There is a direct relation between reduction in agricultural diversity and impoverishment of diets. Genetic erosion in the fields affects not only farmers but consumers as well. Agricultural production models have direct implications for human nutrition and health. "Modern" agriculture and cultivation of few agricultural species favor standardization of dietary habits and loss of cultural appreciation of native species. In the Andes, for instance, many plants traditionally used in the diets of Indigenous populations and local farmers, such as quinoa, amaranth, *chocho*, *kañina*, *viraca*, and *yacón*, are currently being abandoned and replaced by imported species, such as spinach, cauliflower, and leeks, which require far greater quantities of fertilizer to grow. In tropical regions of the Americas, species such as *Portulaca oleracea*, also known as verdolaga or pigweed, grown to make salads with nutritional values close to spinach, and *Tropaeolum majus* (known popularly as "capuchinha"), which were once of great importance to local agricultural systems and food security of rural populations, are used less and less (FAO, 2001). Developing countries are increasingly adopting the simplified diets of the western world, undermining dietary diversity around the world, and conversion from polycultural to monocultural systems usually means that the household must rely on purchased foods that are often unaffordable or inaccessible (Thrupp, 1998)

Diets centered on consumption of plants (fruits, legumes, and vegetables) have been replaced by excessively calorie-rich diets filled with fats yet poor in vitamins, iron, and zinc. Foods are made with increasingly lower numbers of plant species and varieties, and derivatives of corn and soy, for example, are present in most industrialized food products. There are between 250,000 and 420,000 species of superior plants, but only 30 account for 95 percent of human food, of which only seven (wheat, rice, corn, potatoes, manioc/cassava, sweet potatoes, and rye) account for 75 percent of this total. More optimistic estimates, however, indicate that 103 species would be responsible for 90 percent of the foods consumed worldwide, rather than the 20 or 30 most commonly mentioned species (Walter *et al.*, 2005). According to the FAO's *Second Report on the State of the World's Plant Genetic Resources for Food and Agriculture*, published in 2010, fewer than 150 species are under commercial cultivation. A wide range of tuber crops are grown worldwide, but only five species account for the majority of the total world production: potato, cassava, sweet potato, yams, and taro.

More and more, human diets are based on a reduced number of crop varieties and livestock breeds, which compromises health. Diets which are poor in nutritional value and balance are responsible, in part, for the world epidemic of chronic diseases such as obesity, type 2 diabetes, cardiovascular diseases, and some forms of cancer. Often considered diseases of affluence, these are actually the result of oversimplified diets and poor nutrition. Modern western diets focus on high-energy cereals and refined carbohydrates such as maize, wheat, rice, and sugar. Eating a diverse diet by reintroducing agricultural biodiversity can mitigate the effects of simplified western diets. Diversity of diet, founded on diverse farming systems, delivers better nutrition and greater health, with additional benefits for human livelihoods.

According to the World Health Organization (WHO), approximately 177 million children all over the world are threatened by diseases associated with obesity, and the prediction is that 2.3 billion people over the age of 15 will be obese by 2015. Currently, there are 1.5 billion obese people worldwide, while 854 million are undernourished. There are also indirect impacts of poor nutrition on human health: for example, for resource-poor populations in countries faced with the problems caused by HIV/AIDS, the consumption of diverse diets represents an important way of boosting human resistance and tolerance.[4] A balanced diet helps to prevent the illnesses and complications that often occur with HIV infection, for example fungal diseases, herpes, lung infections, tuberculosis, etc. Besides, antiretroviral treatment can be especially successful when access to appropriate nutrition is guaranteed (GTZ, 2009).

Agriculture interacts with the environment in many forms which affect human health. The harmful effects of uncontrolled use of pesticides are well known. In extreme situations, genetic anomalies, tumors, and cancer can arise. WHO estimates that approximately 3 million cases of severe intoxication by pesticides occur, leading to 220,000 deaths each year, 70 percent of which are in developing countries.[5] In addition to poisoning rural workers in direct or indirect contact with these products, contamination of food products also affects consumers. Owing to the risks for human health and the environment, pesticides are subject to legal control in many countries, including Brazil.[6] Environmental changes brought about by irrigation and deforestation encourage development of diseases such as malaria and schistosomiasis.[7]

Agrobiodiversity is directly associated to at least two of the Millennium Development Goals[8]: the eradication of poverty and hunger and the achievement of environmental sustainability. In addition, it is essential for food security, which is achieved "when all people, at all times, have physical and economic access to sufficient, safe and nutritious food to meet their dietary needs and food preferences for an active and healthy life" (FAO, 1996). Among the actions to be taken by national governments to attain food security, recommended by the Voluntary Guidelines on the Right to Food, is the adoption of specific national policies, legal instruments, and supporting mechanisms to prevent the erosion of agrobiodiversity and to ensure the conservation and sustainable use of genetic resources for food and agriculture (Guideline 8D), which is a recognition of the essential importance of agricultural diversity to food security and human health. However, greater efforts are needed to strengthen international and national programs for the conservation and sustainable use of agricultural biodiversity and all of its components.

Notes

1 A pathogen is any organism capable of causing infectious diseases in plants, that is, fungi, bacteria, viruses, nematodes, and protozoa.

2 Although there has been much discussion of these concepts (genetic vulnerability and genetic erosion) since the first report, these definitions have not changed significantly (according to the *Second Report on the State of the World's Plant Genetic Resources for Food and Agriculture*, published in 2010). Genetic erosion can also occur at the level of germplasm collections in genebanks as a result of improper management, especially due to inadequate regeneration procedures.

3 Resilience is the way in which an ecosystem responds to and recovers from disturbance. Resilience to perturbation may be manifested by a smaller drop in productivity, a more rapid recovery, and lower variability over time; all are underpinned by biodiversity, and the risk of simplifying ecosystems is that those systems become more vulnerable to perturbations. Source: Frison *et al.* (2011).

4 According to FAO's *Second Report on the State of the World's Plant Genetic Resources for Food and Agriculture*, published in 2010 (p. 191).

5 In 2008, Brazil became the leading consumer of pesticides. Sales totaled 733.9 million tons and were worth about US$7.1 billion, according to Brazil's National Agricultural Defensive Products Industry (Sindicato Nacional da Indústria de Produtos para a Defesa Agrícola – Sindag). Brazil surpassed the United States,

the largest food producer in the world, which consumed 646 million tons of pesticides in the same period. Source: 'No reino dos agrotóxicos: a Anvisa pode banir 13 pesticidas do Brasil, novo líder mundial de consumo', *CartaCapital*, May 20, 2009, no. 546.

6 Brazilian Law 7802/1989, regulates the use, marketing, transportation, storage, imports, and exports of pesticides.

7 Schistosomiasis, or bilharzia, is a parasitic disease caused by trematode flatworms of the genus *Schistosoma*. Larval forms of the parasites, which are released by freshwater snails, penetrate the skin of people in the water. Source: http://www.who.int/topics/schistosomiasis/en/ (accessed January, 20, 2011).

8 In 2000, the United Nations Millennium Declaration was adopted, committing nations to a new global partnership to reduce extreme poverty and setting out a series of time-bound targets, with a deadline of 2015. The targets have become known as the Millennium Development Goals (MDGs). Other MDGs are achievement of universal primary education; promotion of gender equality and empowerment of women; reduction of child mortality; improvement of maternal health, combat HIV/AIDS, malaria and other diseases; and construction of a global partnership for development.

Bibliography

Bartoletti, S.C. (2002) *Black Potatoes: The Story of the Great Irish Famine, 1845–1850*, Gale, MI.

Bezerra, M.C. and Veiga, J.E. (coords.) (2000) *Agricultura Sustentável*, Ministério do Meio Ambiente (Brazilian Ministry of Environment), IBAMA, and Consórcio MPEG, Brasília.

Bioversity International (2010) 'Agriculture, agricultural biodiversity and sustainability', booklet based on the paper 'Sustainable Agriculture and the Sustainable Use of Agricultural Biodiversity: Concepts, Trends and Challenges', prepared for the 14th meeting of the Convention on Biological Diversity's Subsidiary Body on Scientific, Technical and Technological Advice, held in Nairobi, Kenya.

Ehlers, E. (2008) 'Agricultura sustentável', in *Almanaque Brasil Socioambiental: Uma Nova Perspectiva para Entender o País e Melhorar Nossa Qualidade de Vida*, Instituto Socioambiental, São Paulo, pp. 414–419.

Food and Agriculture Organization (FAO) (1996) *Rome Declaration on World Food Security and World Food Summit Plan of Action.* World Food Summit, November 13–17, 1996, FAO, Rome.

Food and Agriculture Organization (FAO) (2001) 'Seed policy and programmes in Latin America and the Caribbean', in *Proceedings of the Regional Technical Meeting on Seed Policy and Programmes in Latin America and the Caribbean*, March 20–24, 2000, Merida, Mexico (FAO Plant Production and Protection Paper, 164). Plant Production and Protection Division, Seed and Plant Genetic Resources Service, FAO, Rome.

Food and Agriculture Organization (FAO) (2011) *Gardens of Biodiversity: Conservation of Genetic Resources and Their Use in Traditional Food Production Systems by Small Farmers of the Southern Caucus (Armenia, Azerbaijan and Georgia)* (http://www.fao.org/agriculture/gardens_of_biodiversity/en/; accessed April 10, 2011).

Fowler, C. and Mooney, P. (1990) *Shattering: Food, Politics, and the Loss of Genetic Diversity*, University of Arizona Press, Tucson.

Frison, E., Cherfas, J. and Hodgkin, T. (2011) 'Agricultural biodiversity is essential for a sustainable improvement in food and nutrition security', *Sustainability*, vol. 3, pp. 238–253 (www.mdpi.com/journal/sustainability; accessed January 20, 2011).

GTZ (German Technical Cooperation) (2006) 'Agrobiodiversity – the key to food security', sector project 'People, Food and Biodiversity' (Division 45), Germany.

GTZ (German Technical Cooperation) (2009) 'Nutrition security is key in the fight against HIV and Aids', sector project 'People, Food and Biodiversity' (Division 45), Germany.

Ingram, J., Ericksen, P. and Liverman, D. (eds.) (2010) *Food Security and Global Environmental Change*, Earthscan, London.

Jhamtani, H. and Jenny, P. A. (2007) 'Superando a desnutrição com cultivos e sistemas alimentares locais', *Agriculturas: Experiências em Agroecologia*, vol. 4, no. 4, pp. 23–25.

Lawrence, G., Lyons, K. and Wallington, T. (eds.) (2010) *Food Security, Nutrition and Sustainability*, Earthscan, London.

Mooney, P. (1979) *Seeds of the Earth: A Private or Public Resource?*, Canadian Council for International Cooperation and International Coalition for Development Action, Ann Arbor, MI.

Mouillé, B., Charrondière, R. and Burlingame, B. (2010) 'The contribution of plant genetic resources to health and dietary diversity', thematic background study included in the *Second Report on the State of the World's Plant Genetic Resources for Food and Agriculture*, FAO, Rome.

Stern, L., Killough, S., Borja, R., Sherwood, S., Hernidiah, N., Joicey, P. and Berti, P. (2007) 'Trabalhando agricultura e saúde conjuntamente', *Agriculturas: Experiências em Agroecologia*, vol. 4, no. 4, pp. 18–22.

Thrupp, L.A. (1988) 'The political ecology of pesticide use in Central America: dilemmas in banana production of Costa Rica', PhD thesis, University of Sussex, UK.

Thrupp, L.A. (1998) *Cultivating Diversity: Agrobiodiversity and Food Security*, World Resources Institute, Washington, DC.

Walter, B.M.T, Cavalcanti, T.B., Bianchetti, L. and Valls, J.F.M. (2005) 'Coleta de germoplasma vegetal: relevância e conceitos básicos' in Walter, B.M.T. and Cavalcanti, T.B. (eds.) *Fundamentos para a Coleta de Germoplasma Vegetal*, EMBRAPA, Brasilia, pp. 28–55.

Woodham-Smith, C. (1991) *The Great Hunger: Ireland 1845–1849*, Penguin Books, London.

World Health Organization (WHO) (2010) 'Micronutrient deficiencies: vitamin A deficiency, and iron deficiency anaemia' (http: //who.int/nutrition/topics/en; accessed March 15, 2011).

3

AGROBIODIVERSITY AND CLIMATE CHANGE

Agrobiodiversity conservation and sustainable use are essential components of international and national strategies to adapt or mitigate the impacts of global climate change. There is a close relationship between climate change, food security, and agrobiodiversity: it is diversity that makes it possible for species, varieties, and agroecosystems to adapt to changes and variations in environmental conditions. They can cope with future challenges, including those brought about by climate change, only if they have wide genetic variability. Interactions between agrobiodiversity and climate change are many, not least of which is that climate change is reducing the number of species and agricultural ecosystems. On the other hand, agrobiodiversity is also essential to face the impacts caused by global warming.

There is relative international consensus that climate change is a reality. There are diverging opinions among scientists regarding the magnitude, speed, and impacts of global climate change, but very few deny its existence, or that it is a result of human activities. Throughout the planet's history there have been climate fluctuations due to natural causes, such as glaciations. Extinction of dinosaurs and other species is thought to be the result of one such natural climate fluctuation. Nevertheless, these natural climate fluctuations cannot be confused with global climate change, which results from human activities that have occurred since the Industrial Revolution. Burning of fossil fuels, such as coal, oil, and natural gas, by the industry and transportation sectors, accounts for approximately 80 percent (by concentration) of all "greenhouse gases" in the Earth's atmosphere (mainly carbon dioxide, methane, nitrous oxide, etc.). The remaining 20 percent results from inappropriate land use, especially tropical forest burning and clearing (deforestation). Historically, industrialized countries have been responsible for the largest emissions of greenhouse gases (the United States alone is responsible for around 30 percent of global emissions). Currently, however, many developing countries, such as China, India, and Brazil, have also joined the list of great emitters. Brazil is responsible for about 5 percent of global emissions, most of which (61 percent) results from land use change, e.g. deforestation and fires in the Amazon. When forests are cleared and burned, the carbon stored in the plant tissue of the trees is released into the atmosphere in the form of carbon dioxide, one of the main greenhouse gases. In industrialized countries, most emissions come from the burning of fossil fuels, whereas in developing countries most emissions come from land use change (IPAM, 2010).

According to the Fourth Report of the United Nation's Intergovernmental Panel on Climate Change (IPCC),[1] released on November 17, 2007, as a result of increased concentration of greenhouse gases in the atmosphere, the average temperature has risen 0.7°C over the last 100 years, and is expected to increase between 1.4 and 5.8°C over the next century (taking 1990 as the base year). The most likely is an average warming of 2–4°C. This temperature increase will not be homogeneous, with some areas warming more than others. In particular, the polar regions (Arctic and Antarctic) and tropical regions are likely to be affected more strongly. As a result, changes will occur in the distribution patterns and intensity of extreme climate events, such as droughts, floods, hurricanes, and tropical storms, as well as an increase in the level of oceans. Global greenhouse gas emissions due to human activities have grown since pre-industrial times, with an increase of 70 percent between 1970 and 2004, and the global average sea level has risen since 1961 at an average rate of 1.8 (range 1.3–2.3) mm/year and since 1993 at 3.1 (2.4–3.8) mm/year, with contributions from thermal expansion, melting glaciers and ice caps, and the polar ice sheets. In its Fourth Report (2007), the IPCC showed that, considering the period from 1850 to 2005, the final 12 years broke all temperature records (with the exception of 1996). The warmest year was 1998, followed by 2005 (IPCC, 2007).

Some impacts of climate change on biodiversity are irreversible, and the Fourth IPCC Report mentions studies which indicate that approximately 20–30 percent of all studied plant and animal species are at risk of extinction if average global temperatures rise above 1.5–2.5°C (compared with 1980–1999). If the average global temperature increase exceeds 3.5°C, there are projections that 40–70 percent of these species will be at risk of extinction (IPCC, 2007). Other studies show that some butterfly species are already migrating and, in some cases, to areas 95 kilometers north of those they occupied 100 years ago. Further scientific research has found that over 70 species of frogs from tropical America stand to be virtually eliminated by a fungus that thrives on higher temperatures, and the routes of migratory birds are drastically affected (Marengo, 2006). Climate change threatens not only biological diversity, but also the ecological functions of many ecosystems, such as deserts, mangroves, forests and mountains, which will undergo drastic changes in their function and structure. The Arctic, for instance, has lost approximately 7 percent of its ice surface since 1900, and in the spring this loss reaches 15 percent. Africa may lose approximately two-thirds of its productive land by 2025, while Asia and South America may lose one-third and one-fifth, respectively (IPAM, 2010).

Climate change affects the fundamental requirements for health – clean air, safe drinking water, sufficient food, and secure shelter. Many of the major killers, such as diarrheal diseases, malnutrition, malaria, and dengue, are highly climate sensitive. Risks to health range from deaths in extreme high temperatures to changing patterns of infectious diseases. Intense short-term fluctuations in temperature can seriously affect health – causing heat stress (hyperthermia) or extreme cold (hypothermia) – and lead to increased death rates from heart and respiratory diseases. Recent studies suggest that the record high temperatures in western Europe in the summer of 2003 were associated with a spike of an estimated 70,000 more deaths than the equivalent periods in previous years. Levels of pollen and other aeroallergen are also higher in extreme heat. This can trigger asthma, which affects around 300 million people (WHO, 2009a,b).

Climatic conditions affect diseases transmitted through water and by vectors such as mosquitoes. Diarrhea, malaria, and protein–energy malnutrition alone caused more than 3 million deaths globally in 2004, with over one-third of these deaths occurring in Africa. More variable rainfall patterns are likely to compromise the supply of fresh water. Globally, water scarcity already affects 4 out of every 10 people. A lack of water and poor water quality can compromise hygiene and health. This increases the risk of diarrhea, which kills approximately 2.2 million people every year, as well as trachoma (an eye infection that can lead to blindness) and other illnesses (WHO, 2009a,b).

Increasing temperatures and more variable rainfalls are expected to accelerate infectious cycles and to facilitate spatial dispersion of endemic infectious diseases, such as dengue, malaria, leishmaniasis, infectious diarrhea, etc. According to the WHO, malaria kills at least 100,000 people each year, and 50 million cases of dengue occur worldwide, 500,000 of which require hospitalization and 125,000 are fatal. WHO (2009a,b) estimates that by 2080 climate change will increase the number of dengue cases worldwide to 2 billion. Climate change is projected to widen significantly the area of China where the snail-borne disease schistosomiasis occurs.

Climate change is one of the factors responsible not only for the increase in contagious diseases but also for the spread of tropical diseases to other parts of the world. Mosquitoes which transmit malaria, for instance, can already be found in regions where the disease did not previously exist, and Western Nile fever has reached southern Italy. In Europe, there has been dissemination of encephalitis to Scandinavia, caused by the average increase of 3°C in the continent's temperature in the past 40 years, and the tick which causes the disease has migrated to Nordic countries (WHO, 2009a,b).

Rising sea levels – another outcome of global warming – increase the risk of coastal flooding, and can cause population displacement. More than half of the world's population now lives within 60 kilometers of shorelines.

Extreme events caused by climate change (storms, floods, droughts, etc.) provoke physical and psychological traumas (Confalonieri, 2007). They generate environmental migrants, which are populations forced to leave their places of origin because of environmental stresses and events, such as droughts, desertification, sea level rise, hurricanes, cyclones, flooding, tornados, etc. According to the International Organization for Migration (IOM), environmental migrants are "persons or groups of persons who, for reasons of sudden or progressive changes in the environment that adversely affect their lives or living conditions, are obliged to leave their habitual homes, or choose to do so, either temporarily or permanently, and who move either within their territory or abroad." Many human rights organizations argument that "climate refugees" should have the same status as political refugees, which is a very controversial issue.[2]

Poorer classes in developing countries are the most vulnerable, as they have limited resources to adapt to climate change. For instance, the German state of Baden-Württemberg plans to spend US$685 million a year on protection against floods. However, the Special Climate Change Fund, created to assist poor countries in mitigation of global warming effects, has received only US$279 million (approximately half of the amount for the single European state mentioned above). France currently spends more on meteorological monitoring systems than all of sub-Saharan Africa (made up of 47 countries). Holland has 32 times more weather stations per 10,000 square kilometers than Africa (UNDP, 2007/2008).

Agriculture is one of the activities most affected by climate change, as it depends directly on temperature and rainfall conditions. Higher temperatures in tropical and subtropical areas, which include most developing countries, such as Brazil, will affect agricultural production directly. Estimates point to developing countries losing 9 percent of their capacity for agricultural production by 2080 if climate change is not controlled. Latin America is one of the regions where the impacts on agriculture are expected to be greatest: productive potential is estimated to drop 13 percent, second only to Africa (17 percent), and by more than in Asia (9 percent) and the Middle East (9 percent). Corn production in Latin America is predicted to fall by 10 percent by 2055 and in Brazil by 25 percent, which would increase hunger among populations that depend on this crop for subsistence. A study of *in situ* conservation of wild relatives of cultivated plants, carried out by the United Nations Environment Program (UNEP) in partnership with Bolivian institutions (Centro de Investigaciones Fitoecogenéticas de Pairumani and Museo de Historia Natural), estimates that within 10 years wild relatives of manioc/cassava (such as *Manihot tristis*) and peanut (such as *Arachis duranensis*) may be threatened with extinction in Bolivia, a country in which 43 percent of the population depends on agriculture for survival, but only 3 percent of the country's area is planted (Zapata Ferrufino *et al.*, 2008). Another study, conducted by researchers from Stanford University, found that southern Africa may lose 30 percent of its main agricultural product – corn – in the next two decades, and southern Asia over 10 percent of its corn and rice harvests (Lobell *et al.*, 2008). In developed countries, the opposite trend is predicted: agricultural production should grow by 8 percent, as climate change should make growth cycles of crops longer and increase rainfall in regions with higher latitudes (UNDP, 2007/2008). Losses in agriculture tend not only to increase hunger, but also to aggravate inequalities between rich and poor countries and internal inequalities in poorer countries. Climate change will impact productivity of species which are important food sources in poorer areas of the world, such as large parts of Asia, sub-Saharan Africa, the Caribbean, and Central and South America, where 95 percent of all undernourished people in the world live. Most of the world's rural poor live in areas that are resource poor, highly heterogeneous and risk prone, often located in arid or semiarid zones, and in mountains and hills that are ecologically vulnerable (Altieri and Koohafkan, 2008).

Brazil has a high level of vulnerability to climate change.[3] With over 8,000 kilometers of coastline, an increase in the level of oceans will affect not only coastal cities (where the largest part of the Brazilian population lives), but also oceanic islands, beaches, marshlands, etc. Economic activities such as fishing, tourism, and ports stand to be directly affected, as well as important Brazilian ecosystems. There are predictions of eastern Amazon "savannization" and increase in desertification of the northeastern semiarid regions of Brazil. Studies show that the Amazon and the northeast of Brazil are "climatic change hot spots," and represent the Brazilian regions that are most vulnerable to climate change (Carneiro, 2008). Destabilization of the rainfall patterns in the Amazon would affect not only the local climate, but also the entire region of the Prata River Basin, where some of the largest South American cities are located.

A study conducted by EMBRAPA[4] (Brazilian agricultural research corporation) and the University of Campinas (Unicamp, in the Brazilian state of São Paulo), found that climate change will severely impact Brazilian agriculture and have major consequences for the geographic distribution of crops (Assad and Pinto, 2008). Temperature increases (predicted by the IPCC in 2007) will significantly reduce the areas favorable to the cultivation of crops such as soybean, cotton, rice, beans, sugarcane, coffee, sunflower, manioc/cassava, and maize (together, these crops account for 86.17 percent of the Brazilian cultivated area). The study considered two scenarios predicted by IPCC: the most pessimistic, which estimates a temperature increase of between 2°C and 5.4°C by 2100 (A2 scenario), and a more optimistic scenario, which estimates a temperature increase of between 1.4°C and 3.8°C by 2100 (scenario B2). The most severely affected crop would be soybean (the main Brazilian export crop), which, according to the study, may suffer, by 2070, a reduction of 41.39 percent in the area favorable to its cultivation (in the worst scenario). However, according to the study, soybean may suffer, as early as 2020, a reduction in the area favorable to its cultivation of 21.62 percent (B2 scenario) to 23.59 percent (scenario A2).

According to the study carried out by EMBRAPA and Unicamp (Assad and Pinto, 2008), the crop that would be the next most affected by climate change in Brazil is Arabica coffee. The area favorable to the cultivation of Arabica coffee (inside the Brazilian territory) would, in the worst scenario (A2), be reduced by 9.48 percent by 2020, by 17.1 percent by 2050, and by 33 percent by 2070. According to the more optimistic scenario (B2), the area reduction would be 6.75 percent by 2020, 18.3 percent by 2050, and 27.39 percent by 2070. Maize, rice, beans, cotton, and sunflower would also suffer significant production decreases in the northeastern region of Brazil. On the other hand, there would be an increase in the area favorable to the cultivation of manioc (cassava) in Brazil (by 7.29 percent by 2050 and by 16.61 percent in 2070, in scenario B2, and by 13.48 percent in 2050 and by 21.26 percent in 2070 in scenario A2). This increase would be particularly marked in the south of Brazil because of a decrease in areas subject to snow frost, and in the Amazon region, owing to a decrease in surplus water. However, temperature increases may lead to significant reduction in manioc/cassava production areas in the northeastern, semiarid regions of Brazil, where this crop is very important for local food security. Temperature increases may also extend sugarcane cultivation areas to southern regions of Brazil in the next 10–20 years. According to the study, sugarcane's cultivation area may increase from 6 million hectares (currently) to 17 million in 2020 (scenario B2).

Other consequences of climate change for Brazilian agriculture include displacement of perennial crops, for example oranges, southward, in search of cooler temperatures (Marengo, 2006). Furthermore, the largest producer of apples in Brazil has already announced that it has decided to take its orchards up the mountain. The plantations, which currently occupy the western state of Santa Catarina (southern Brazil), will move to São Joaquim, in the highlands of the state. The reason is the increasingly warm winters. Between 1960 and 2009, the Company for Agricultural Research and Rural Extension of Santa Catarina (EPAGRI) recorded an increase of 1.3°C in the monthly average temperature in January.[5]

In another study, EMBRAPA (Brazilian agricultural research corporation) researcher Raquel Ghini found that climate change provokes significant alterations in the occurrence and severity of plant diseases. New climate and soil conditions may result in infestations of many plagues and diseases, owing to its effects on pathogen–host relations and the effect of carbon dioxide (CO_2) on plant diseases and microorganisms. She cites correlations found between El Niño effects and epidemics of potato late blight and tobacco blue mold in Cuba and rust on wheat, in northern China and the midwestern United States (Ghini, 2005). A study performed by the

University of Illinois revealed that higher concentrations of carbon CO_2 in the atmosphere make soy plantations more vulnerable to insect attacks. Soy plants submitted to high levels of CO_2 not only produce more carbohydrates, which attract insects, but also lose their ability to synthesize a chemical substance that acts as a natural defense mechanism against insects, according to the study (Agência Estado, 2008).

In the Upper Rio Negro region, in the Brazilian Amazon, Indigenous peoples are particularly concerned with the impacts of climate change on natural cycles and processes that affect some of their main activities, such as fishing, hunting, tilling the soil for crops cultivation, preparing seedbeds, cultivating and harvesting, and navigation. In this region, there is a large Indigenous population that practices traditional, low-technology agriculture, which is highly vulnerable to rainfall variations. Both extreme increases and decreases in humidity can also impact Indigenous agricultural practices. The Baniwa[6] Indigenous people have been closely monitoring the phenological behavior of the tree species found in their forests, and of maniocs/cassava and other cultivated plants, as well as changes in the behavior of fish, all very important to their food security. To understand what happens in the world, the Baniwa use several natural indicators, such as the times of the year when plants flower and fruit, summer rainy season, *piracema* time (when fish swim to the headwaters of rivers to reproduce), lunar cycles, and constellations. According to the oral accounts of the older Baniwa people, many natural phenomena no longer happen at the right time. There has been a decrease in fish each year and *piracema* does not happen with the same strength as it used to. Terrestrial animals, such as wild pig, monkeys, pacas, carará, and mutum, are becoming scarcer. Rains and droughts no longer occur at the right time, impacting especially manioc/cassava and peppers cultivations. When there is too much rain, the soil becomes too humid, killing new plantings. Many plants are flowering and fruiting at different (strange) times, and one typical example, often cited by the elderly Baniwa, is the *umari* plant (of the Icacinacea family). This can be a consequence of temperature rise, which increases the accumulated thermal unit (a measure of the cumulative effect of temperature over time) of the plant, and makes it reproduce out of season. Abnormal reproduction of fish may be also influenced by higher water temperature, which causes the reproductive organ of the fish to be ready (to reproduce) too soon, as a higher temperature accelerates its development. In this case, fish reproduce out of season, and when *piracema* time arrives, it does not have the same strength (Cardoso, 2008; Silva *et al.*, 2010). High water temperature also has direct effects on fish physiology, as it reduces the ability of the fish to obtain enough oxygen to breathe. Amazonian waters have an extremely rich biodiversity, holding the greatest wealth of species of freshwater fish in the world. It has been estimated that 3,000 fish species exist in the Amazon region. Fish consumption in the Amazon is among the highest in the world, and commercial fishing is an extremely important economic activity, amounting to around 173,000 tons per year (Zuanon, 2008).

The impact of climate change on Amazonian tropical palm trees – essential for food security and the survival of all Indigenous and traditional communities who live in Amazonia – has also been assessed by some studies (Miles *et al.*, 2004; Clement *et al.*, 2008). Miles *et al.* (2004) have used the Hadley Centre model[7] to evaluate the impacts of expected global climate change on a representative sample of 69 individual angiosperm species in Amazonia. The impacts were simulated from 1990 to 2095. In the resulting simulations, 43 percent of all species became nonviable by 2095 because their potential distributions had changed drastically. Widely distributed species (such as *pupunha*, *tucumã*, and *buriti*) with high tolerance to environmental variation exhibited the least response to climate change, and species with narrow ranges and short generation times (such as *piaçaba* and *açaí chumbo*) the greatest response to climate change. Climate changed most in northeast Amazonia while the best remaining conditions for lowland moist forest species were in western Amazonia (Miles *et al.*, 2004). Tropical rainforests in Amazonia are estimated to contain at least 12 percent of all species of flowering plants in the world.

One of the strategies proposed by scientists to deal with climate change is development of agricultural systems and varieties that are adapted to extreme climate events, such as droughts and floods. In order to do this, it is fundamental that the genetic diversity of agricultural species and varieties and their wild relatives be tapped. All plants domesticated by humans originated, at some point in time, from their wild relatives, which are sources of genes for the development of new varieties adapted to adverse socioenvironmental conditions. Wild relatives developed resistance to droughts, floods, and extreme heat and cold. When cultivated plants are attacked by a given pest or disease, or start suffering the effects of climate change, farmers and geneticists need to turn to their wild relatives in search of genes that are resistant to this type of stress.

A study carried out by two research centers connected with the Consultative Group on International Agricultural Research and made public on May 22, 2007 (International Biodiversity Day), found that, by 2055, 61 percent of 51 wild species of peanuts and 12 percent of 108 wild potato species analyzed by the study may become extinct as a result of climate change. Of the 48 species of cowpeas, two would be threatened with extinction (Jarvis *et al.*, 2008). The coordinator of the study, agronomist Andy Jarvis, explains that survival of wild relatives of many species, not only peanuts, potatoes and cowpeas, would be threatened even if the most conservative estimates were taken into account regarding the magnitude of global climate change. According to Jarvis, the vulnerability of a wild plant to climate change depends on its capacity for adaptation, and a form of adaptation by plants to climate change is migration to regions with milder temperatures. Researcher Annie Lane, from the Bioversity International Agricultural Research Center, who participated in the study, points out that

> Geneticists will need, more than ever before, wild varieties in order to develop new agricultural varieties which can adapt to climate change. However, it is precisely due to climate change that we are at risk of losing a great part of these genetic resources at the moment in which they are most needed in order to maintain agriculture.[8]

Annie Lane failed to add that not only geneticists and plant breeders, but also traditional and local farmers depend on broad genetic heterogeneity in order to face the challenges brought on agriculture by global climate change. Agrobiodiversity is important for all forms of agricultural production. Both agribusiness, which is highly dependent on homogeneous varieties improved by geneticists, and traditional and local agricultural systems, which make use of local seeds selected and improved by farmers themselves, need genetic diversity. However, poor farmers in developing countries are especially vulnerable to climate change because of their geographic exposure, low incomes, and greater reliance on agriculture (Altieri and Koohafkan, 2008).

Another study, by the Stanford University Center for Science and Environmental Policy and some North American research institutions, made public on May 2, 2007, evaluated the impact of climate change on rice plantations in Indonesia. Agriculture in Indonesia is already under strong influence from rainfall variations caused by monsoons and climate fluctuations. The study focused mainly on Bali and Java, important rice-growing regions, and came to the conclusion that the probability of delays in rain of more than 30 days (seriously damaging agriculture) should increase from the current 9–18 percent to 30–40 percent by 2050, that is, more than double. The study predicts that Asian farmers will face more intense and frequent droughts and floods (Naylor *et al.*, 2007). The impact of climate change on maize, a crop that is fundamental for food security of American and African populations, was also evaluated in a study by the International Tropical Agriculture Center (CIAT) and the International Livestock Research Center (ILRI). Results indicated an average decrease of 10 percent in corn productivity by 2055 (CGIAR, 2007).

What is the best way to deal with the effects of climate change on rice and corn crops, which are so fundamental to food security of Asian, American, and African populations? Among the solutions pointed out by scientists, the following stand out: diversification of agricultural production and development of agricultural varieties which are more resistant to droughts and higher temperatures. In both cases, the diversity of species and varieties of cultivated plants – agrobiodiversity – are a crucial instrument against climate change.

The International Rice Research Institute (IRRI), with headquarters in the Philippines, started a research program in 2006 aimed at the development of rice varieties that are tolerant to higher temperatures and extreme climate events, in addition to using higher levels of CO_2 to increase agricultural productivity (Brahic, 2006).

Other scientists have defended the need for agricultural research to give more priority to increasing plant resilience than to increasing productivity, in light of global climate change. Martin Parry, one of the current directors of the IPCC, and William Dar, director-general of the International Center for Research on Semi-arid Tropics, in a workshop held in India in 2007, argued that agricultural research should be reoriented to adaptation to environmental stress, such as higher temperatures and water scarcity, resulting from climate variations (Padma, 2007).

Some civil society organizations have, nonetheless, accused the world's largest agricultural biotechnology companies of filing multigenome patents in pursuit of exclusive control over "climate-ready" crops, to monopolize a potentially lucrative market for gene-altered crops that can handle global warming (Weiss, 2008). According to a report by the Canadian organization ETC Group,[9] 262 patent families (subsuming 1,663 patent documents worldwide) published between June 2008 and June 2010 make specific claims to abiotic stress tolerance (such as drought, heat, flood, cold, and salt tolerance) in plants. These claims extend, in many cases, to multiple traits in scores of genetically modified crops and even to the harvested food and feed products. According to the report, just six corporations (DuPont, BASF, Monsanto, Syngenta, Bayer, and Dow) and their biotech partners (Mendel Biotechnology and Evogene) control 201 or 77 percent of the 262 patent families (both issued patents and applications). Three companies – DuPont, BASF, and Monsanto – account for 173 or 66 percent, and the public sector has only 9 percent. The report says that the dramatic upsurge (between June 2008 and June 2010) in the number of patents related to "climate-ready," genetically engineered crops is a bid to control the world's food security, ease public acceptance of genetically engineered crops, and increase farmers' dependence on genetically modified crops (ETC Group, 2010a).

The government of Norway, in partnership with the international organization Global Crop Diversity Trust, has, in anticipation that in the worst-case scenario global warming could result in natural disasters or even wars, built the largest seedbank in the world, in one of the coldest regions of the planet: a cavern in a mountain in the Arctic, near the city of Longyearbyen, in the Svalbard Archipelago, a region in Norway that is in total darkness for three months every year (it is the "polar night"). The temperature in the seedbank reaches approximately –18°C, and the area's natural permafrost,[10] the snow and ice that cover the mountain for the greater part of the year, helps to keep temperatures very low. The seedbank was inaugurated on February 28, 2008, with storage capacity for 4.5 million seed samples. It was conceived to keep them viable for a long period of time, so that in the event of natural catastrophe or extreme climate variations, food production can be restarted in any part of the planet. Occasional loss of seeds in *ex situ* collections can also be replaced with samples stored in the seedbank.

The Svalbard Global Seed Vault is designed to store duplicates of seeds from seed collections around the globe. It is not a genebank, but a free-of-charge safety storage, and the seeds in the vault can only be accessed when the original seed collections have been lost for any reason. The depositors of seeds (countries or institutions) still own and control access to their seeds and, therefore, access to seeds by third parties is subject to the depositors' consent. The vault's location takes into account all known scenarios for rising sea levels caused by global climate changes. The facility has also been located so deep inside the mountain that any possible changes to Svalbard's climate, as known about today, will not affect the efficacy of the permafrost. The facility consists of three separate underground chambers. Each chamber has the capacity to store 1.5 million different seed samples. From the opening of the Svalbard Global Seed Vault in February 2008 until February 2011, more than 600,000 unique seed samples had arrived at the Vault from seedbanks all over the world.[11] The Svalbard Global Seed Vault celebrated its third anniversary on February 24, 2011, with the arrival of seeds of rare lima beans, blight-resistant cantaloupe, and progenitors of antioxidant-rich red tomatoes from Peru and the Galapagos Islands.[12]

According to Cary Fowler, Executive Director of the Global Crop Diversity Trust, there are over 1,500 seedbanks worldwide, but only 35–40 percent meet international standards. "The Svalbard seedbank will serve as a backup for other collections throughout the world," explains Cary Fowler. "It is the world's best freezer." According to Fowler, even the most severe climate change will not have strong impacts on the seedbank, as it is located on the coldest part of the mountain, which is one of the coldest places in the world. The bank has become known as "the new Noah's Ark" (Shanahan, 2006; Qvenild, 2008). By its entrance, there is a large metallic sculpture by Norwegian artist Dyveke Sanne, visible from kilometers away, which shines during summer nights, and lightens, with optical fibers, the long Arctic winters. According to Hawtin and Fowler (2011), the cost of construction of the vault, some US$9 million, was funded entirely by the government of Norway, which owns the facility (but not the seeds stored within it) and is responsible for maintaining and administering it. Under the terms of a tripartite agreement between the Norwegian government, the Global Crop Diversity Trust, and the Nordic Genetic Resource Center (NordGen), responsibility for managing the vault lies with NordGen, overseen by an International Advisory Council. All the material in the vault is maintained under "black box" conditions, that is, with ownership and access rights to the material remaining with the depositor. This means that seed packages and boxes sent for storage cannot be opened or sent to anyone except the original depositor and that the responsibility for testing material and for any subsequent regeneration remains with the depositor (Hawtin and Fowler, 2011).

Norwegian legislation forbids entry of genetically modified seeds into the country, as well as the deposit of genetically modified seeds in Svalbard, and some scientists believe that the Svalbard collections may be used in the future for comparison with contaminated seeds in the countries of origin. However, according to a recent Svalbard statement, Norwegian gene technology legislation was formulated before the Svalbard Global Seed Vault was set up, and therefore fails to take into account the vault's special status, or the low risk (of contamination) related to handling seeds in sealed packaging. Until changes can be made to the rules or exemptions can be provided from them, long-term storage of the seeds of genetically modified organisms (GMOs) will not be allowed in the vault. "However, if it becomes apparent at a later date that modification of the facilities to include GMO seeds can be necessary in order to comply with the purpose of the vault, Norway will consider revising its policy and rules," according to the Norwegian government.

The Svalbard seedbank is, however, as is any initiative for *ex situ* conservation of agrobiodiversity, only a partial solution, as a large part of the genetic diversity is maintained on-farm, and has undergone serious erosion. Initiatives and resources aimed at *in situ* and on-farm conservation of agrobiodiversity have, so far, been insufficient. The government of Norway itself, after building the "Noah's Ark," announced that, starting in 2009, it would donate the amount correspondent to 0.1 percent of all its seed sales to initiatives aimed at promoting on-farm conservation and management of agrobiodiversity, through the benefit-sharing fund of FAO's International Treaty on Plant Genetic Resources for Food and Agriculture (ITPGRFA).

Keeping agricultural varieties stored deep-frozen in germplasm banks is only a part (albeit an important one) of agrobiodiversity conservation and climate change adaptation strategies. On-farm and *in situ* conservation of agrobiodiversity are also essential components of such strategies. Climate change requires not only that genetic resources be conserved, but also that they keep adapting to climate change. Plants, animals and ecosystems have the capacity to adjust to changes in factors such as heat, drought, or salinity, but adaptation is a dynamic process brought about through an organism's interaction with its environment, on farmers' fields and considering the wide agroecological variation of sites (GTZ, 2006; Kotschi, 2007). Standardized and genetically homogeneous varieties, produced and distributed on a large scale, do not address local agroecological variations, and do not offer site-specific solutions.

On one hand, poor farmers who live in developing countries (in Africa, Asia, and Latin America) are more vulnerable to climate change, but, on the other hand, they have been continuously (and historically) coping with extreme weather events and climatic variability. Therefore, they have developed and/or inherited complex farming systems that have the potential to bring solutions to many uncertainties brought by climate change, as Altieri and Koohafkan (2008) point out. According to these authors, these farming systems have been managed in ingenious ways, allowing small farming families to meet their subsistence needs in the midst of environmental variability without depending much on modern agricultural technology.

Altieri and Koohafkan (2008) also point out that generations of farmers and herders have developed complex, diverse, and locally adapted agricultural systems, which enable them to generate sustained yields, despite marginal land endowments, climatic variability, and low use of external inputs. One of the salient features of these traditional farming systems is their high degree of biodiversity, in particular the plant diversity in the form of polycultural and/or agroforestry patterns. Many of these agricultural systems offer examples of adaptation measures that can help reduce the impact of climate change on agriculture, such as multiple cropping, wild plant gathering, home gardening, use of local genetic diversity, soil organic matter enhancement, etc. According to Altieri and Koohafkan (2008), "the diversity of these (traditional) agricultural systems and the creativity and knowledge of family farmers and Indigenous communities are assets of great value for solving the daunting problems affecting agriculture in the 21st century."

Notes

1 The Intergovernmental Panel on Climate Change (IPCC) is the leading international body for the assessment of climate change. It was established by the United Nations Environment Program (UNEP) and the World Meteorological Organization. IPCC is a scientific body, and it reviews and assesses the most recent scientific, technical, and socioeconomic information produced worldwide relevant to the understanding of climate change. Source: http://www.ipcc.ch/organization/organization.shtml (accessed January 20, 2010). The IPCC Fifth Assessment Report is under way. On June 23, 2010, IPCC announced that 831 experts had been selected to participate in the report preparation. For more information on experts' meetings, see https://www.ipcc-wg1.unibe.ch/meetings/meetings.html (accessed March 2, 2011).

2 Although widely used, the term "climate refugee" is very controversial. The UN Convention relating to the Status of Refugees (known as the Geneva Convention, approved in 1951) defines a refugee as "a person who, owing to a well-founded fear of being persecuted for reasons of race, religion, nationality, membership of a particular social group, or political opinion, is outside the country of nationality, and is unable to or, owing to such fear, is unwilling to avail him/herself of the protection of that country." The Convention makes

no reference to environmental refugees. Some organizations understand that referring to environmental migrants as refugees might weaken the protection of political refugees, and while political refugees cannot turn to their own government for support, environmental migrants often can (see Panda, 2010).

3 On December 29, 2009, the Brazilian government took a historic step by establishing its National Policy on Climate Change, through Federal Law No. 12,187. According to this law, Brazil will adopt measures to reduce its projected emissions between 36.1 percent and 38.9 percent by 2020 (this equates to a 17 percent reduction in its emissions compared with 2005 levels). The National Policy on Climate Change is divided into five sector plans: (1) 80 percent reduction in deforestation in the Amazon; (2) 40 percent reduction in deforestation in the Cerrado (the Brazilian savanna); (3) actions for the energy sector; (4) for agriculture and livestock; and (5) for industry.

4 EMBRAPA (Brazilian agricultural research corporation) created a climate change platform, aimed at defining its strategies to cope with the impacts of climate change on Brazilian agriculture. For more information see http://www.cnptia.embrapa.br/content/plataforma-de-mudancas-climaticas-estrutura-projetos-de-pesquisa-210808.html (accessed January 10, 2009).

5 Source: 'As armas da ciência frente às mudanças climáticas na agricultura' (2009) (http://www.macroprograma1.cnptia.embrapa.br/climapest/noticias/as-armas-da-ciencia-frente-as-mudancas-climaticas-na-agricultura; accessed February 15, 2009).

6 The Baniwa (Indigenous people) live on the borders of Brazil with Colombia and Venezuela, in villages located on the banks of the Içana River and its tributaries (Cuiari, Aiari, and Cubate), as well as in communities on the Upper Rio Negro/Guainía and in the urban centers of São Gabriel da Cachoeira, Santa Isabel, and Barcelos (in the Brazilian state of Amazonas). The Kuripako, who speak a dialect of the Baniwa language and are kin of the Baniwa, live in Colombia and on the upper Içana (Brazil). Both groups are highly skilled in the manufacture of arumã (aririte) basketry, an age-old art that was taught to them by their creator heroes and which is being commercialized today in Brazilian markets. Recently, they have also become noted for their active participation in the Indigenous movement in the Rio Negro region. This movement includes a cultural complex of 22 different Indigenous peoples who are articulated through a network of trade and are very similar in their social organization, material culture, and cosmovision. The two basic subsistence activities of the Baniwa are agriculture and fishing, which have equal and complementary cultural and economic importance. Baniwa knowledge of the forests is extensive. Every man knows where to find the best lands for crops cultivation and gardening, where to look for fruits, and where to hunt game. Since colonial times, the name "Baniwa" has been used to refer to all peoples who speak Arawakan languages who live along the Içana River and its tributaries. The Baniwa more frequently use as collective self-designations the names of their phratries such as Hohodene, Walipere-dakenai, or Dzauinai. Source: Instituto Socioambiental (http://pib.socioambiental.org/en/povo/baniwa; accessed January 12, 2011).

7 The Hadley Centre second generation coupled ocean-atmosphere GCM (HADCM2) scenario assumes an annual 1 percent increase in atmospheric CO_2 content with effects mitigated by sulfate forcing.

8 'Climate change threatens wild relatives of key crops', *Bioversity International News*, May 18, 2007 (http://www2.bioversityinternational.org/Themes/Crop_Wild_Relatives/index.asp; accessed December 10, 2007).

9 Source: ETC Group (2010a).

10 Permafrost is the type of soil found in the Arctic region. It is made up of earth, ice, and permanently frozen rocks.

11 Source: http://www.regjeringen.no/en/dep/lmd/campain/svalbard-global-seed-vault.html?id=462220 (accessed February 2, 2011).

12 Source: 'Rare, unique seeds arrive at Svalbard Vault, as crises threaten world crop collections' (2011) (http://www.sciencedaily.com/releases/2011/02/110224201859.htm; accessed February 28, 2011). The Peruvian desert lima bean variety (stored in the Norwegian seed vault) is on the verge of extinction and was rescued by the Colombia-based International Center for Tropical Agriculture (CIAT), as well as other lima beans and relatives that grow in very dry or high-altitude locations. In total, CIAT's new shipments include 3,600 bean and forage samples collected from 94 countries, including Afghanistan, Nepal, Yemen, Vietnam, and Zimbabwe. The new accessions also include Agricultural Research Service–US Department of Agriculture (USDA) donations of soybeans collected by USDA researchers in China in the 1920s, and seed collections of *Solanum chilense* and *Solanum galapagense*, wild relatives of the tomato whose genetic material was used by breeders at USDA and the University of California, Davis, to create tomatoes high in lycopene (an antioxidant) and beta-carotene (a source of vitamin A). Other US shipments included seeds for disease-resistant varieties of spinach, maize, and cantaloupe.

Bibliography

Agrobiodiversity Platform (2009) Climate Change Project, 'Coping with climate change: the use of agrobiodiversity by Indigenous and rural communities' (http://www.agrobiodiversityplatform.org/blog/wp-content/uploads/2009/12/PAR_briefing_final.pdf; accessed January 15, 2010).

Altieri, M. and Koohafkan, P. (2008) 'Enduring farms: climate change, smallholders and traditional farming families', Third World Network, Penang, Malaysia (Environment & Development Series no. 6) (http://www.fao.org/nr/water/docs/Enduring_Farms.pdf; accessed January 15, 2010).

Agência Estado (2008) 'Alto nível de CO_2 deixa soja vulnerável a insetos', March 25, 2008 (www.estadao.com.br/geral/not_ger145508,0.htm; accessed March 26, 2008).

Assad, E. and Pinto, H.S. (coords.) (2008) 'Aquecimento global e cenários futuros da agricultura brasileira', Brazilian Agricultural Research Corporation (EMBRAPA) and University of Campinas/Center for Meteorological and Climatic Research Applied to Agriculture (Cepagri) (http://www.agritempo.gov.br/climaeagricultura/CLIMA_E_AGRICULTURA_BRASIL_300908_FINAL.pdf; accessed January 7, 2011).

Borron, S. (2006) 'Building resilience for an unpredictable future: how organic agriculture can help farmers adapt to climate change', FAO/Sustainable Development Department (ftp://ftp.fao.org/docrep/fao/009/ah617e/ah617e.pdf; accessed January 7, 2011).

Brahic, C. (2006) 'Urgent need for rice that tolerates climate change', Science and Development Network, March 29, 2006 (http://www.scidev.net/en/news/urgent-need-for-rice-that-tolerates-climate-chan.html; accessed January 25, 2011).

Brown, M. and Funk, C. (2008) 'Food security under climate change', *Science*, vol. 319, no. 5863, pp. 580–581.

Campbell, A., Kapos, V., Scharlemann, J.P.W., Bubb, P., Chenery, A., Coad, L., Dickson, B., Doswald, N., Khan, M.S.I., Kershaw, F. and Rashid, M. (2009) *Review of the Literature on the Links between Biodiversity and Climate Change: Impacts, Adaptation and Mitigation* (Technical Series no. 42), Secretariat of the Convention on Biological Diversity, Montreal.

Campbell-Lendrum, D., Corvalán, C. and Neira, M. (2007) 'Global climate change: implications for international public health policy', *Bulletin of the World Health Organization*, vol. 85, no. 3, pp. 161–244 (http://www.who.int/bulletin/volumes/85/3/06–039503/en/index.html; accessed January 7, 2011).

Cardoso, J. (2008) 'Um olhar de índio Baniwa sobre a mudança climática', in *Rio Negro, Manaus e as Mudanças no Clima*, Instituto Socioambiental, São Paulo, pp. 15–23.

Carneiro, A. (2008) 'Introdução', in *Rio Negro, Manaus e as Mudanças no Clima*, Instituto Socioambiental, São Paulo, pp. 9–14.

Centre National de Ressources en Agriculture Biologique (ABioDoc) (2008) 'Agriculture biologique et changement climatique, Contribution de l'agriculture biologique et de nos choix alimentaires à l'effet de serre' (CD), ABioDoc, Lemdes, France.

Clement, C., Miranda, I.P., Desmoulière, S. and Oliete, I. (2008) 'Possíveis respostas das palmeiras da bacia do rio Negro em dois cenários de mudanças climáticas ao longo do século 21', in *Rio Negro, Manaus e as Mudanças no Clima*, Instituto Socioambiental, São Paulo, pp. 25–29.

Confalonieri, U. (2007) 'Mudança climática global e saúde', *ComCiência: Revista Eletrônica de Jornalismo Científico*, no. 85 (www.comciencia.br/comciencia/handler.php?section=8&edicao=22&id=237; accessed January 30, 2008).

Consultative Group on International Agricultural Research (CGIAR) (2007) *Global Climate Change: Can Agriculture Cope?* (www.cgiar.org/impact/global/cc_mappingthemenace.html; accessed January 25, 2011).

ETC Group (2010a) 'Gene giants stockpile patents on climate-ready crops in bid to become biomassters' (http://www.etcgroup.org/upload/publication/pdf_file/FINAL_climate-readyComm_106_2010.pdf; accessed January 15, 2010).

ETC Group (2010b) 'The new biomassters: synthetic biology and the next assault on biodiversity and livelihoods' (www.etcgroup.org/upload/biomassters_0.pdf; accessed January 15, 2010).

Food and Agriculture Organization (FAO) (2010) The Hague Conference on Agriculture, Food Security and Climate Change (31 October to 5 November 2010), 'Climate-smart" agriculture: policies, practices and financing for food security, adaptation and mitigation' (http://www.fao.org/docrep/013/i1881e/i1881e00.htm; accessed January 15, 2011).

Food and Agriculture Organization (FAO) and Bioversity International (2008) 'Climate change and biodiversity for food and agriculture' (www.fao.org/fileadmin/user_upload/foodclimate/HLCdocs/HLC08-bak-3-E.pdf; accessed January 15, 2010).

Food and Agriculture Organization (FAO) and Bioversity International (2008) 'Climate change adaptation and mitigation in the food and agriculture sector' (www.fao.org/fileadmin/user_upload/foodclimate/HLCdocs/HLC08-bak-1-E.pdf; accessed January 15, 2010).

Fujisaka, S., Williams, D. and Halewood, M. (eds.) (2009) "The impact of climate change on countries' interdependence on genetic resources for food and agriculture', FAO, Commission on Genetic Resources for Food and Agriculture (Background Study Paper no. 48) (ftp://ftp.fao.org/docrep/fao/meeting/017/ak532e.pdf; accessed January 15, 2011).

Ghini, R. (2005) *Mudanças Climáticas Globais e Doenças de Plantas*, EMBRAPA Meio Ambiente, Jaguariúna, São Paulo.

Global Crop Diversity Trust (2007) 'Engineers begin critical "cooling down" of Arctic Doomsday Seed Vault for deep-freeze and 24-hour polar night', *Global Seed Vault News*, Oslo, November 16, 2007 (http://www.croptrust.org/documents/Press%20Releases/SGSV%20Cooling%20Down%20Release_Final.pdf; accessed January 30, 2011).

Goloubinoff, M., Katz, E. and Lammel, A. (1997) *Antropología del Cima en el Mundo Hispanoamericano* (Tomo 1), Abya Yala, Quito, Ecuador.

Gonzales, T. (2011) 'Peruvian Indigenous Andean-Amazonian communities assessment and strategies to tackle climate change', paper presented at the International Expert Group Meeting on Indigenous Peoples, Marginalized Populations and Climate Change: Vulnerability, Adaptation and Traditional Knowledge, Mexico City, Mexico, July 19–21, 2011.

GTZ (German Technical Cooperation) (2006) 'Agrobiodiversity and climate change – a complex relationship', sector project "People, Food and Biodiversity" (Division 45), Germany.

Guedes, I. (ed.) (2010) *Mudanças Climáticas Globais e a Produção de Hortaliças*, EMBRAPA, Brasília.

Hawtin, G. and Fowler, C. (2011) 'The Global Crop Diversity Trust: an essential element of the Treaty's Funding Strategy' in Frison, C., López, F. and Esquinas-Alcázar, J.T. (eds.) *Plant Genetic Resources and Food Security: Stakeholder Perspectives on the International Treaty on Plant Genetic Resources for Food and Agriculture*, Earthscan, London.

Instituto de Pesquisa Ambiental da Amazônia (IPAM) (2010) 'Perguntas e respostas sobre aquecimento global' (http://www.ipam.org.br/biblioteca/livro/Perguntas-e-respostas-sobre-Aquecimento-Global/572; accessed January 10, 2011).

Instituto Socioambiental (2011) 'Indigenous Peoples of Brazil', (http://pib.socioambiental.org/en; accessed January 12, 2011).

Intergovernmental Panel on Climate Change (IPCC) (2002) *Climate Change and Biodiversity* (IPCC Technical Paper, V), IPCC, Geneva (www.ipcc.ch/pdf/technical-papers/climate-changes-biodiversity-en.pdf; accessed November 15, 2007).

Intergovernmental Panel on Climate Change (IPCC) (2007) *Climate Change 2007*: Synthesis Report. Contribution of Working Groups I, II and III to the Fourth Assessment Report of the Intergovernmental Panel on Climate Change, Geneva (http://www.ipcc.ch/publications_and_data/publications_ipcc_fourth_assessment_report_synthesis_report.htm; accessed January 20, 2009).

Jarvis, A., Lane, A. and Hijmans, R. (2008) 'The effect of climate change on crop wild relatives', *Agriculture, Ecosystems & Environment*, vol. 126, pp. 13–23.

Jarvis, A., Upadhyaya, H., Gowda, C.L.L., Aggarwal, P.K., Fujisaka, S. and Anderson, B. (2010) 'Climate change and its effect on conservation and use of plant genetic resources for food and agriculture and associated biodiversity for food security',

FAO, Rome (http://www.fao.org/docrep/013/i1500e/i1500e16.pdf; accessed January 10, 2010).

Kotschi, J. (2007) 'Agricultural biodiversity is essential for adapting to climate change', *Gaia – Ecological Perspectives for Science and Society*, vol. 16, no. 2, pp. 98–101.

Lobell, D. and Burke, M. (eds.) (2010) *Climate Change and Food Security: Adapting Agriculture to a Warmer World*, Springer, New York.

Lobell, D., Burke, M., Tebaldi, C., Mastrandrea, M., Falcon, W. and Naylor, R. (2008) 'Prioritizing climate change adaptation needs for food Security in 2030', *Science*, vol. 319, no. 5863, pp. 607–610.

Local Initiatives for Biodiversity, Research and Development (LI-BIRD) and Platform for Agrobiodiversity Research (PAR) (2009) 'Climate change and agrobiodiversity in Nepal: opportunities to include agrobiodiversity maintenance to support Nepal's National Adaptation Programme of Action (NAPA)' (www.agrobiodiversityplatform.org/blog?getfile=3537; accessed 10 January 10, 2010).

Magrin, G. and Gay Garcia, C. (coords.) (2007) 'Latin America', in *Climate Change 2007: Impacts, Adaptation and Vulnerability*, Contribution of Working Group II to the Fourth Assessment Report of the Intergovernmental Panel on Climate Change, Cambridge University Press, Cambridge, UK.

Marengo, J.A. (2006) *Mudanças Climáticas Globais e Seus Efeitos Sobre a Biodiversidade: Caracterização do Clima Atual e Definição das Alterações Climáticas Para o Território Brasileiro ao Longo do Século XXI*, Ministério do Meio Ambiente, Brasília.

Miles, L., Grainger, A. and Phillips, O. (2004) 'The impact of global climate change on tropical forest biodiversity in Amazonia', *Global Ecology and Biogeography*, vol. 13, pp. 553–565 (http://www.geog.leeds.ac.uk/projects/rainfor/papers/Miles%20L%20et%20al%20%20GEB%202004.pdf; accessed January 16, 2010).

Moreira, P. and Baniwa, D. (2011) 'Linking global climate change policies to local reality of Indigenous peoples: a preliminary assessment of climate change impacts and needs for adaptation from Baniwa based on the Içana River, Brazilian Amazon, a case study', paper presented at the International Expert Group Meeting on Indigenous Peoples, Marginalized Populations and Climate Change: Vulnerability, Adaptation and Traditional Knowledge, Mexico City, Mexico, July 19–21, 2011.

Moutinho, P., Santilli, M., Schwartzman, S. and Rodrigues, L. (2005) Why ignore tropical deforestation? A proposal for including forest conservation in the Kyoto Protocol. *Unasylva*, vol. 222, pp. 27–30.

Naylor, R., Battisti, D.S., Vimont, D.J., Falcon, W.P. and Burke, M.B. (2007) 'Assessing risks of climate variability and climate change for Indonesian rice agriculture', *Proceedings of the National Academy of Sciences of the United States of America*, vol. 104, no. 19, pp. 7752–7757 (http://www.ncbi.nlm.nih.gov/pmc/articles/PMC1876519/; accessed January 25, 2011).

Padma, T.V. (2007) 'Crop research must switch to climate adaptation', Science and Development Network, November 23, 2007 (http://www.scidev.net/en/news/crop-research-must-switch-to-climate-adaptation.html; accessed January 25, 2011).

Panda, A. (2010) 'Climate refugees: implications for India', *Economic & Political Weekly*, vol. XLV, no. 20, pp. 76–79.

Panduro Meléndez, R. (2011) 'Los campesinos Amazónicos–Andinos y su "conversación" com el cambio climático en la región San Martin', paper presented at the International Expert Group Meeting on Indigenous Peoples, Marginalized Populations and Climate Change: Vulnerability, Adaptation and Traditional Knowledge, Mexico City, Mexico, July 19–21, 2011.

Parry, M., Evans, A., Rosagrant, M.W. and Wheller, T. (2009) *Climate Change and Hunger: Responding to the Challenge*, World Food Programme, Rome.

Qvenild, M. (2008) 'Svalbard global seed vault: a Noah's Ark for the world's seeds', *Development in Practice*, vol. 18, no. 1, pp. 110–116.

Rosenzweig, C., Karoly, D., Vicarelli, M., Neofotis, P., Wu, Q., Casassa, G., Menzel, A., Root, T.L., Estrella, N., Seguin, B., Tryjanowski, P., Liu, C., Rawlins, S. and Imeson, A. (2008) 'Attributing physical and biological impacts to anthropogenic climate change', *Nature*, vol. 453, pp. 353–357.

Royal Society (2008) Science Policy Section, 'Biodiversity–Climate interactions: adaptation, mitigation and human livelihoods', Royal Society, London (www.royalsociety.org/WorkArea/DownloadAsset.aspx?id=5557; accessed January 30, 2008).

Sahai, S. (2010) (ed.) National Conference on Climate Change and Food Security (21–24 April, 2001, New Delhi, India): *Recommendations* (http://www.genecampaign.org/climate-change/cc-recom.pdf; accessed January 25, 2011).

Santilli, M., Moutinho, P., Schwartzman, S., Nepstad, D., Curran, L. and Nobe, C. (2005) 'Tropical deforestation and the Kyoto Protocol: an editorial essay', *Climatic Change*, vol. 71, pp. 267–276.

Shanahan, M. (2006) 'Arctic cave to safeguard global crop diversity', Science and Development Network, London, January 13, 2006 (http://www.scidev.net/en/news/arctic-cave-to-safeguard-global-crop-diversity.html; accessed January 25, 2011).

Silva, A.L, Fontes, O., Cardoso, J. and Rodrigues, I.L. (2010) 'Visões Baniwa sobre as mudanças climáticas'. in Cabalzar, A. (org.) *Manejo do Mundo: Conhecimentos e Práticas dos Povos Indígenas do Rio Negro, Noroeste Amazônico*, Instituto Socioambiental, São Paulo, Federação das Organizações Indígenas do Rio Negro, São Gabriel da Cachoeira, Amazonas, Brazil, pp. 67–75.

Sthapit, B., Padulosi, S. and Mal, B. (2009) 'Role of on-farm/in situ conservation and underutilized crops in the wake of climate change', http://www.bioversityinternational.org/index.php?id=19&user_bioversitypublications_pi1%5BshowUid%5D=4768; accessed January 10, 2010).

Ulloa, A. (ed.) (2011) *Perspectivas Culturales del Clima*, ILSA and Universidade Nacional de Colombia, Bogotá, Colombia.

United Nations Development Programme (UNDP) (2007/2008) Relatório de desenvolvimento humano 2007/2008: 'Combatendo a mudança climática: solidariedade humana num mundo dividido' (http://www.pnud.org.br/rdh/; accessed January 30, 2011).

Weiss, R. (2008) 'Firms seek patents on climate-ready altered crops', *Washington Post*, May 13, 2008.

World Health Organization (WHO) (2009a) 'Protecting health from climate change: connecting science, policy and people' (http://www.who.int/world-health-day/toolkit/report_web.pdf; accessed January 15, 2010).

World Health Organization (WHO) (2009b) '10 facts on climate change and health' (http://www.who.int/features/factfiles/climate_change/facts/en/index.html; accessed January 15, 2010).

World Health Organization (WHO) (2010) 'Climate change and health' (http://www.who.int/mediacentre/factsheets/fs266/en/; accessed January 15, 2010).

Zapata Ferrufino, B., Atahuachi, M. and Lane, A. (2008) 'The impact of climate change on crop wild relatives in Bolivia', *Crop Wild Relative*, no. 6, pp. 22–23 (http://intranet.iucn.org/webfiles/doc/SSC/Gen_docs/e_bulletin_/CWR_6_online_.pdf; accessed February 14, 2009).

Zuanon, J. (2008) 'Peixes, pesca e clima na Amazônia: um ensaio sobre os efeitos das mudanças climáticas globais sobre os recursos pesqueiros na região do rio Negro, Amazonas, Brasil', in *Rio Negro, Manaus e as Mudanças no Clima*, Instituto Socioambiental, São Paulo, pp. 25–29.

4

SEED LAWS

The paradigms of industrial agriculture, traditional/local agricultural systems, and agrobiodiversity

Seeds – we shall use this term in a broad sense, to include all plant propagating materials[1] – contain the entire life of a plant and constitute the basis of agrobiodiversity. The impact of legal systems on agricultural diversity cannot be understood without an analysis of regulations on production, marketing, and utilization of seeds. Seed laws not only impact agricultural systems, but are also directly linked to public policies related to sustainable development, food and nutritional security, social equity, and cultural diversity. Seed laws must, therefore, contemplate the diversity of agricultural systems and social stakeholders involved in food production.

Seed laws aim to ensure the identity and quality of plant propagating materials, regulating how they are produced, used, and marketed. Seed laws must not be confused with intellectual property (IP) laws or plant breeders' rights, which grant ownership rights over new plant varieties and aim to stimulate innovations in plant breeding. However, seed laws and IP rights over plant varieties follow similar logics and share common concepts. Seed laws were adopted as a "legal support" (or component) to the so-called process of "modernization" or "industrialization" of agriculture, whose main paradigms were productivism, standardization of agricultural products, and fragmentation of the many stages involved in agricultural production. This new industrial paradigm promoted high-yield, homogeneous and stable plant varieties, highly dependent on external inputs. Analyzing the historical development of formal/commercial seed systems, Niels Louwaars (2007) explains that they emerged in industrialized countries in the second half of the nineteenth century and rapidly developed after the reinvention of Mendel's laws of heredity[2] in the early twentieth century. They were enhanced by the discovery of the phenomenon of heterosis[3] and the subsequent introduction of hybrid varieties of maize. In the post-Green Revolution period, in the 1960s and 1970s, many developing countries also started to promote formal/commercial seed systems, with wide support from agricultural development programs funded by international organizations. However, formal/commercial seed systems were never able to replace (totally) farmers' seed systems and their locally adapted seeds.

Biologist and science historian Christophe Bonneuil (see Bonneuil *et al.*, 2006) draws attention to the role, in this industrial agriculture model, of what he calls a "fixist (or static) paradigm of variety," which considers a genetically homogeneous and stable plant variety as the "most perfect form of variety." Bonneuil *et al.* (2006) cite a revealing article published in 1944 by the influential French biologist Jean Bustarret, who was one of the main directors of INRA (Institut National de la Recherche Agronomique, the French national agricultural research institute) between 1949 and 1973, and exerted strong influence on plant breeding programs. In this article, called "Variétés et variations," Bustarret distinguishes pure-bred/lineage varieties (*variété-lignée pure*), F1 hybrids, and clones, which are genetically homogeneous, and population varieties (*varietés de pays*), which are genetically heterogeneous, and holds that agricultural research should focus on the first three types of variety. He argues that genetic homogeneity ensures predictability and stability of agronomic and technological values of plant varieties. Population varieties have two disadvantages, according to Jean Bustarret (1944): owing to their genetic heterogeneity, they are "much harder to describe and characterize" than pure-lineage varieties and they are "susceptible to variations over time and space." The "perfect" uniform and stable plant variety became then the paradigm to be adopted and promoted by seed laws.

Bustarret (1944) tended to disregard the role of farmers in the development of local varieties, and considered the latter the result of "natural selection" only. He did not view farmers as active actors in plant development, and drew a clear distinction between professional scientists (the "innovators") and farmers, whom he considered to be merely the users and beneficiaries of science. Bustarret supported the introduction of the criteria of homogeneity, stability, and "distinctive characteristics" as mandatory requirements for plant varieties to be included in the official catalogue[4] and to be legally marketed. Such criteria excluded most local varieties and landraces, and ignored the evolution of plant varieties over time and space and the social, cultural and environmental contexts in which they developed. They were biased towards intensive and large-scale agricultural models, and reduced the range of genetic diversity available to farmers.

Bonneuil *et al.* (2006) explain that the introduction of essays/trials to assess the "agronomical and technological" value of plant varieties also contributed to narrow genetic diversity in farming systems. Such trials evaluated only some characteristics, notably yield and productivity, and did not take into account the diversity of environments, as they were conducted in artificial environments, and made intensive use of chemical pesticides and fertilizers. Furthermore, they were increasingly conducted in laboratories and agronomical research stations, and not on farmers' fields. The evaluation of the "agronomical and technological" value of plant varieties without the participation of farmers and without considering social, cultural, and environmental factors tends to exclude any variety that is not uniform and not adapted to the industrial agriculture model. It reduces the availability of varieties particularly adapted to social, economic, and ecological conditions, especially in marginal areas.

"Modernization" of agriculture promoted the idea that both plant breeding and seed production should be activities performed by specific professional sectors (plant breeders, agronomists, etc.). Farmers were seen as mere producers and consumers of seeds and other industrially produced agricultural inputs. This conception denied the role of farmers as innovators and holders of knowledge and practices that are crucial for the sustainability of agroecosystems and for on-farm conservation of agrobiodiversity. Seeds and varieties developed and produced by farmers, adapted to local conditions, were more and more replaced by "modern" uniform varieties, and agricultural knowledge was increasingly produced off-farm, away from farmers. Official policies did not, however, stop farmers from innovating, selecting, and producing their own seeds, developing new plant varieties and exchanging seeds and agricultural knowledge among themselves, through social networks.

Mainstream conceptions – that the homogeneous and stable variety is "perfect" and the most appropriate for any farming system and that only professional scientists are capable of innovating in agriculture – were the basis for seed laws passed in the post-Green Revolution period. Seed laws attempt to "modernize" agriculture through artificial legal impositions, ignoring the social, cultural, and economic reality of farming systems in developing countries. They do not recognize the existence of complex and diversified local seed systems (also called farmers' systems), which predominate in most developing countries. Formal and local seed systems follow distinct logics and dynamics, and they cater to the needs of different agricultural models.

In spite of having their own peculiarities in each country, seed laws tend to follow a linear approach, influenced by a seed system development paradigm that is well described by Niels Louwaars (2007):[5] laws and public policies must favor the development of "modern" and commercial seed systems, in which private enterprises play a central role in seed production and sales and from which government can withdraw. Governmental policies must encourage private investments (from national and foreign companies) in seeds systems, adopting legal and economic (such as tax benefits) measures to encourage the private sector to take over seed improvement, production, distribution, and sales. The main idea is that seed systems must "evolve" from traditional/local seed systems to "modern" seed systems, which use "modern" technologies and produce uniform high-yielding plant varieties. Farmers' seed systems are considered to be inferior, providing poor-quality seeds, and not warranting major support or investments. Seed laws should, therefore, support the "formal" seed systems and eliminate the "informal" ones (according to such an approach). Niels Louwaars (2007) criticizes this linear approach, and shows that it bypasses the capacities of farmers to deal with seeds. It is built around the idea that scientific knowledge can solve the problems of farming by breeding "improved" varieties and providing "high-quality seeds." Farmers are positioned as recipients of this technology, who only need to be persuaded to adopt the new seeds, and such an approach excludes support for the improvement of other types of seed supply even when farmers are capable of producing seeds of good quality. In addition, there is an assumption that seed systems can or should develop along the same lines for all crops and for all seed users. According to Niels Louwaars (2007), such a seed production model may be successful for a limited number of major (high-value) crops, such as hybrid cereals and industrial crops, such as cotton and soybean, and for certain types of (commercial) farmers. But not for all crops and not for all farmers. Farm-produced seeds are by far the most important source for small-scale farmers in low-input agriculture in developing countries, especially for crops such as Indigenous vegetables, manioc/cassava, yam, sweet potato, and many pulses, important for home consumption and the local market (Louwaars, 2007).[6]

Most seed laws are biased toward the so-called "formal" seed system, and do not take into consideration the role of "local" seed systems (also called "informal"[7]), which are managed and controlled by farmers themselves, and involve the production, multiplication, distribution, exchange, improvement, and conservation of seeds (at the local level). The term "formal" (or "conventional"), when applied to seed systems, emphasizes that seed production, development, and distribution are operated by public and private specialists and institutions, such as agronomical research institutions, plant breeders, producers, processors, traders, and seed certifiers, whose activities are regulated by standard technical norms and methodologies. Formal seed systems operate at the national and international levels, distributing seeds of uniform high-yielding varieties at large scale, and the operators of such systems usually have no commercial interest in distributing seeds of locally adapted varieties in small quantities, to specific geographical regions.

Seed laws must take into consideration the different needs of agricultural systems. They must recognize the value of different seed systems for different situations, and the important role of farmers' systems in the conservation and sustainable use of plant genetic resources for food and agriculture. Seed regulations (on variety release, evaluation, and registration and on seed quality control, production, storage, and distribution) must be reviewed and adjusted to the need to promote on-farm and *in situ* agrobiodiversity conservation, and this is a legally binding obligation of all contracting parties to the ITPGRFA (according to Article 6). Seed regulations must support the development and maintenance of diverse farming systems (and, for that purpose, they must consider the need of diversified seed systems), increase the range of genetic diversity available to farmers (as carriers of genetic diversity, seeds must be diversified), and promote the expanded use of local and locally adapted crops, varieties and underutilized species.

Ricardo Bocci, from Associazione Italiana per l'Agricoltura Biologica (AIAB), Véronique Chable, from Institut National de la Recherche Agronomique (INRA), and Visser (2002) explain the current situation very clearly: seed rules and regulations were devised after World War II and responded to well-defined needs – to increase agricultural production, foster varietal innovation in the seed industry, and provide farmers with good-quality seeds. Some of these requirements are no less important today, but additional ones have emerged and therefore alternatives must be explored to respond to old objectives. The conservation of agricultural biodiversity and marketing of varieties specifically adapted to particular conditions (such as organic or low-input farming, in both developed and developing countries) are important issues that seed laws and regulations must address.

Seed laws in Latin American countries

Walter de Boef (2007) explains that, in local/informal systems, farmers produce their own seeds and are directly involved in both the cultivation and improvement of plant varieties. Farmers control plant genetic resources in an integrated manner and with different goals. According to Boef (2007), farmers' management and selection of varieties, combined with natural processes such as genetic mutation and cross-breeding between varieties and their wild relatives, as well as the influence of the natural environment (including abiotic agents, such as temperature, humidity, and soil type, and biotic agents, such as diseases, pests, and symbionts), determine a system of "continuous crop evolution." Local systems maintain genetic diversity in farmers' fields, as well as plant varieties adapted to specific local conditions, which formal systems cannot do or are not interested in doing. It is also local systems that produce seeds in remote and hard-to-access areas, which formal systems cannot reach. Heterogeneity of seeds and varieties produced by local systems is, on the other hand, what makes them more flexible and capable of adapting to social and environmental changes. Additionally, for low income farmers, the possibility of eliminating the costs of commercial seeds also plays an important role in the choice of local seeds (Boef, 2007).

Local systems are widely predominant in developing countries, where most farmers reproduce their own seeds (of both local and commercial varieties). According to the *Second Report on the State of the World's Plant*

Genetic Resources for Food and Agriculture, launched by FAO in 2009 and published in 2010, many country reports indicated that "informal": seed systems remain a key element in the maintenance of crop diversity on-farm, and can account for up to 90 percent of seed movement. Although seed exchange can take place over large distances, in many cases it appears to be more important locally, especially within traditional farming systems. In Peru, for example, between 75 and 100 percent of the seed used by farmers in the Aguaytia Valley is exchanged within the community, with little going outside (FAO, 2010). In Nepal, formal systems contribute less than 5 percent of the main crop seeds, the rest being produced by farmers themselves (Joshi, 2000), and in 2005 the Seed Act of 1988 was changed in favor of participatory plant breeding. The changes included the introduction of farmers' perception data, organoleptic taste data, and data from participatory assessment as criteria for variety registration; national listing (registration) of landraces and local crop varieties, including farmers' varieties; and provisions for production and marketing of farmers' varieties – both notified and non-notified (Shrestha, 2009).

FAO estimates that approximately 75 percent of the seeds used by farmers in Latin American and Caribbean countries are supplied through local systems (which FAO calls "informal systems"). Local systems include methods such as retaining seed on-farm from previous harvests to plant the following season and farmer-to-farmer seed exchange networks. Despite having been left out of governmental or donor efforts geared toward improving the seed sector over the past three decades, local seed systems still prevail in Latin American and Caribbean countries (FAO, 2001). Most investments made by governmental, bilateral, and multilateral organizations are focused on formal seed systems.

FAO (2001) points out that the level of reliance of farmers (in Latin American and Caribbean countries) on informal seed systems varies depending on country, farming community, and crop. Major reasons for the use of informal seed systems include the fact that formal seed systems often do not produce seeds of the local varieties that are important to farmers, perhaps because they are considered to be economically nonviable, especially in the case of vegetatively propagated crops. It is difficult, under the prevailing conditions of most countries in the region, for the formal seed sector to reach every farmer in the region. Besides, most improved varieties produced by formal seed systems are targeted to commercial farmers in well-endowed areas where rainfall, irrigation, and other agricultural inputs are easily available, rather than to resource-poor farmers working in marginal lands and those living in more remote areas.

FAO (2001) alerts Latin American countries to the need to protect the interests of small farmers, especially those living in harsh environments, as their main subsistence crops are unlikely to attract private investors' interests. FAO (2001) also points out that in local systems farmers share, exchange, or sell, at low prices, seeds to other farmers, and the advantages brought by low prices, adaptability, and easy access tend to compensate for any qualitative differences between these and commercial seeds. For such reasons, FAO (2001) recommends:

> It is of paramount importance for countries in the Latin America and Caribbean region to create conditions conducive to the development of both the formal system – public and private – and informal seed supply systems. *It is advisable for governments in the region to not only recognize the importance of informal seed systems, but also to introduce the necessary policies and actions to stimulate their growth* (emphasis added).

FAO has also introduced the concept of "Quality Declared Seed System," which makes use of resources already available in seed production organizations, especially in Asia and Africa, to increase the availability of high-quality seed for agricultural communities (a semiformal system).[8]

Trust and reciprocity relationships among farmers are also important factors to explain the predominance of local seed systems in Latin American countries. Lone Badstue (2007) performed an interesting study in the central valleys of Oaxaca, in Mexico – a center of genetic diversity for maize – focusing on the importance of social relations in the exchanges of seeds and the central role played by trust and reciprocity in traditional seed systems. Many farmers in the central valleys of Oaxaca consider it far more risky to buy seeds in a store than to obtain them in their community, where people know each other and are responsible for the quality of seeds they donate, exchange, or sell. Badstue found that farmers have little trust in salespersons in agricultural stores, believing that the latter will attribute any problems with seeds to the failure of the farmer to sow them properly or to the lack of irrigation; thus, farmers prefer to trust other farmers living in the same or in other local communities (Badstue, 2007).

Despite the wide recognition of the relevance of farmers' seed systems, Latin American countries have been increasingly adopting seed laws aimed at promoting formal seed systems, which leave little (if any) legal space for farmers' and local seed systems. Despite varying considerably in each country, such laws tend to favor the growth of the private seed sector, and establish mandatory seed registration and certification requirements that can be met only by the large seed industry. With a few exceptions, farmers' seed exchanges and local seed sales are outlawed, and strong penalties are imposed upon those who violate seed laws. Some examples of this trend are the new Mexican seed law (Ley sobre Producción, Certificación y Comercio de Semillas), published in June 15, 2007, which replaced the seed law of 1991; the Peruvian seed law (Ley General de Semillas) No. 27262, published in May 13, 2000, regulated by Supreme Decree 026-2008-AG; the

Ecuadorian seed law (Codificación de la Ley de Semillas), approved in 2004; the Venezuelan seed law (Ley de Semillas y Material para la Reproducción Animal e Insumos Biológicos), approved in 2002; and the Costa Rican seed law (No. 6289, of 1978), which is being completely revised by the National Congress (Expediente 16098 – Reforma a la Ley Nacional de Semillas). Chile is also revising its current seed legislation (Decreto Ley No. 1764, del Ministerio de Agricultura: normas para la investigación, producción y comercio de semillas) to impose stricter rules on registration and certification of seeds.[9] Brazil also enacted a new seed law in 2003, but it is among the few Latin American countries that have ensured (limited) legal space for farmers' seed systems, which will be discussed next.

The Brazilian seed law and traditional, local, and creole plant varieties

Brazilian agriculture is characterized by a sharp duality of agricultural models: family farming (also called peasant, small-scale, or traditional farming) and agribusiness. The coexistence of these two agricultural models creates divergent political, social, and economic interests that have important consequences for the definition of agricultural policies and regulations, including seeds laws. Historically, two very distinct models of agricultural production have developed. (In fact, there are not just two distinct models, but a large plurality of agricultural models in Brazil, but we are going to describe the two main broad categories.)

1 Family farming takes a large diversity of forms and expressions, such as Indigenous, traditional, and agroecological agricultural systems. Family farming produces most of the food consumed nationally, and it is one of the main driving forces in local economies, accounting for 87 percent of manioc/cassava consumed in the country, 70 percent of beans, 59 percent of pork and 50 percent of poultry, 58 percent of milk, 46 percent of maize, 38 percent of coffee, 34 percent of rice, and 21 percent of wheat. Soybean is the crop that involves less participation of family farming in its production (16 percent).

   Family farming employs 74.4 percent of rural labor, despite occupying only 24.3 percent of the total arable land in Brazil, according to the Secretariat of Family Farming, of the Brazilian Ministry for Agrarian Development.[10] According to Law No. 11.346/2006, which establishes the national policy on family farming, the following criteria determine which farmers are considered to be "family farmers" or "rural family entrepreneurs": their properties may not have more than four fiscal modules (they vary from one region to another), they must be explored mainly by farmers themselves, and their families and their main sources of income must originate from their rural properties. The definition also includes Indigenous and other traditional communities, such as artisanal fisherfolk, rubber tappers, and other extrativists. According to the National Program to Strengthen Family Farming (Programa Nacional de Fortalecimento da Agricultura Familiar – PRONAF), "traditional farmers" constitute approximately 15 percent of the total Brazilian agricultural population. Traditional, family, and agroecological farmers are the main producers and users of local, traditional, and creole varieties.

2 Agribusiness is aimed mainly at exporting agricultural products, generating income, and increasing the surplus in the Brazilian trade balance. Agribusiness accounts for approximately one-third of the Brazilian GNP (Gross National Product) and soybean, the country's main agricultural export product, occupies 45 percent of all arable land in the country. From April 2010 to March 2011, the value of exports of soybean and its products increased by 26.3 percent, according to the Brazilian Ministry of Agriculture.[11] Agribusiness is based on large monocultural crop systems, with intensive use of chemicals, fertilizers, and other external inputs, and produces commodities whose prices are set by the international market (soybean, corn, wheat, cotton, coffee, etc.).

Owing to this duality (plurality) of agricultural models, Brazil has two ministries responsible for agricultural and agrarian development policies: (1) the Ministry of Agriculture, Livestock and Food Supply (Ministério da Agricultura, Pecuária e Abastecimento – MAPA), whose mission is to stimulate the growth of agricultural production and agribusiness; and (2) the Ministry of Agrarian Development (Ministério do Desenvolvimento Agrário – MDA), responsible, among other things, for the implementation of the National Program to Strengthen Family Farming (Programa Nacional de Fortalecimento da Agricultura Familiar – PRONAF). Although both ministries are part of the federal administration, they frequently promote contradictory and conflicting policies, as they represent the political and economic interests of different stakeholders involved in agricultural production.

Such conflicts of interests are also reflected in the Brazilian seed law. Currently, Federal Law No. 10,711, of August 5, 2003,[12] regulates the National Seed System, which is aimed at "ensuring the identity and quality of plant propagating materials, produced, sold and used in the national territory." Despite being a legal instrument aimed at regulating the Brazilian "formal" seed system, it creates some legal space for farmers' varieties, which are called "local, traditional or creole." Such legal space was incorporated into the law as a result of strong pressure from farmers' and civil society organizations during discussions and public hearings held in the National Congress.

Some studies carried out in Brazil show the magnitude of the social, economic, and cultural importance of "local" (or "informal") seed systems. According to the Brazilian Seed Association (Associação Brasileira de Sementes e Mudas – ABRASEM, 2008), which represents the largest producers of seeds in Brazil, the proportions of seed used by farmers in the 2006/2007 harvest that came from the "formal" system were as follows: cotton 49 percent, rice 43 percent, beans 15 percent, maize 85 percent, soybean 50 percent, sorghum 74 percent, and wheat 71 percent. This means that seeds produced by local systems accounted for 51 percent of cotton, 57 percent of rice, 85 percent of beans, 15 percent of corn, 50 percent of soybean, 26 percent of sorghum, and 29 percent of wheat. In the 2007/2008 harvest, the proportion of seeds produced by "formal" systems decreased in the case of nearly all crops (except soybean and sorghum). According to ABRASEM (2008a), the percentages of seeds produced by the formal sector that were used in the 2007/2008 harvest were as follows: cotton 44 percent, rice 40 percent, beans 13 percent, corn 83 percent, soybean 54 percent, sorghum 88 percent, and wheat 66 percent. In other words, local systems are responsible for the supply of the majority of seeds for Brazilian crops, and the use of seeds produced by the formal system has been decreasing. Local seed systems are especially important for traditional, family, and agroecological farmers.

Next we will see what the Brazilian seed law says about local varieties and what are the main bottlenecks and obstacles for the enforcement of its provisions:

Legal definition of local, traditional, or creole cultivars

The Brazilian seed law determines that "local, traditional or creole" cultivars are:

> varieties developed, adapted or produced by family farmers, agrarian reform settlers or Indigenous peoples, with well established phenotypical characteristics, that are recognized by their respective communities and which, according to the Ministry of Agriculture, and considering also social, cultural and environmental describers, are not characterized as substantially similar to commercial cultivars.
>
> (Article 2, XVI, of Law 10,711/2003)

In spite of recognizing local seeds in an inclusive manner (they comprise *all* varieties developed, adapted, or produced by such farmers), the law leaves it up to the Ministry of Agriculture to define which plant varieties are not "substantially similar to commercial cultivars" and therefore can be considered to be local. There is no consensus on what must be the genetic distance between local and commercial varieties. This raises questions especially in relation to improved/commercial varieties that have been further adapted and developed by local farmers. Even though the intent of the law is, clearly, to include not only local/traditional plant varieties and landraces, but also improved/commercial varieties that have become locally adapted, it is not clear how the difference between them will be determined.

Many family farmers argue that it should be up to local farming communities (with the participation of official agencies) to define the criteria for identification and characterization of varieties that were developed, produced, or adapted to local and specific social and environmental conditions, as well as criteria to set them apart from commercial cultivars. The seed law requires that social, cultural, and environmental describers be taken into consideration, rather than just agronomical and botanical describers, but does not specify what such describers are, making them also subject to different interpretations. The Ministry of Agriculture has not yet published any regulatory instruments on local varieties, and it remains unclear whose responsibility it is to decide which varieties can be considered "local, traditional or creole" and "not substantially similar to commercial cultivars."

Laure Emperaire (2008) points out that the notion of local variety, or local cultivar, varies according to the cultural context in which it is used. She gives manioc/cassava as an example: for the geneticist, a manioc/cassava variety, which is a plant of vegetative propagation, is a clone, that is, the variety is made up of genetically identical individuals. For the local farmer, a variety is constituted of individuals with morphological traits considered sufficiently similar (among them) and different from others to constitute a management unit and receive a name. The local notion of variety can be very different from the one accepted by legal instruments. Nivaldo Peroni (2007) explains that, among the *caiçaras* (traditional communities living in the Brazilian Atlantic forest; *caiçaras* are artisanal fishermen and agriculturists of mixed African, European, and Indigenous descent), 58 local names were identified for manioc/cassava varieties, which correspond both to varieties with the same name but different genotypes and to varieties with the same genotype but different names. According to him, this happens because farmers often disregard small morphological variations in manioc/cassava varieties, and identify them solely by the most outstanding traits, making it relatively common to find varieties which, in reality, belong to the same genotype family and have some degree of genetic differentiation, but high morphological similarity.

Exemption from official registration for local, traditional, or creole cultivars

The Brazilian seed law also sets forth that registration of local, traditional or creole cultivars used by family farmers, agrarian reform settlers or Indigenous peoples in the National Registry of Cultivars (Registro

Nacional de Cultivares) is not mandatory (Article 11, paragraph 6, of Law 10,711/2003). Such exemption recognizes the specificities of local cultivars, and the difficulties of meeting the requirements set by the National Registry of Cultivars (especially homogeneity and stability). However, as it is not clear what varieties may be considered "local, traditional or creole," there are doubts also as to which varieties are not required to be registered officially in order to be produced or sold.

Exemption from official registration for family farmers, agrarian reform settlers, and Indigenous peoples

According to the Brazilian seed law, family farmers, agrarian reform settlers and Indigenous peoples who multiply seeds for distribution, exchange or sales among themselves do not have to be registered in the National Seeds and Seedlings Registry (Registro Nacional de Sementes e Mudas, RENASEM), in accordance with Article 8, paragraph 3, of Law 10,711/2003. This means that as long as distribution, exchange, and trade of seeds take place among family farmers, agrarian reform settlers, and Indigenous peoples themselves, there is no need for them to be registered. There are two official registries in Brazil: one is the National Registry of Cultivars (Registro Nacional de Cultivares), mentioned above, where plant varieties are registered in order to be produced and sold, and the second one is the National Seeds and Seedlings Registry (Registro Nacional de Sementes e Mudas, RENASEM), where all (natural or legal) persons who produce and sell seeds must be registered. According to the seed law, local, traditional, or creole cultivars do not have to be registered in the National Registry of Cultivars, and family farmers, agrarian reform settlers, and Indigenous peoples (who grow seeds for distribution, exchange, or sale among themselves) do not have to be registered in National Seeds and Seedlings Registry.

However, Decree 5,153 of 2004, which regulated the seed law, created the following restriction: when seeds are distributed by farmers' *organizations*, they can take place only among members of such farmers' organizations. That is, family farmers, agrarian reform settlers and Indigenous peoples, when acting as *individuals*, can distribute, exchange, or sell seeds among themselves; however, their *organizations* can distribute seeds only among farmers who are members of these specific organizations, and farmers' *organizations* cannot sell seeds (without being registered) in any circumstance. Many farmers' organizations have argued that this restriction is illegal, as the decree, an administrative act, creates restrictions that do not exist in the law. They also argue that such restriction violates the constitutional principle of freedom of association. The seed law states that family farmers, agrarian reform settlers, and Indigenous people can distribute, exchange, and sell seeds among themselves, and does not make any restriction in relation to the possibility of these activities being carried out through farmers' organizations. Thus, the legality of the decree is being questioned by several farmers' organizations and civil society organizations.

Prohibition of restrictions on inclusion of local varieties in public programs

The seed law (Article 48) prohibits any restrictions on the inclusion of seeds of local, traditional, or creole cultivars in publicly funded seed programs. Such legal provision was an important achievement for local farmers, as the previous Brazilian seed law (6,570 of 1977) did not acknowledge local seeds, which were treated merely as "grains." This made it difficult for public policies to support initiatives aimed at recovering, improving, and reintroducing local, traditional, or creole seeds developed by farmers' organizations. Legal recognition made possible government support for various projects and initiatives developed by nongovernmental organizations (NGOs) and farmers' organizations. It also allowed the creation, in 2003, of an important program called Programa de Aquisição de Alimentos (PAA, Food Acquisition Program), which is developed by CONAB (a public agency that is responsible for ensuring food supply and distribution). Through this program, the federal government buys agricultural products from family farmers and agrarian reform settlers, for reasonable prices, and distributes them among community kitchens and restaurants, public elementary schools and daycare centers, hospitals, and charitable institutions. The same program also buys seeds of local varieties (mainly of food crops, such as beans, maize, and rice) from family farmers and agrarian reform settlers to distribute them among other farmers.[13]

In the 2004/2005 and 2005/2006 harvests, many farmers who used local cultivars and made use of public rural credit and insurance lost their crops as a result of a severe drought in the center-south region of Brazil. These farmers faced several difficulties in accessing rural insurance because they had used local, traditional, and creole cultivars. According to the official rules, rural insurance can be accessed only by farmers who used cultivars included in the climatic risk agricultural zoning made by the Ministry of Agriculture, and only registered cultivars can be included in the agricultural zoning. However, according to the seed law, local, traditional, or creole cultivars do not have to be registered, and an impasse was created, as there was a contradiction between the seed law provisions on local varieties and agricultural zoning established by the Ministry of Agriculture, which allowed only the use of registered cultivars. In the 2004/2005 harvest, the federal government authorized insurance coverage for farmers who planted cultivars not included in the farming zoning established by the Ministry of Agriculture. In the 2005/2006 harvest, the same thing happened. On July 18, 2006, the Ministry of Agrarian Development created a national register of organizations which "develop recognized work with the recovery, management and/or conservation of local, traditional or creole cultivars." Currently, such register is regulated by Administrative Directive No. 51, of October 3, 2007, edited by the Ministry of Agrarian Development. In order to be registered, organizations must exist legally for at least two years and describe at least two of their activities involving recovery, management, and/or conservation of local, traditional, or creole cultivars. Registered organizations must state which local, traditional, or creole cultivars they have been working with, their basic characteristics, and the geographical regions to which they are adapted, as well as appoint technical experts responsible for such information. This register was established to facilitate insurance coverage (in case of losses resulting from climatic events) for farmers (and their organizations) who use traditional, creole, or local cultivars. The organizations and the local cultivars that these organizations work with, and not farmers themselves, are the ones registered. This register must not be confused with the other two registries described above (National Registry of Cultivars and National Seeds and Seedlings Registry), as they were created with different purposes.

Once the organizations working with local, traditional or local cultivars are registered (*as organizations*), they can also register the cultivars that they work with. According to Administrative Directive 51/, 2007, local, traditional, or creole cultivars (1) must have been developed, adapted, or produced by family farmers, agrarian reform settlers, or traditional or Indigenous communities; (2) must have phenotypical characteristics which are well established and recognized by these communities; (3) must have been used by farming communities for more than three years; (4) cannot be genetically engineered or contain any transgenes (only non-GMO cultivars can be registered); and (5) cannot be hybrids developed out of the control of local farming communities. Directive 51/2007 also establishes that local, traditional, or creole cultivars, owing to their nature and tradition, constitute a social and cultural heritage of local communities, and cannot be appropriated through patents or other IP rights.

Some farmers' organizations have criticized the registration because they believe that it may "freeze" local seeds, which are characterized by their evolution through time and space. These varieties are essentially dynamic, subject to continuous evolution and transformation processes. In addition, different varieties may bear the same name in distinct regions, and the same variety may have distinct names in the same or different places, as they are constantly exchanged among local communities. The Ministry of Agrarian Development considers, however, that registration is necessary not only to meet the demands for rural insurance, but also to identify experiences involving local, traditional, and creole cultivars and to develop public policies targeted to them.

Despite providing some legal space for local, traditional, and creole cultivars, and for the farming systems that need them, the Brazilian seed law was conceived mainly to regulate the formal seed system. It contemplates local seed systems in some specific provisions, but establishes strict conditions for seed production and marketing that can only be met by large seed industries. Small seed producers faced enormous difficulties in meeting the requirements established by the seed law, as they are excessively costly for small-scale seed production targeted to local markets. The seed law not only benefits formal seeds systems, but it also favors large seed companies, by imposing heavy bureaucratic conditions that only they can fulfill. The impact on agrobiodiversity is perverse: seeds of varieties that are adapted to specific social and environmental conditions are no longer produced (or used) and only commercial/uniform varieties are produced and sold on a large scale. The costs/burdens involved in the maintenance of the technical infrastructure required by the seed law are not economically feasible for small seed enterprises. They can be compensated only through sales of large quantities of seeds, across the national territory.

Another difficulty is that, for a cultivar to remain registered in the National Registry of Cultivars, there must be at least one maintainer/provider who is responsible for making available a minimum supply of the cultivar's propagating material. If, for any reason, such minimum of seeds can no longer be supplied, the cultivar is withdrawn from registry, and can no longer be produced or sold. Protected varieties can be registered only by plant breeders (who own the varieties). The registration of cultivars in the public domain, on the other hand, may be requested by any person capable of making available a minimum supply of the cultivar's propagating material. When registered cultivars fall into the public domain, seed companies no longer have an interest in keeping them on the market, as they will not receive any royalties for their use. Thus, farmers lose access to these plant varieties, and can no longer use them. Access to plant varieties is determined by the commercial interests of large private seed companies, and not by their importance to farmers.

Another negative aspect of the Brazilian seed law (from the point of view of small-scale and local farmers) is how it regulates the use of farm-saved seed. According to Article 2, XLIII, farm-saved seed is defined as "the amount of plant propagating material saved by the farmer, at each harvest, for seeding or planting exclusively in the next harvest, in his own property, or in any other property which he owns, in accordance with the parameters set by the National Registry of Cultivars." The use of farm-saved seed is widespread not only in Brazil and other Latin American countries but also in industrialized countries such as France, Germany, and the United States, especially regarding self-pollinating species. Saving of seeds for sowing in the next harvest is a tradition among most family farmers and includes different crops. Legal safeguarding of this practice is essential for local systems and for the conservation of the diversity of species, varieties, and agroecosystems.

The Brazilian seed law allows the use of farm-saved seed, but restricts this practice to the next harvest and limits the quantity of seeds that can be saved. If the seed law aims to ensure the identity and quality of seeds produced and used in the country, it does not make sense to impose any restrictions on farmers' rights to save part of their seeds, at each harvest, for sowing in future harvests. After all, if farmers selected some of their seeds to be used in future harvests, no one knows better than them the quality of such seeds. Fortunately, Decree 5153/2004, which regulates the seed law, sets forth, in its Article 115, that the above-mentioned conditions (for the use of farm-saved seed) *do not apply* to family farmers, agrarian reform settlers, and Indigenous people who multiply seeds for distribution, exchange, or trade among themselves. Other small-scale farmers, however, will have to deal with such legal restrictions on the use of farm-saved seed.

Seed laws must adopt one of the following options: either they must regulate only formal seed systems, leaving local and farmers' seed systems out of their scope, or they must regulate formal and local seed systems differently, considering their specificities and distinct characteristics. If only seeds of commercial uniform varieties are allowed to be produced, the use of locally adapted crops will be greatly discouraged and may even be illegal in some circumstances. Seeds determine, in large part, the agricultural model to be adopted and, if only seeds of uniform high-yielding varieties are made available, this will be the only farming system imposed on all farmers, with great losses for agrobiodiversity, farmers, and consumers. The main justification for a strict control over the use, production, and trading of seeds has been the risk of diseases – especially dissemination across different regions, and the need to insure the genetic purity and capacity for germination and vigor of seeds. This is a solid argument, and it certainly must be taken into consideration. However, local seed systems use mainly locally adapted varieties, which are distributed and sold at the local level, and generally do not involve distribution over large distances or across different geographical regions. It is important to evaluate what benefits such strict controls really bring to farmers, and compare them with the difficulties that they create for them.

Furthermore, seed laws were made for long supply chains, in which production, distribution, and trading of seeds involve many intermediaries between producers and consumers, in which seed consumers (farmers) have no direct relations with producers (large seed producers and traders), as Anvar (2008) explains. The distance between seed producers and consumers is enormous, and no trust, collaboration, or reciprocity exists between them. Strict regulations, established for long supply chains, are not justified, however, when seeds are produced and sold at the local level and farmers have direct access to seed producers. These rules are even less justifiable when farmers produce their own seeds or receive them from other local farmers, through their social networks or through short supply chains, such as direct sales. If, originally, the intent of seed laws was to avoid the dissemination of poor-quality seeds, such original purposes have been overstepped. New priorities, such as agrobiodiversity conservation and use, must be taken into consideration and determine the revision of seed laws.

Seed laws and the limited legal space left for local systems are obstacles to the adoption of a "sustainable" agricultural model. Agrobiodiversity is an essential component of sustainable agricultural systems, and each agro-ecosystem has its distinct characteristics, requiring specific solutions and adaptation to social and environmental conditions. The choice of seeds is determined not only by agronomical, but also social, cultural, and environmental factors. In order for farmers to be able to choose their seeds, public policies must promote widespread diversification of seeds and leave more legal space for local systems, rather than impose, artificially, a single system.

The European directives on conservation varieties, the Italian regional laws, and seed laws in Switzerland and Norway

In the European Union (EU), only certified seeds of officially registered varieties can be marketed commercially,[14] although local, small-scale, non-commercial exchange of planting material remains quite common. A recent research revealed that traditional varieties and landraces of horticultural crops, legumes, and grains are still extensively planted by farmers and gardeners throughout Europe and they are often found in the home gardens of rural households. Invaluable diversity of traditional varieties of many crops, especially of fruits and vegetables but also of maize and wheat, is still available, even in countries where modern commercial varieties dominate the seed systems, crop fields, and commercial orchards (FAO, 2010). Locally adapted varieties, old landraces, and mixed populations play an important role in organic and low-input agriculture in European countries. However, strict EU seed regulations create several difficulties for farmers and gardeners in accessing traditional and local varieties, which leads to genetic erosion and loss of agrobiodiversity.

On June 20, 2008, the European Commission published Directive 2008/62, which provided for a certain flexibility in the registration and marketing of *agricultural* landraces and varieties that are "naturally adapted to the local and regional conditions and threatened by genetic erosion," called "conservation varieties."[15] On November 26, 2009, a second directive (Directive 2009/145) established special rules for the registration and marketing of *vegetable* landraces and varieties which "have been traditionally grown in particular localities and regions and are threatened by genetic erosion, and of vegetable varieties with no intrinsic value for commercial crop production but developed for growing under particular conditions"; these are also called "conservation varieties." EU directives must be transposed to national legal systems, in order to be implemented in each country. In transposing EU directives, countries have some flexibility to adapt them to their local conditions.

Directive 2008/62 considers that (1) in order to ensure *in situ* conservation and the sustainable use of plant genetic resources, landraces and varieties that are naturally adapted to local and regional conditions and threatened by genetic erosion (conservation varieties) must be grown and marketed even where they do not comply with the general requirements as regards the acceptance of varieties and the marketing of seeds and seed potatoes; and (2) in order to achieve that objective, it is necessary to provide for derogations as regards the acceptance of conservation varieties, for inclusion in the national catalogues of varieties of agricultural plant species as well as for the production and marketing of seeds and of seed potatoes of those varieties. The directive allows EU countries to adopt derogations of seeds regulations to include such "conservation varieties" in their national catalogues of varieties of agricultural plant species, and to allow production and marketing of seeds of those varieties. EU countries may adopt specific provisions on the criteria of distinctness, stability, and uniformity, as long as minimum standards are observed, that is, if the uniformity level is established on the basis of off-types, a population standard of 10 percent must be applied (that is, the variety must be 90 percent uniform). A landrace is defined as "a set of populations or clones of a plant species which are naturally adapted to the environmental conditions of their region." Genetic erosion is defined as "loss of genetic diversity between and within populations or varieties of the same species over time, or reduction of the genetic basis of a species due to human intervention or environmental change."

According to Directive 2008/62, to be included in national catalogues (so that they can be produced and marketed), the region or regions of origin (where the varieties have historically been grown and are naturally adapted to) of conservation varieties must be identified. When the regions of origin are located in more than one country, there must be an agreement between them. Seeds of conservation varieties can be produced only in their regions of origin. In exceptional situations (owing to specific environmental problems), countries may approve additional regions for seed production. However, these seeds (produced in additional regions) can only be used in their regions of origin. Marketing of seeds of conservation varieties can take place only in their regions of origin, but countries may approve additional regions in their own territories, if those regions are comparable to the regions of origin as regards the natural and seminatural habitats of those varieties.

The directive also sets limits on the quantity of seeds of conservation varieties which can be marketed: they cannot exceed 0.5 percent of the seed of the same species used in the country in one growing season, or the quantity necessary to sow 100 hectares, whichever is the greater quantity. For some species – peas (*Pisum sativum*), wheat (*Triticum* spp.), barley (*Hordeum vulgare*), maize (*Zea mays*), potatoes (*Solanum tuberosum*), canola (*Brassica napus*), and sunflower (*Helianthus annuus*) – that percentage cannot exceed 0.3 percent or the quantity necessary to sow 100 hectares, whichever is the greater quantity. Furthermore, the total quantity of seed of conservation varieties marketed in each country cannot exceed 10 percent of the seed of the species concerned used yearly in the country. In cases where this leads to a quantity lower than that necessary to sow 100 hectares, the maximum amount of seed of the species used yearly in the country may be increased to reach that necessary to sow 100 hectares.

Varieties that are already listed in the European common catalogue of varieties of agricultural plant species (as a variety other than a conservation variety), or which were deleted from the common catalogue within the last two years, cannot be accepted for inclusion in the national catalogues (as conservation varieties). The same applies to varieties protected by IP rights or when applications for their protection have already been filed (but are still pending).

According to the directive, no official examination will be required (for the inclusion of conservation varieties in national catalogues) if there is sufficient information on: the description of the conservation variety and its denomination; the results of unofficial tests; knowledge gained from practical experience during cultivation, reproduction and use; and other information, in particular from the plant genetic resource authorities or from organizations recognized for this purpose by the EU member states. In addition, the conservation variety may have more than one denomination/name, if the names concerned are historically known. This is a derogation from the general principle of seed regulations that each variety must have *one* denomination/name.

Directive 2009/145 regulates conservation varieties of *vegetable* species. It reproduces most of the rules established by Directive 2008/62, but adds a new category: vegetable varieties with no intrinsic value for commercial crop production but developed for growing under particular conditions. According to Directive 2009/145, in addition to the general aim of protecting plant genetic resources, the particular interest of preserving such varieties (with no intrinsic value for commercial crop production but developed for growing under particular conditions) lies in the fact that they are apt to be grown under particular climatic, soil, or agro-technical conditions (such as manual care, repeated harvesting). Therefore, countries may also adopt special rules for their production and marketing, especially regarding distinctness, stability, and uniformity. For varieties developed for growing under particular conditions, "quantitative restrictions must take the form of requiring seed to be marketed in small packages, and the relatively high cost of the seed sold in small packages having the effect of a quantitative limitation."

Directives 2008/62 and 2009/145 resulted mainly from the efforts and public campaigns promoted by European civil society organizations, working in networks, such as the Réseau Semences Paysannes, in France, the Red de Semillas Resembrando e Intercambiando in Spain, and Rete Semi Rurali in Italy, among others. Nevertheless, most of these organizations have found that the final texts of these directives impose too many geographic, quantitative, and packaging restrictions. They call attention, for instance, to the fact that if only 10 percent of off-types are allowed (to meet the uniformity requirement for conservation varieties), this excludes any population reproduced through successive multiplications and not from a return to the basic lines, particularly in the case of cross-pollinating species. They argue that strict uniformity standards should not be applied, since the objective of registering conservation varieties is to promote the sustainable use of diversity and that *identifiability*, and not uniformity, must be the primary concern. Moreover, if varieties must be "historically" or "traditionally" grown in an identified area, this precludes recent farmer selection or evolution of traditional varieties.

The European Commission has funded a program called "Farm seed opportunities: opportunities for farm seed conservation, breeding and production," which brings together a consortium of 12 organizations comprising research institutions and organic and seed associations in six European countries, and started its activities in 2007. This program was conceived to support the implementation of seed regulations and to propose several regulation scenarios aiming at conserving agrobiodiversity and promoting on-farm conservation and participatory plant breeding. The program aims to contribute to the recognition of the role of (European) farmers in conserving genetic diversity through the use of landraces and the breeding of new varieties (i.e., farmers' varieties). It shows the variability of agricultural systems in different European countries, and supports the diversification of seed laws. This program considers the European directives as important steps toward meeting the objectives of agrobiodiversity conservation, but calls attention to some of the bottlenecks of the European directives, such as:

1 To be listed in catalogues, conservation varieties must be at risk of genetic erosion, but this would be very difficult to define and express in numerical terms. The variability of each local variety would have to be known – and these are often fairly heterogeneous populations – to estimate the risk of intravarietal erosion. Besides, if the seed of a conservation variety is sold in conformity with all the rules, can it still be considered at risk of genetic erosion?

2 Linking a conservation variety to a certain region of origin may be complicated in some circumstances, since plants have always traveled with humans. The dynamic of crops is not taken into account, and the processes of varietal adaptation and evolution are not considered. Moreover, many old varieties, which may be thought to be local varieties, originate from elsewhere. The program has concluded that legal space is still missing in Europe for nonuniform varieties, mainly populations, or farmers' varieties, which could be very relevant for the survival of food tradition and organic and low-input farming systems (FAO, 2010).

Italian regional laws on conservation varieties

On April 6, 2007, Italy approved a national law (No. 46[16]) establishing a national registry on conservation varieties, and it was implemented through a decree issued by the Ministry of Agriculture, Food and Forestry on April 18, 2008.[17] Italy was the only country in Europe that had a national legislation on conservation varieties when the European Directive 2008/62 was approved (on June 20, 2008). Italian regional and national regulations influenced some provisions of the European directives (Bocci, 2009). In 2008, Italy adopted the National Agrobiodiversity Plan (Piano Nazionale sulla Biodiversità di Interesse Agricolo), and on October 29, 2009, Italy approved Legislative Decree No. 149,[18] aimed at implementing the European Directive 2008/62. The Italian Ministry of Agriculture, Food and Forestry also enacted a decree on December 17, 2010,[19] setting operational guidelines for its implementation. Many Italian regions already had regional laws aimed at safeguarding local plant varieties and animal breeds when the Italian national regulations were approved in 2007 and 2008. In fact, their main objective was to give a national legal framework for already existing Italian regional laws. Tuscany was the first Italian region to approve a law (Legge regionale 50, of July 16, 1997), which was replaced by the current Tuscan law (Legge regionale 64, of November 11, 2004), aimed at protecting and enhancing local breeds and varieties of interest to agriculture, husbandry, and forestry. Tuscany's law was followed by similar initiatives[20] by the regions of Lazio,[21] Umbria,[22] Friuli-Venezia Giulia,[23] Marche,[24] Emiglia-Romagna,[25] and Basilicata.[26]

Italian regional laws were developed within the scope of policies aimed at promoting rural development and safeguarding local agroecosystems and typical and regional products. They are aimed at safeguarding and promoting autochthonous genetic resources, especially those at risk of erosion. In some cases, only animal and plant varieties of agricultural interest are considered (Lazio, Umbria, and Marche), and in others forestry resources are also included (Tuscany and Friuli-Venezia Giulia). The definition of authochthonous breeds and varieties include those that are originally from the regional territory; those that, although not originally from the regional territory, have lived within it for more than 50 years; and those originally from the regional territory and no longer present on it, but conserved elsewhere (Bertacchini, 2009). Enrico Bertacchini points out, that the concept of autochthony (in Italian regional laws) "emerges as being broad and especially dynamic," and is "adaptable and elastic to shifts in local farming methods" (Bertacchini, 2009).

These laws establish regional registers (*repertori/registri*), subdivided into animal section and plant section, in which local species, varieties, breeds, populations, cultivars, landraces, and clones are registered. *Ex situ* germplasm collections are also established in some regions, and they are called Banche Regionali delle Varietà e Razze Locali/Autoctone (ALPA, 2009). The establishment of regional networks for the conservation and security of local genetic resources (Rete di Conservazione e Sicurezza) is another important instrument. Municipalities, public and private organizations, universities and research institutions, genebanks and farmers (individually and in their associations) participate in these regional networks. Farmers who promote *in situ*/on-farm conservation of local varieties and breeds and participate in such networks are called *agricoltori/conservatori custodi* (guardian or steward farmers). These networks are responsible for *in situ*/on-farm conservation of local varieties and for multiplying and distributing them. Technical-Scientific Committees evaluate the genetic materials listed in regional registers.

The main objectives of Italian regional laws are to gather knowledge, document, classify and safeguard local plant and animal heritage, to allow the exchange of local genetic materials among farmers, to stimulate their use and on-farm conservation and to give a referential legal framework for such activities (ALPA, 2009). According to Bertacchini (2009), these networks put a variety of actors in touch with each other who are interested in the protection and sustainable use of local genetic resources.

"Steward" farmers may save and locally exchange a small quantity of seeds of local varieties, which is determined when they register to participate in networks. Farmers, research centers, and agrarian universities provide regional development agencies with samples of their genetic materials, for conservation and multiplication purposes. Laws also include the establishment of regional plans and guidelines aimed at conserving and protecting local genetic heritage. These plans and guidelines include support and incentives for farmers to participate in conservation networks, multiply and distribute seeds of local varieties, and disseminate information about them, among other activities. These plans often include measures aimed not only at producing but also at encouraging consumption of local and typical products associated with environmental and cultural values. In Tuscany, for instance, many initiatives are aimed at promoting direct relationships between producers and consumers (the "short supply chain") and agrotourism (in rural territories), making small farmers exempt from a number of legal obligations imposed on large farmers.[27] A recent survey conducted in Tuscany found that direct sales and short supply chains play important roles in promoting more diversified agricultural systems and maintaining agrobiodiversity (Naziri, 2009).

Regional Italian laws tend to establish a distinction between property rights over plants or animals and the (collective) rights of local communities over genetic heritage, that is, between the biological resource in its material dimension (the plant or animal) and its immaterial dimension (its genetic makeup). Ownership of the plant or animal does not exclude recognition of the collective rights of local communities over genetic resources. Bertacchini (2009) explains that Italian regional laws do not contemplate the institution of any form of individual exclusive rights over the plant variety. The person who suggests that a variety be registered enjoys no exclusive right to the variety involved. The register accrues collective benefits for the community as a whole.

The Basilicata law, for example, sets forth, in its Article 7, that, in spite of property rights over plants and animals being included in the regional registries, the genetic heritage of these plants and animals is subject to collective rights held by local communities which conserve them, and with whom benefits deriving from their use must be shared. These collective rights are inalienable, and farmers are entitled to exchange regionally seeds and other plant propagating materials. Local varieties and breeds are not, however, "common heritage of humanity," as pointed out by Antonio Onorati (2005), but rather a (collective) cultural and environmental heritage of local communities which conserved and developed them in their territories.

Another common aspect of these laws is that local genetic resources – or their parts and components – cannot be appropriated through IP rights. There are, however, some differences among regional laws on this aspect. Lazio's law expressly affirms that plant and animal genetic resources belong to local communities (and, therefore, cannot be appropriated through patents or plant breeder's rights). Other regional laws, such as those in Emiglia-Romagna and Tuscany, set forth that any claims of patents or plant breeders' rights over varieties essentially derived from varieties included in regional registers depend on prior authorization from the regional government. So far, however, no such authorizations have been granted by Emiglia-Romagna or Tuscany.

Collective rights over local varieties are largely inspired by an Italian legal concept known as *usi civici* ("civic use rights"), which comprehend several forms of collective rights. *Usi civici* include both the rights of collectivities to use natural resources in private properties for certain purposes (hunting, fishing, timber extraction, pasture, use of water and forest resources, etc.), which are called *usi civici sulla proprietà privata*, as collective property rights over lands and to the collective use of their natural resources (*terre civiche e proprietà collettive*). Civic use rights are indivisible and inalienable, and collective lands (*terre civiche*) are destined for agricultural and forest management activities.

Usi civici were regulated by Law 1766, of 1927, and, after the publication of Law 431/1985, which protects cultural and environmental heritage (known as *Legge Galasso*), *terre civiche* were also recognized by Italian environmental law for their importance for the conservation of natural resources and as a model of shared territorial management. Originally, *usi civici* were collective rights over material heritage, but regional Italian laws have been inspired by this juridical institution to regulate immaterial heritage (genetic information from plants and animals) as well.

Some interesting characteristics of Italian agriculture and farming systems may explain why Italian regions have taken so many initiatives to safeguard and promote the use of local plant varieties and animal breeds. Bocci and Tiberio (2009) explain that as far as agriculture is concerned, Italy holds a position in between tradition and modernity. Despite significant decreases in farm numbers over the last few decades, Italy still holds the third place in Europe, in terms of number of farmers, after Romania and Poland. Italian agriculture is dominated by small farms: farms of less than 10 hectares account for 85 percent of the total. Farms of more than 50 hectares represent only 2.2 percent of the total in numerical terms and account for only 5.6 percent of the overall utilized arable land. In fact, the average size of farms in Italy is sensibly smaller than the average in the EU area and in line with the size of farms in the new EU countries of eastern Europe. This means that in Italy the average farm size is 7.4 hectares; in France that is sevenfold higher (48.6 hectares) and in the United Kingdom nearly eight times greater (55.6 hectares). An interesting fact that enables a better understanding of the Italian farming system is the workers' average age: in Italy only 3.5 percent of farmers are younger than 35 years compared with a European average of 6.9 percent, and the number of farmers older than 64 years is 41.4 percent. The largest portion of Italian agrobiodiversity and associated traditional knowledge is preserved by small farms, not listed as "enterprises," and managed by people older than 65 years. Furthermore, for the 551 species of cultivated plants that have been recorded in northern and central Italy, Italian farmers informally use no fewer than 10,672 vernacular names to refer to them (Bocci and Chiari, 2009). That is, local farming systems are still very important in Italy, and they explain the need to guarantee legal space for local varieties and landraces.

The Swiss seed law

Despite not being a member of the EU, Switzerland adopted the European legislation on production and marketing of seeds, which is extremely restrictive (with exceptions for "conservation varieties"). Switzerland belongs to the "European Seed Area," but established different regulations for certain seeds and plants marketed within its territory. The Swiss law[28] only permits sales of certified seeds and plants of varieties registered in the official catalogue, in homogeneous lots, and with officially sealed and labeled packaging. However, it established some exemptions for:

- seeds and plants of local varieties (*variétés du pays*), defined as plant populations of the same species that resulted from mass and natural selection, in the context of traditional agriculture of a specific region, and which can comprise several types of plants presenting morphological and physiological differences;
- seeds and plants of ancient varieties (*anciennes variétés*), defined as those that have been removed from the official catalogue for more than two years;
- ecotypes of forage plants (*ecotypes de plantes fourragères*), defined as populations of plants of the same species resulting from natural selection in the particular ecological conditions of a certain region, and which are composed of various types of plants, presenting morphological and physiological differences among them.

The Swiss seed law defines as niche varieties (*variétés de niche*) all local varieties, ancient varieties, and ecotypes of forage plants and any other varieties that do not have to meet the requirements for registration in the catalogue (with the exception of genetically modified varieties). Seeds of local varieties, ancient varieties, and ecotypes of forage plants can be sold in the Swiss territory (without the need to be registered and certified), as long as they have an unofficial label, of a different color (from the label of registered and certified seeds), containing the following warning: "non-certified seeds of authorized niche variety" (Article 29). Sales of these seeds are also subject to quantitative limitations and some regulations regarding sampling and weight of the lots. The Swiss Ministry of Agriculture requires, for the description of local varieties, information on their origin, region, specific value, and use (Anvar, 2008).

The Swiss law permits farmers to choose between homogeneous commercial varieties and local varieties, and to make a conscious and deliberate choice. The legal space given to local varieties has made possible important initiatives for *in situ* conservation of agrobiodiversity, such as the network of collections of varieties that are rare and threatened with extinction, maintained by farmers and gardeners (amateur or professional) in all of Switzerland, which is managed by the ProSpecie Rara Foundation (Fondation Suisse pour la Diversité Patrimoniale et Génétique liée aux Végétaux et aux Animaux[29]). This network, which involves approximately 2,500 people and institutions, is responsible for *in situ* conservation of about 900 varieties of legumes and vegetables and 1,800 fruit and 700 berry varieties. These farmers use the ProSpecie Rara seal of quality to identify their products.

Gardens, orchards, and farms where the varieties (and breeds) are kept are visited by approximately 300,000 people each year.

The Norwegian seed law

Norway is not a member of the EU, but it is a member of the European Economic Area Agreement and has to implement all EEA-relevant EU directives. The EU directives on variety release and seed marketing are EEA relevant and are thus implemented in Norway. As a result of introduced laws from EU, variety release and seed marketing regulations have prohibited seed exchange/sale among farmers and the marketing of varieties that are not on the official catalogues of crop varieties from 2004 and until 2010. In 2010, however, Norway introduced new seed regulations based on the EU Directives on Conservation Varieties. The new regulations allow farmers and gardeners to exchange, give away, or sell seeds on a noncommercial basis. Norwegian farmers are allowed to exchange seed of any variety (whether it is protected by plant breeders' rights, registered in the official registry of plant varieties, or not registered at all) among themselves. Farmers may also sell small quantities of seed of any variety except for those that are protected by plant breeders' rights, provided it is on a noncommercial basis (selling in small quantities without active marketing is regarded as noncommercial). There is one exception: potato. Norwegian farmers are not allowed to distribute any seed potato because of the high risk of disease transmission. The new regulations also allow for the release of plant varieties that are considered conservation varieties, following the EU rules but interpreting them less strictly than may be the case in other European countries. The new Norwegian rules also allow farmers to establish authorized seed shops for conservation varieties with simple procedures and lower requirements than for other seed shops. This is meant to enable the marketing of conservation varieties on a commercial scale. The customary use of farmers to select seeds for use the next season is quite widespread in Norway. For this reason, and to ensure a fair balance between farmers' and breeders' rights, Norway is a member of the Union for the Protection of New Varieties of Plants (UPOV) based on its 1978 Act, and not on its 1991 Act.[31]

According to the 2010 Report of the Farm Seed Opportunities Program,[30] Norway has few farmers left (around 45,000 farms), and a small but growing minority of these are engaged in diversity farming. They normally use older varieties that are more genetically heterogeneous and thus easier to adapt to their specific growing conditions and develop further according to their own preferences. The majority of these farmers, who probably do not number more than around 100 people, are small-scale organic, particularly biodynamic, farmers. They need varieties that are not adapted to the use of fertilizers and pesticides. There is also an expanding niche market in terms of demand for better nutrition and old traditional food. A kind of informal seed system is still in place: landraces are used to some extent, old commercial varieties are frequently used, farmers exchange seeds among themselves and across borders, some farmers get seeds from Nordgen, the Nordic genebank, and farmers develop the varieties from season to season by selection.

Notes

1 To propagate or reproduce a plant means to multiply it or generate another plant. There are two main methods of propagating plants: sexual propagation, which is the natural process a plant uses to make seeds with or without human intervention; and asexual reproduction, which is broken into various subcategories of techniques. Asexual propagation, sometimes called vegetative propagation, can be done by cloning, using cuttings, dividing, layering, or grafting. In other words, plant propagating materials include not only (botanical) seeds but also roots, tubers, bulbs, rhizomes, etc. We distinguish "seeds" from "grains" because of their use and destination: seeds are used to generate other plants, whereas "grains" are directly consumed by humans or animals, or transformed into flour, oils, agrofuels, etc.

2 Austrian botanist and monk Gregor Mendel formulated the laws of heredity in 1865, but his findings were ignored until early in the twentieth century, when he became known as the "father of genetics." Mendel experimented on different pea varieties, demonstrating the existence of hereditary units (now known as genes) responsible for transmission of dominant and recessive traits.

3 Heterosis (or hybrid vigor) can be defined as an increase in yield and productivity arising from cross-breeding among genetically different varieties.

4 In France, the first law to establish quality control for seeds was passed in 1905. In 1932, an official seed catalogue was created for some species and varieties, starting with wheat and followed by potatoes, rye, oats, and corn. This catalogue, however, was, at first, optional. Inserting a variety in the catalogue was not mandatory in order to trade it. In the early 1960s, garden varieties were included in the official catalogue. In 1942, the Comité Technique Permanent de la Sélection was created, and became responsible for management of the official catalogue of cultivated seeds and varieties. A 1949 decree forbade trading of any seeds of agricultural varieties not included in the official catalogue. Inclusion in the official catalogue (for marketing purposes) is mandatory for most cultivated species. In 1997, France created an annex to its national catalogue, listing varieties used by "amateur" gardeners for noncommercial purposes (many people in France, and Europe as a whole, dedicate time and effort to gardens as a leisure activity). In 1996, the European Community created a common catalogue (for member countries), which currently has 32,000 varieties. European Directive 2002/53 makes it mandatory for genetically modified varieties which are listed in the European common catalogue to be clearly identified as such. Sources: www.droit-et-semence.blogspot.com and Anvar (2007).

5 According to Niels Louwaars (2007), these policies were highly influenced by a seed system development paradigm published by J. Douglas (1980) *Successful Seed Programs: A Planning and Management Guide*, Westview Press Boulder, Boulder, CO. There are different levels of state intervention in seed production and sales, through regulation. In the United States, for instance, seed certification is voluntary, and releasing varieties is the responsibility of each

company. Seed laws regulate only requisites for seed certification. This system reflects belief that the market itself will eliminate low-quality seed producers. In Europe, on the other hand, most countries demand seed registration and certification prior to production and marketing. China, for example, left seeds developed by farmers out of the coverage of its new seed law. The Indonesian seed law regulates the formal system, but does not include in its scope local seeds traded and exchanged locally. In other countries (such as Cameroon, Nigeria, and Senegal), only traded seeds have to be registered and certified. There are also countries in which mandatory registration and certification applies to only some agricultural species and/or varieties (Zambia, Malawi, India, Bangladesh). In other countries, rules apply only to certified seeds, in order to ensure that only effectively certified seeds are sold as such, excluding local seed systems. Source: Louwaars (2005).

6 It is interesting to point out that production of seeds by farmers themselves is also significant in industrialized countries. European seed producers estimate that approximately 50 percent of seeds used in cultivation of the main grain crops are produced by farmers, and, in some countries of southern Europe, such as Italy and Greece, only 10 percent of the seeds (of grains) are bought by farmers. In France, 50 percent of the seeds of self-pollinating agricultural species, such as wheat, are produced by farmers, and in Germany it is estimated that this figure reaches 46 percent. In Portugal, there are estimates that this number is 75 percent, and in Spain the figure is estimated to be 88 percent. European farmers maintain the traditional practice of reserving part of the harvest for seeding in the next year. Even in the United States the average use of seeds produced in the formal system, in the period between 1986 and 1997, was 37 percent for wheat, 78 percent for cotton, and 81 percent for soy, with 100 percent for corn owing to the use of hybrids. See Toledo (2002), Kastler (2005), and Louwaars (2007).

7 Although frequently called "informal" seed systems, farmers' and local seed systems also have their "formalities," that is, they follow their own rules, customs, and practices, which are rooted in the culture of local communities. As the term "informal" may have, in some contexts, a negative connotation, we prefer to use the terms local or farmers'. Almekinders and Louwaars (1999) use the term "farmers' seed systems" to make clear that the ones operating this system are the farmers themselves.

8 FAO (2003) Plant Production and Protection paper no. 185, on "Quality Declared Seed System."

9 In Argentina, Ley de Semillas y Creaciones Fitogenéticas no. 20247 was approved in 1973, but it was modified in 1983, 1989, and 1991. In Guatemala, the regulation called Acuerdo Gubernativo, approved in 1961, introduced "normas reglamentarias para la producción, certificación y comercialización de semillas agrícolas y forestales."

10 http://www.mda.gov.br/portal/noticias/item?item_id=3594546 (accessed April 10, 2011).

11 http://www.agricultura.gov.br/portal/page/portal/Internet-MAPA/pagina-inicial/comunicacao/noticias/noticia-aberta?noticiaId=31693 (accessed April 10, 2011).

12 This law replaced the previous seed law (No. 6507, of 1977), which, in turn, revoked the first Brazilian seed law, No. 4727, of 1965. The seed law currently in force, no 10,711, of 2003, is regulated by Decree 5153/2004.

13 Source: http://www.conab.gov.br/conabweb/agriculturaFamiliar/paa_o_que_e.html (accessed April 10, 2011).

14 Since 1966, it has been mandatory in the EU that, for varieties to be legally marketed, they be recorded in the official register. To be registered, they must

meet distinctness, uniformity and stability (DUS) requirements and, in the case of agricultural species, each new variety must have a superior value for cultivation and use (VCU). The European Catalogue is based on the national catalogues of EU members, and lists those varieties whose seeds are marketable within the EU. The following European legislation governs the marketing of seeds and plant propagating material within the EU: 66/401/EEC, 66/402/EEC, 2002/53/EC, 2002/54/EC, 2002/55/EC, 2002/56/EC, 2002/57/EC, 68/193/EC, 1998/56/EC, 92/33/EEC, 92/34/EEC, and 1999/105/EC. However, the European Commission has decided to evaluate and review the legislation on the marketing of seeds and plant propagating material, under the framework of its Better Regulation policy, with the aims of simplifying and improving existing regulation in different fields. Many EC directives on the marketing of seeds and plant propagating material date back to the 1960s and 1970s. Source: http://ec.europa.eu/food/plant/propagation/evaluation/index_en.htm (accessed April 15, 2011).

15 A previous European directive (1998/95), of December 14, 1998, had already opened the legal possibility that member states define specific conditions for the marketing of conservation varieties.

16 Law 46/2007 was aimed at converting into law, with some modifications, Decree Law 10, of February 15, 2007. Law 46/2007 has also modified Law 1096, of November 25, 1971 (which regulates production and marketing of seeds in Italy). Law 1096/1971 was regulated by Presidential Decree 1065/1973.

17 Published on May 26, 2008.

18 Published on October 31, 2009.

19 Published on February 17, 2011.

20 Italy is divided into 20 regions, which have considerable administrative and legislative autonomy. Each region is divided into provinces, which, in turn, are subdivided into municipalities (*comuni*).

21 Legge regionale No. 15/ 2000.

22 Legge regionale No. 25/ 2001.

23 Legge regionale No. 11/ 2002.

24 Legge regionale No. 12, of June 6, 2003.

25 Legge regionale No. 1, of January 29, 2008.

26 Legge regionale No. 26, of October 14, 2008.

27 Sources: www.arsia.toscana.it/network/ and http://filieracorta.arsia.toscana.it/ (accessed March 10, 2010). See also Bazzanti *et al.* (2006).

28 Swiss Ordinance of December 7, 1998 (sur les semences et les plants des espèces de grandes cultures et de plantes fourragères et des cultures maraîchères), which was modified by the Ordinance of June 7, 2010, in force since July 1, 2010 (Articles 2, 27, and 29). Available at: www.admin.ch/ch/f/rs/916_151_1/index.html (accessed April 18, 2011).

29 www.prospecierara.ch.

30 http://www.prodinra.inra.fr/prodinra/pinra/data/2011/04/PROD20101e4b818_20110406034307470.pdf (accessed April 10, 2011).

31 Source: Regine Andersen, personal communication, April 26, 2011.

Bibliography

Almekinders, C. and Louwaars, N. (1999) *Farmers' Seed Production. New Approaches and Practices*, Intermediate Technology Publications, London.

Associazione Lavoratori Produttori dell'Agroalimentare (ALPA), Rete Semi Rurali, and Centro per lo Sviluppo Agricolo e Rurale (2009) 'Farmers as stewards of agrobiodiversity in agriculture', informative campaign (www.alpainfo.it/progettieuropei/20092010/Home_Page_files/ALPA_opuscolo_Biodiversita_1.pdf; accessed April 10, 2011).

Anvar, S.L. (2007) 'Les indicateurs de biodiversité: de l'importance du contexte réglementaire', *Le Courrier de l'environnement de L'INRA*, no. 54, pp. 9–18.

Anvar, S.L. (2008) 'Semences et droit: l'emprise d'un modèle économique dominant sur une réglementation sectorielle', PhD thesis, Université de Paris I Panthéon-Sorbonne, France.

Associação Brasileira de Sementes e Mudas (ABRASEM) (2008a) *O Mercado de Sementes no Brasil*, Brasília.

Associação Brasileira de Sementes e Mudas (ABRASEM) (2008b) *Semente: Inovação Tecnológica. Anuário 2008*, Brasília.

Badstue, L. (2007) 'Confiança mútua como base para a aquisição de sementes', *Agriculturas: Experiências em Agroecologia*, vol. 4, no. 3, pp. 18–21.

Bazzanti, N., Turchi, R. and Bartoli, M. (2006) *La Tutela e la Valorizzazione del Patrimonio di Razze e Varietà Locali in Toscana*, ARSIA (Agenzia Regionale per lo Sviluppo e l'Innovazione nel Settore Agricolo-Forestale), Florence.

Benedetti, A. (2009) 'Gli usi civici nella storia e nella legislazione italiana', *Geopunto*, vol. 6, no. 5, pp. 11–22 (www.georoma.it/geopunto/geopunto6/usi%20civici.pdf; accessed April 10, 2011).

Bertacchini, E. (2009) 'Regional legislation in Italy for the protection of local varieties', *Journal of Agriculture and Environment for International Development*, vol. 103, no. 1–2, pp. 51–61.

Bocci, R. (2009) 'Seed legislation and agrobiodiversity: conservation varieties', *Journal of Agriculture and Environment for International Development*, vol. 103, no. 1–2, pp. 31–49.

Bocci, R. and Chable, V. (2008) 'Semences paysannes en Europe: enjeux et perspectives', *Cahiers Agricultures*, vol. 17, no. 2, pp. 216–222.

Bocci, R. and Chable, V. (2009) *Farm Seed Oportunities: Seed Regulation Scenarios for Europe* (http://www.farmseed.net/home/resources/publication; accessed March 10, 2011).

Bocci, R. and Chiari, T. (2009) (eds.). *The Sustainable Use of Agrobiodiversity in Italy*. Report on Case Studies on Article 6 of the International Treaty on Plant Genetic Resources for Food and Agriculture, Istituto Agronomico per l'Oltremare, Florence, Italy.

Bocci, R., Chable, V., Kastler, G. and Louwaars, N. (2010) 'Policy recommendations: set of recommendations on farm conservation strategy, the role of innovative market mechanisms, legislative framework for landraces, conservation varieties and amateur varieties in Europe'. Farm Seed Opportunities: Opportunities for farm seed conservation, breeding and production, May 31, 2010 (http://www.prodinra.inra.fr/prodinra/pinra/data/2011/04/PROD20101e4b818_20110406034307470.pdf; accessed April 10, 2011).

Boef, W.S. de (2007) 'Uma perspectiva de sistemas aproximando agricultores e pesquisadores no manejo comunitário da agrobiodiversidade', in Boef, W.S. de, Thijssen, M.H., Ogliari, J.B. and Sthapit, B. (orgs.) *Biodiversidade e Agricultores: Fortalecendo o Manejo Comunitário*, L & PM, Porto Alegre, Brazil, pp. 59–66.

Bonneuil, C. and Thomas, F. (2009) *Gènes, Pouvoirs et Profits: Recherche Publique et Régimes de Production des Savoirs de Mendel aux OGM*, Éditions Quae, Versailles, France.

Bonneuil, C., Demeulenaere, E., Thomas, F., Joly, P.B., Allaire, G. and Goldringer, I. (2006) 'Innover autrement? La recherche face à l'avènement d'un nouveau régime de production et de régulation des savoirs en génétique végétale', in Gasselin, P. and Clèment O. (coords.). *Quelles variétés et semences pour des agricultures paysannes durables?, Les Dossiers de L'Environnement de l'INRA*, no. 30, pp. 29–52.

Branca, G. and Perrone-Pacifico, C. (2003) *Le Terre Collettive del Lazio: un'Analisi del Possibile Ruolo delle Proprietà Collettive nelle Politiche per lo Sviluppo Rurale Regionale*, Università della Tuscia, Viterbo, Italy.

Bustarret, J. (1944) 'Variétés et variations', *Annales Agronomiques*, vol. 14, pp. 336–362.

Carraro, I. (2005) 'A Empresa de Sementes no Ambiente de Proteção de Cultivares no Brasil', PhD thesis, Faculdade de Agronomia Eliseu Maciel, Universidade Federal de Pelotas, Rio Grande do Sul, Brazil.

Chable, V. and Kastler, G. (2006) 'Maintien, re-découverte et création de la diversité cultivée pour l'agriculture biologique', *Alter Agri*, no. 78, pp. 13–17.

Deléage, E. (2004) *Paysans, de la Parcelle à la Planète*, Syllepse, Paris.

Deverre, C. and Kastler, G. (2006) 'Semences, ressources génétiques et droit', in Gasselin, P. and Clèment O. (coords.) *Quelles variétés et semences pour des agricultures paysannes durables?, Dossiers de L'Environnement de L'INRA*, no. 30, pp. 167–168.

Emperaire, L. (2008) 'O manejo da agrobiodiversidade: o exemplo da mandioca na Amazônia', in Bensusan, N. (org.) *Seria Melhor Mandar Ladrilhar? Biodiversidade: Como, Para Que e Por Quê*, Universidade de Brasília, IEB, Peirópolis, Brasília, pp. 337–352.

Espinosa Calderon, A., Tadeo Robledo, M., Turrent Fernández, A., Gomez Montiel, N., Sierra Macias, M., Palafox Caballero, A., Caballero Hernández, F., Valdivia Bernal, R. and Rodriguez Montalvo, F.A. (2007) 'Variedades mejoradas disponibles y abastecimento de semillas ante la Nueva Ley de Semillas en México', X Congreso Internacional de Ciencias Agrícolas, Mexicali, Baja California, 18 y 19 de octubre de 2007, Universidad Autónoma de Baja California Instituto de Ciencias Agricolas.

Fernandez-Conejo, J. (2004) *The Seed Industry in U.S. Agriculture: an Exploration of Data and Information on Crop Seed Markets, Regulation, Industry Structure, Research and Development* (Agriculture Information Bulletin, 786), United States Department of Agriculture, Economic Research Service, Washington, DC.

Food and Agriculture Organization (FAO) (1998) *Seed Production and Improvement: Assessment for Sub-Saharan Africa*, Plant Production and Protection Division, Seed and Plant Genetic Resources Service, Rome.

Food and Agriculture Organization (FAO) (2001) 'Seed policy and programmes in Latin America and the Caribbean, in *Proceedings of the Regional Technical Meeting on Seed Policy and Programmes in Latin America and the Caribbean*, March 20–24, 2000, Merida, Mexico, Plant Production and Protection Division, Seed and Plant Genetic Resources Service, Rome.

Food and Agriculture Organization (FAO) (2010) Commission on Genetic Resources for Food and Agriculture. *Second Report on the State of the World´s Plant Genetic Resources for Food and Agriculture*, FAO, Rome (http://www.fao.org/docrep/013/i1500e/i1500e.pdf; accessed February 20, 2011).

Fulciniti, L. (2000) *I Beni d'Uso Cívico*, CEDAM, Padova.

Galluzzi, G., Eyzaguirre, P. and Negri, V. (2009) 'Uncovering European home gardens: their human and biological features and potential contribution to the conservation of agrobiodiversity', in Bailey, A., Eyzaguirre, P. and Maggioni, L. (eds.) *Crop Genetic Resources in European Home Gardens. Proceedings of a Workshop*, 3–4 October 2007, Ljubljana, Slovenia, Bioversity International, pp. 8–17.

Gisselquist, D (1999) 'Regulatory issues', in Wood, D. and Lenné (eds.) *Agrobiodiversity: Characterization, Utilization and Management*, Cabi Publishing, New York, pp. 409–423.

GRAIN (2005a) 'Africa's seed laws: red carpet for the corporations', *Seedling*, July, pp. 28–35 (http://www.grain.org/seedling/?id=342; accessed April 10, 2011; seed laws in selected African countries can be consulted at http://www.grain.org/seedling_files/SL_Africa_maps.pdf; accessed April 10, 2011).

GRAIN (2005b) Latin America: the mantra of privatization, *Seedling*, July 2005 (http://www.grain.org/seedling_files/seed-05–07.pdf; Latin American seed laws can be consulted at: http://www.grain.org/go/seedlaws; accessed April 10, 2011).

GRAIN, with contributions from Dr. Devinder Sharma (2005c) 'India's new Seed Bill' (http://www.grain.org/seedling/?id=338, july 2005; accessed 10 April 2011; to consult seed regulation and certification laws in Southeast Asian countries: http://www.grain.org/seedling_files/SL_Asia_maps.pdf; accessed April 10, 2011).

Guei, R. (coord.) (2010) *Promoting the Growth and Development of Smallholder Seed Enterprises for Food Security Crops: Case studies from Brazil, Côte d'Ivoire and India* (http://typo3.fao.org/fileadmin/templates/agphome/documents/PGR/PubSeeds/seedSynthesis_book7.pdf; accessed 10 April, 2011).

Joshi, K.D (2000) 'Strengthening the farmers' seed system in Nepal', *Biotechnology and Development Monitor*, no. 42, pp. 15–17.

Kastler, G. (2005) 'Seed laws in Europe: locking farmers out', *Seedling*, July 2005 (http://www.grain.org/seedling/?id=343; accessed April 10, 2011).

Kastler, G. (2009) 'Les variétés de conservation momifiées contre la volonté unanime des deputés', Réseau Semences Paysannes, Brens, France (www.semencespaysannes.org/varietes_conservation_momifiees_contre_volont_115-actu_64.php#date64; accessed April 10, 2011).

Lenné, J.M. and Wood, D. (2011) *Agrobiodiversity Management for Food Security: A Critical Review*, Cabi International, Wallingford, UK.

Louwaars, N. (ed.) (2002) *Seed Policy, Legislation and Law: Widening a Narrow Focus,* Food Products Press, Binghamton, NY.

Louwaars, N. (2005) 'Biases and bottlenecks: time to reform the south's inherited seed laws? *Seedling*, July 2005 (http://www.grain.org/seedling_files/seed-05-07-2.pdf; accessed April 10, 2011).

Louwaars, N. (2007) 'Seeds of confusion: the impact of policies on seed systems', PhD dissertation, Wageningen, the Netherlands.

Louwaars, N. (2010) 'Seed systems and plant genetic resources for food and agriculture: thematic background study', in *Second Report on the State of the World's Plant Genetic Resources*, FAO, Rome.

Londres, F. (2006) *A Nova Legislação de Sementes e Mudas no Brasil e seus Impactos sobre a Agricultura Familiar*, AS-PTA, Rio de Janeiro, Brazil.

Lopez Noriega, I. (2009) 'European legislation in support of home gardens conservation', in Bailey, A., Eyzaguirre, P. and Maggioni, L. (eds.) *Crop Genetic Resources in European Home Gardens, Proceedings of a Workshop*, 3–4 October 2007, Ljubljana, Slovenia, Bioversity International, pp. 62–69.

Lorenzetti F. and Falcinelli M. (2009) 'La salvaguardia delle risorse genetiche delle piante e l'utilizzazione delle varietà da conservazione nella attuale disciplina sementiera', *Dal Seme*, no. 49.

Martínez Morales, C.S. (2007) 'Estudio comparativo sobre el sistema juridico legal y marco de politicas que inciden en la cadena de produccion y difusion de semillas de granos basicos en Centroamerica', Programa Colaborativo de Fitomejoramiento Participativo en Meso América, Managua, Nicaragua.

Moreschini, I. (2007) 'Gli usi civici nella regione Lazio', *Rivista di Diritto Pubblico Italiano, Comunitario e Comparato*, vol. 5, no. 12.

Naziri, D. (2009) 'Direct sale as a means for promoting the sustainable use of plant genetic resources: the case of the Tuscany Region', *Journal of Agriculture and Environment for International Development*, vol. 103, no. 1–2, pp. 65–68.

Neate, P. and Guei, R. (coords.) (2010) *Promoting the Growth and Development of Smallholder Seed Enterprises for Food Security Crops* (http://typo3.fao.org/fileadmin/templates/agphome/documents/PGR/PubSeeds/seedpolicyguide6.pdf; accessed April 10, 2011).

Onorati, A. (2005) 'Collective rights over farmers' seeds in Italy', *Seedling*, July 2005, pp. 17–21 (http://www.grain.org/seedling/?id=340; accessed April 10, 2011).

Osman, A. and Chable, V. (2009) 'Inventory of initiatives on seeds of landraces in Europe', *Journal of Agriculture and Environment for International Development*, vol. 103, no. 1–2, pp. 95–130.

Paoloni, L. (2005) *Diritti degli Agricoltori e Tutela della Biodiversità*, G. Giappichelli, Torino, Italy.

Peroni, N. (2007) 'Manejo e domesticação de mandioca por caiçaras da Mata Atlântica e ribeirinhos da Amazônia', in Boef, W.S. de, Thijssen, M.H., Ogliari, J.B. and Sthapit, B. (orgs.) *Biodiversidade e Agricultores: Fortalecendo o Manejo Comunitário*, L & PM, Porto Alegre, Brazil, pp. 234–242.

Perelmuter, T. (2009) 'Modificaciones a las Leyes de Semillas en Argentina y México', paper prepared for delivery at the 2009 Congress of the Latin American Studies Association, Rio de Janeiro, Brazil, June 11–14, 2009.

Red de Semillas 'Resembrando e intercambiando (2008), Sevilla, Spain: Análisis de la nueva directiva sobre variedades de conservación', *Cultivar Local*, no. 22, pp. 3–5 (www.redsemillas.info/wp-content/uploads/2009/02/cultivar-local-n-22-diciembre-08.pdf; accessed April 10, 2011).

Réseau Semences Paysannes (2010) 'The co-evolution of the European legislative landscape and industrial strategies of appropriation of our vital resources', 5th meeting of the European Seed Network Let's Liberate Diversity!, Volkshaus, Graz, Austria (http://bienscommuns.org/blog/wp-content/uploads/2010/03/Revision-of-EU-seeds-laws-1.pdf; accessed April 10, 2011).

Ribeiro, S. (2007) 'Nueva ley de semillas contra los campesinos' (http://www.iade.org.ar/modules/noticias/article.php?storyid=1984; accessed April 10, 2011).

Riesco, A. (2002) Annual Report for the Project: Strengthening the Scientific Basis of in situ Conservation of Agricultural Biodiversity: Peru Country Component, IPGRI, Rome. In *Second Report on the State of the World's Plant Genetic Resources for Food and Agriculture*, FAO, Rome (http://www.fao.org/docrep/013/i1500e/i1500e.pdf; accessed February 20, 2011).

Toledo, A. (2002) 'Saving the seed: Europe's challenge', *Seedling*, April (www.grain.org/seedling/seed-02-04-2-en.cfm; accessed April 10, 2011).

Schutter, O.D. (2009) *Seed Policies and the Right to Food: Enhancing Agrobiodiversity and Encouraging Innovation*. Report of the Special Rapporteur on the right to food, United Nations (http://www.srfood.org/images/stories/pdf/officialreports/20091021_report-ga64_seed-policies-and-the-right-to-food_en.pdf; accessed April 10, 2011).

Shrestha, P. (2009) 'Implication of seed policy and laws for conservation of local crop varieties in Nepal' (http://www.farmseed.net/home/resources/publication/Pratap%20Shrestha.pdf; accessed April 10, 2011).

Toledo, A. (2002) 'Saving the seed: Europe's challenge'. *Seedling*, April 2002 (http://www.grain.org/seedling/?id=191; accessed August 15, 2008).

Vásquez Vásquez, F.J. (2006) 'Análisis de la Ley de Semillas y propuesta para sua actualización', Law Dissertation, Universidad de San Carlos de Guatemala, Facultad de Ciencias Jurídicas y Sociales.

Veteläinen, M., Negri, V. and Maxted, N. (eds.) (2009) *European Landraces: On-Farm Conservation, Management and Use*, Bioversity International, Rome.

Visser, B. (2002) 'An agrobiodiversity perspective on seed policies', in Louwaars, N. (ed.) *Seed Policy, Legislation and Law: Widening a Narrow Focus*, Food Products Press, The Haworth Press, New York, pp. 231–245.

Useful web sites

www.semencespaysannes.org
www.gentechnikfreie-saat.de
www.redsemillas.info
www.redandaluzadesemillas.org
www.redextremeñadesemillas.es
www.semirurali.net
www.semionline.croceviaterra.it
www.agricolturacontadina.org
www.aiab.it
www.farmseed.net
www.seed-sovereignty.org
www.liberate-diversity-hungary2011.org/coordination-en/better-regulation

5

THE CONVENTION FOR THE PROTECTION OF NEW VARIETIES OF PLANTS AND THE UPOV SYSTEM

The protection of intellectual property rights over plant varieties

History

It was the European plant breeders who took the initiative to propose, in the 1950s, the creation of an international regime to protect IP rights over plant varieties. The International Convention for the Protection of New Varieties of Plants was adopted in 1961, and it established the so-called plant breeders' rights (PBRs) or plant variety rights (PVRs). UPOV is the French acronym for Union Internationale pour la Protection des Obtentions Végétales (International Union for the Protection of New Varieties of Plants), which is the name of the organization that the International Convention (called the UPOV Convention) established. The need to establish IP rights for plant varieties was conceived and developed when plant breeding[1] became an economically promising activity and started to attract the interest and investments of private companies, which pushed for proprietary rights over plant varieties. IP protection gives breeders exclusive (monopolistic) rights over the production and selling of newly developed plant varieties for a certain period of time, and it is intended to promote the growth of private investment in plant breeding (of commercial crops, such as wheat, corn, soybean, coffee, etc.).

The UPOV system is aimed at protecting innovations made by professional plant breeders from public and private research institutions, through methods and techniques that are considered "scientific," and usually have as a target the development of plant varieties that are high yielding, genetically homogeneous and stable (after repeated cycles of propagation), and well adapted to the large-scale, industrialized agricultural model. It is based on the assumption that professional plant breeders are the only ones capable of innovating in agriculture. Farmers' innovations, knowledge, practices, and techniques are not considered by the UPOV system as worthy of legal protection, and farmers are regarded as mere users of technologies developed by professional breeders. The UPOV system denies farmers' role as innovators and holders of knowledge and practices which are essential for agricultural development, and fails to recognize that the vast agricultural diversity would not exist were farmers not able to innovate and develop new varieties and agricultural systems.

Historically, plant breeding and seed production have been carried out by farmers, who continue to select and improve their varieties, even if using different methods and techniques than those of professional breeders, especially in developing countries. Although "scientific" plant breeding did not emerge until relatively recently – particularly at the beginning of the twentieth century, after the rediscovery of Mendel's principles of genetics and heredity – plant improvements made by farmers have taken place throughout past millennia. However, an IP regime for plant varieties was considered to be necessary only when farming activities were separated from crop selection and improvement, and plant breeding became a "profession" over which only professional and technical experts have control.

Plant breeders have always argued that, because seeds are self-replicable, farmers can reproduce them easily. Therefore, seeds of newly created plant varieties do not need to be bought, they can be simply reproduced, which would prevent breeders from profiting from their newly developed plant varieties. Plant breeders identified "self-replicability," an inherent and essential characteristic of seeds, as the principal "barrier" preventing them from achieving adequate remuneration for their work. The establishment of IP rights over newly developed varieties aimed to ensure breeders' exclusivity in the production and sales of such plant varieties, so that they would be financially compensated for their professional activities in plant improvement.[2]

When European breeders conceived the UPOV system (an alternative or *sui generis* system), they wanted to keep plant breeding out of the plant patent system established in the United States, because they considered it to be inadequate for plant variety protection. Thus, they developed a *sui generis* IP rights system, whose main principles were established during the International Conference for Protection of New Plant Varieties, held in Paris in 1957, by the French government, at the request of plant breeders' associations.[3] Until then, the United States was the only country to grant patents for plant varieties, by means of the Plant Patents Act of 1930, which was especially designed to protect vegetatively reproduced ornamentals and fruit varieties. Later, Japan, Australia, and New Zealand also allowed plant varieties to be patented, but most countries use a *sui generis* system of plant variety protection, based on the UPOV system. In most countries (Europe,[4] Canada, and Latin American countries), patents are not granted to plant and animal varieties, nor to essentially biological processes for the production of plants and animals. They are, however, granted to genetically modified organisms (GMOs) and nonbiological processes.

For European breeders, plant breeders' rights should grant free access to genetic resources and, at the same time, protect breeders' innovations. To develop new plant varieties, breeders need to have access to the widest genetic diversity possible. Therefore, plant breeders' rights laws should allow free access to new plant varieties not for reproduction and marketing (which would violate breeders' rights), but to be used as an initial source of variation for the purpose of creating other varieties. Free access to genetic resources, including those protected by IP rights, is essential for breeders, as innovation in plant breeding is essentially cumulative.

Thus, a fundamental characteristic of plant breeders' rights (in contrast to patent laws) is that they maintain the so-called "breeders' exemption" (or "privilege"), which means that IP rights over plant varieties do not prevent breeders from using any genetic material as an initial source of variation in the development of new plant varieties. The authorization from the initial breeder (holder of IP rights) is not required for such purposes. In addition, the right to freely use protected varieties is not limited to the stage of research aimed at creating new varieties, but extends to sales of the newly developed variety. If a breeder creates a new plant variety (even if he or she accessed protected varieties as a source of genetic variation), and this new variety becomes distinguishable from others by at least one important trait, its production and sales do not need the prior consent of the original plant breeder. The same is not true for patent law: any new invention that incorporates a patented invention requires authorization from the original inventor, as research exemptions in patent law are much more restricted.

Another peculiarity of plant breeders' rights is that they do not protect plant breeding processes – such as cross-breeding, backcrossing, etc. – which are universal and can be freely used by all.[5] Plant breeders' rights cover the propagating material of the protected plant variety, which must be sufficiently homogeneous and stable, as well as distinguished from other plant groupings by at least one characteristic. The protection granted by the breeder's right is therefore not limited to the innovation that differentiates a plant variety from others or to the gene or group of genes that were introduced, but it covers the complete living organism, with its entire set of characteristics, some new, some old. The protection covers not only the new element that was added to the variety, but the entire new variety, and, thus, the idea of maintaining free access to the plant variety, including to its new components, was defended from the start by European breeders, with very few diverging opinions. It was further agreed that, in the plant variety protection *sui generis* system, the protected variety would not be "described," as commonly happens with patented inventions, but deposited, as a live and concrete object, for evaluation (Hermitte and Kahn, 2004, p. 74).

Plant breeders' rights also safeguard (implicitly) the so-called "farmers' privilege," which I would rather call "farmers' rights."[6] Plant breeders' rights have historically recognized the rights of farmers to save seeds of protected varieties to resow their fields. Farmers can save and reproduce seeds (of protected varieties) for their own use. As this has always been a traditional and universal practice, widely adopted by farmers all over the world, the group of European breeders (gathered in Paris, in 1957, during the International Conference for Protection of New Plant Varieties) imposed no restriction on this practice, recognized as customary law, nor did it oblige farmers to pay royalties to the IP title-holder. According to Hermitte and Kahn (2004, p. 74), breeders were accustomed to having to "tolerate" farmers' custom of saving seeds for their own use. The UPOV Convention approved in 1961 makes no reference to farmers' rights, stating only that the breeders' rights covers only commercial production, offering for sale, and marketing, which has been interpreted as permission for farmers to re-use saved seeds in future harvests.

The UPOV Convention: main concepts

The International Convention for Protection of New Plant Varieties was signed on December 2, 1961, in Paris (and became known as the UPOV Convention), but came into force in 1968. It was initially signed by five countries (France, Holland, Germany, Belgium, and Italy), all wealthy and industrialized. The main points of the UPOV Convention are:

1. The breeders' right is, as is true of all IP rights, exclusive and temporary. Exclusivity extends to production for commercial purposes, offering for sale, and marketing of the (reproductive or vegetative) propagating material of the plant variety.
2. The breeder's authorization is not necessary for use of the plant variety as a source of variation in plant breeding (the so-called "breeder's exemption"). A legal distinction is established between two objects that in reality are only one: the variety as an "invention," protected by exclusive proprietary rights of the breeder, and the variety as a genetic resource, and as raw material/input for the development of new plant varieties, which is free of any right and accessible (Hermitte and Kahn, 2004, p. 25).
3. To be protected a plant variety must be sufficiently homogeneous (or uniform) considering the particular features of its sexual reproduction or vegetative propagation. It must also be stable in its essential characteristics, that is, it must maintain these characteristics throughout repeated successive cycles of propagation or, when the breeder has defined a particular cycle of reproduction or multiplication, at the end of each cycle.

   The homogeneity and stability conditions for the protection of plant varieties reflect a "static" conception of plant varieties, which does not take into account their evolution through time and space, nor the social, cultural, and environmental contexts in which they have developed. Homogeneous and stable plant varieties are well adapted to the logic of IP rights, which require precise and well-defined objects to be protected; this does not occur with genetically heterogeneous varieties, which are dynamic and too "fluid" (subject to change) to be owned by anyone. Such requirements exclude local, traditional, and creole varieties from any kind of protection, as they rarely meet such legal conditions.

4 To be protected, a plant variety must be clearly distinguishable, by one or several important traits, from any other variety whose existence is well known at the time of the filing of the application for protection, that is, what is essential for protection is the difference in relation to what is already known. The characteristics that distinguish the variety must allow precise recognition and description.

5 Protection is granted to the breeder's right independently of the origin (artificial or natural) of the initial variation from which the variety has resulted. In other words, protection extends not only to new plant varieties created through classical plant breeding, but also to varieties whose improvement was based on the discovery and selection of mutants or variants found in a population of cultivated plants. According to the UPOV Convention (1991 Act), the "breeder" is "the person who bred, or discovered and developed, a variety."

6 To be protected, a plant variety must be new (novel), that is, the variety must not have been offered for sale or marketed for longer than one year, with the consent of the breeder, in the territory of the state where the application (for a breeder's right) has been filed. Furthermore, it must not have been offered for sale or marketed, with the agreement of the breeder, in the territory of any other states for longer than six years in the case of vines, forest trees, fruit trees, and ornamental trees, or for longer than four years in the case of all other plants.

7 The variety must have a denomination that enables its identification. The variety's denomination must not mislead or cause confusion concerning the characteristics, value, or identity of the variety or the identity of the breeder. It must be different from any denominations of existing varieties of the same botanical species or closely related species.

The UPOV Convention, signed in 1961 and revised in 1972,[7] 1978,[8] and 1991, remained as an instrument adopted only by rich countries[9] until the Trade-Related Aspects of Intellectual Property Rights (TRIPS) Agreement was signed in 1994,[10] and became one of the main pillars of the World Trade Organization (WTO). Developing countries (led by Brazil and India) initially resisted the inclusion of IP among the issues to be addressed by the international trade system, claiming that WIPO[11] (World Intellectual

Property Organization) was the UN specialized agency for IP (Roffe, 2008, p. 50). In WIPO, the developing countries acted in blocks, and this strengthened them politically, whereas in the General Agreement on Tariffs and Trade (GATT)/WTO several commercial issues were addressed at the same time, which made block voting difficult and divided developing countries, which bargained bilateral commercial advantages, weakening their common position. The initial resistance by developing countries was thus overcome with the concession of benefits by developed countries for agriculture, textiles, and tropical products (Andersen, 2008, p. 150), as well as threats of commercial sanctions. The position that ended up prevailing was the one defended especially by the United States, which represented the interests of multinational companies in the pharmaceutical, chemical, and biotechnology areas. The United States argued, among other things, that WIPO did not have an effective dispute settlement mechanism. The US position was supported by the EU and Japan, and IP rights ended up being incorporated into one of the WTO agreements (TRIPS).

The Agreement on Trade-Related Aspects of Intellectual Property Rights (TRIPS) of the World Trade Organization (WTO)

The TRIPS Agreement was an important landmark for IP rights, as it broke with several principles adopted thus far by international conventions (such as the 1883 Paris Convention for the Protection of Industrial Property and the 1886 Berne Convention for the Protection of Literary and Artistic Works, administered by WIPO). The negotiations involving IP were shifted from WIPO (where they had been traditionally addressed) to international trade organizations (GATT/WTO). Furthermore, the adoption of the TRIPS Agreement became a necessary condition for countries to become members of the WTO. Countries applying for membership had to accept all agreements in the WTO "package," with no exceptions.

The TRIPS Agreement of the WTO brought some important differences, such as:

1 The TRIPS Agreement required that patents be granted to inventions, whether products or processes, in all fields of technology, as long as they are new, involve an inventive step, and are capable of industrial application. Before the TRIPS Agreement, countries could exclude some industrial or technological sectors from patentability, according to their own national development strategies. Pharmaceutical, food, and chemical products, for instance, were excluded from patentability by many countries, such as Brazil.[12]
2 After the TRIPS Agreement, countries were required not only to recognize and protect IP rights but also to establish mechanisms (through

administrative, civil, and criminal procedures) aimed at enforcing these rights at the domestic level.

3 The TRIPS Agreement states that IP rights are subjected to all principles that govern the international trade system (GATT/WTO), which means applying these principles to relationships among member countries. Among these, the main novelty (in relation to the Paris and Berne Conventions) is the principle of the "most favored nation" treatment, which sets forth that any advantages or privileges granted by a WTO member to a national of another country (whether or not a WTO member) are automatically and unconditionally extended to all other WTO members. By this principle, countries cannot be treated differently over the protection of IP (Roffe 2008, p. 54).
4 WTO principles of multilateral prevention and settlement of disputes between member countries also became applicable to IP rights. This included allowing the possibility of commercial retaliation, including "cross-retaliation." In other words, in the event of noncompliance by members of their obligations under TRIPS, a retaliation may be adopted by a complaining party in a different sector from that in which the infraction was made: goods, services or IP rights (Roffe 2008, p. 52). A violation of IP rights by one country can lead another country to retaliate in a different sector, such as goods or services.

According to the TRIPS Agreement, member countries do not have the obligation to implement, in their national legislation, more extensive protection than is required in the Agreement. They may, however, extend protection for IP above the minimum standards established in the TRIPS Agreement, and this has occurred through several bilateral and regional free trade agreements signed between the United States or the EU and developing countries. Known as "TRIPS-plus," these agreements impose obligations not included in TRIPS, such as mandatory adoption of the 1991 Act of the UPOV Convention and the obligation to patent plants, animals, and biotechnological inventions. Some examples are the free trade agreements between the United States and EU and Chile, Colombia, Ecuador, Peru, Singapore, and Sri Lanka, among others; and the free trade agreements between the EU and Algeria, Egypt, Korea, Syria, etc., which set forth such "additional" obligations, imposing stricter restrictions on the limited flexibility allowed by the TRIPS Agreement (Rajotte, 2008, pp. 142–145) and, consequently, preventing developing countries from adopting legal systems that are more coherent with their social, environmental, cultural, and economic needs, in favor of the interests of the global biotechnology market.[13]

In 2004, a free trade agreement involving the United States and Central American countries (Costa Rica, El Salvador, Guatemala, Honduras, and Nicaragua) and the Dominican Republic also imposed obligations on these countries to adopt the 1991 UPOV Convention and to ratify the Budapest

Treaty on the International Recognition of the Deposit of Microorganisms for the Purposes of Patent Procedure, which contains some regulations that are contrary to the Convention on Biological Diversity. More recently, in August, 2008, Colombia, Ecuador, and Peru, which are members of the Andean Community (a South American economic bloc made up of these countries, in addition to Bolivia[14]) decided that Peru could adopt its own legislation on IP in order to adapt to the free trade agreement between Peru and the United States. Bolivia voted against this measure because Decision 486 of the Andean Community of Nations sets forth a common regime of IP for Andean countries, and the decision to allow each country to choose its own rules weakens this common regime.

According to Article 27 of the TRIPS Agreement, countries must grant patents on products and processes in all fields of technology, with no discrimination regarding the place of invention and whether products are imported or locally produced. Before the TRIPS Agreement,[15] countries could establish the duration of patents in their territories, but the TRIPS Agreement states that the duration of patents cannot be less than 20 years, and member countries can exclude patentability only in the exceptional cases listed in Article 27. Article 27.3(b) contains the provisions that are more directly related to plant genetic resources:[16]

> Article 27.3 Members may also exclude from patentability:
> (b) plants and animals other than microorganisms, and essentially biological processes for the production of plants and animals, other than non-biological and microbiological processes. However, Members shall provide for the protection of plant varieties either by patents or by an efficient *sui generis* system, or by any combination thereof.

Member countries may exclude from patentability plants and animals (but not microorganisms), as well as essentially biological processes for the production of plants and animals (but not nonbiological and microbiological processes). However, member countries are obliged to protect plant varieties, and have to choose between a patents system, a *sui generis* system, or a combination of both. (The TRIPs Agreement does not define the term "*sui generis* system," but it is understood as a special, one of a kind, system; at no point in the text, however, is there any indication that the UPOV system would be the appropriate *sui generis* system.) As the TRIPS Agreement does not explain what exactly can be considered a *sui generis* system, WTO member countries are not obliged to become UPOV members, nor must they adopt any national laws in accordance with the UPOV Convention. Nonetheless, the International Union for the Protection of New Plant Varieties (UPOV) has always upheld the position that the adoption of its Convention is the most efficient form of protecting plant varieties.

Thus, many countries have chosen to follow the UPOV Convention model, through the adoption of national laws "inspired" by the UPOV model, even if they do not necessarily adopt the UPOV Convention. There are a number of reasons for such a strategy, including the fact that national laws inspired by the UPOV model are likely to be accepted more easily by the TRIPS Agreement Council than any other *sui generis* system and that UPOV and WIPO offer technical and administrative assistance for the development of these laws, which is rare in countries that choose other models of *sui generis* systems.

Up until 1998, it was still possible to become a UPOV member by adopting the 1978 UPOV Act. Since then, countries wishing to become members of this organization must adopt the 1991 UPOV Act, which requires much stricter protection for breeders' rights and is far more restrictive regarding farmers' rights. When the TRIPS Agreement came into effect, in 1996 (Article 65.1), the 1978 UPOV Act was still in force, as the 1991 UPOV Act only came into force in 1998. Brazil was one of the countries that joined the UPOV Convention under the 1978 Act, after approving a national IP law to protect plant varieties (Law 9456, of April 24, 1997).

US patents on plant varieties: utility patents and plant patents

The United States was the first country to adopt legislation that permits patenting of plant varieties, through its Plant Patents Act, adopted in 1930.[17] Plant patents can only be granted, however, for species of vegetative and asexual propagation[18] (except for edible tubers, such as potatoes), and are applied mainly to ornamentals, fruit, and forest trees. Keeping plants that undergo sexual reproduction or are propagated by tubers out of the US plant patents system reflected the dominant conception (when the Plant Patents Act was adopted, in 1930) that such plants were not sufficiently identifiable, uniform, and stable to justify patent protection. The Plant Patents Act was introduced primarily to benefit the horticulture industry.

In 1952, the United States published the Patent Act, which extended protection, by means of "utility patents," to other agricultural inventions, such as farming machinery and equipment, agrochemicals, etc. This law established a broad definition of what can be protected by a patent, and opened the way for patenting of biotechnological innovations and GMOs. In 1970, the United States published the Plant Variety Protection Act (PVPA), which applies only to sexually reproducing species. This law recognizes the breeder's exemption and farmers' rights to save seeds to be replanted on their lands, as well as to sell saved seeds to other farmers (both the buyer and the seller must be farmers whose primary farming occupation is the growing of crops for sale for other than reproductive purposes), but such farmers' rights were restricted by an amendment passed in 1994.

The 1994 amendment to the PVPA (which came into force in April, 1995) adapted the US legislation to the UPOV Convention (1991 Act), enabling the United States to ratify the Convention. This amendment extended IP protection to tuber-reproduced plants (such as potatoes) and to the first generations of hybrids (called F1). It also prohibited the sale of any farm-saved seed without the permission of the variety owner (the holder of IP right). The storage of seeds for planting in farmers' own lands was still allowed, but the sales of seeds of protected varieties became illegal. The amendment made it illegal, under any circumstances, to sell farm-saved seed for planting purposes of varieties protected by the PVPA after April 4, 1995. The 1994 amendment also extended protection to "essentially derived varieties" and to harvested material of the variety. The length of protection was increased from 18 to 20 years for most plants and to 25 years for trees, shrubs, and vines.

The PVPA grants the breeders' rights over varieties that are new, distinct, genetically uniform (homogeneous), and stable throughout successive generations. While "utility patents" apply to any invention that fulfills the patenting requirements (novelty, inventive step, and industrial application), the Plant Patents Act and the PVPA are aimed specifically at the protection of plant varieties. Table 5.1 summarizes the plant variety protection forms in the US model.

US legislation does not regulate the interactions among the three legal instruments (Plant Patents Act, PVPA, and Utility Patent Act), and it is possible to obtain protection for the same plant under both a utility patent and a plant patent, and the same variety can also be covered by several patents, granted to different holders. This overlap leads to countless legal disputes over these rights. Many decisions of the US courts have thus impacted the IP system, among which some of the most important, from a standpoint of plant variety protection, are these paradigmatic precedents:

Table 5.1 Forms of IP protection for plant varieties in the United States

Patent Act (utility)	*Plant Patents Act*	*Plant Variety Protection Act*
Protection for inventions in any technological area, as long as the requirements of novelty, inventive step and industrial application are met. Used mainly for biotechnological inventions	Protects only species of vegetative and asexual propagation (except for edible tubers, such as potatoes). Used mainly for ornamentals, fruit, and forest trees	Protects only species which undergo sexual reproduction. After the 1994 amendment was passed, tuber-propagated plants (such as potatoes) and first generations of hybrids (F1) were also covered by protection. The United States ratified the 1991 UPOV Convention.

Source: US Plant Patents Act, Plant Variety Protection Act, and Patent Act (http://www.patentlens.net/daisy/bios/1234#limited_types; accessed December 10, 2010).

- The US court ruling in the case of Diamond *vs.* Chakrabarty, in 1980, allowed (on a count of five votes in favor and four against) grant of patents for GMOs (products of biotechnology). In the case taken to court, the discussion revolved around whether a genetically modified bacterium (capable of breaking down crude oil and used in treating oil spills) should be considered a nonpatentable product of nature or a human invention, eligible for patent protection. The US Patent and Trademark Office had denied an application for a patent on the bacterium, under the argument that neither the Plant Patents Act of 1930 nor the PVPA of 1970 mentioned patents for bacteria, and a new law was thus necessary to make any other living organism eligible for patents. Ananda Chakrabarty, a microbiologist working for General Electric, appealed to the Patent Office Board of Appeals, which upheld the patent examiner's decision, and then Chakrabarty appealed to the US Court of Customs and Patent Appeals, which also confirmed the Board's decision. Chakrabarty then appealed to the US Supreme Court, which considered that the bacterium could be patented because it had distinctive characteristics, and was not a naturally occurring microorganism.
- The US Patent and Trademark Office's Board of Patent Appeals and Interferences[19] ruling in the Hibberd case, in 1985, about the possibility of granting a utility patent for a plant variety. Hibberd and his team of researchers had developed a corn variety that was rich in tryptophan (an amino acid), and claimed a patent on it, which was denied by the patent's examiner because he understood that only a breeder's right could be granted (in accordance with the PVPA) and not a utility patent (in accordance with the Utility Patent Act). To the patent examiner, the PVPA is more specific than the Utility Patent Act and is thus the only alternative for plant protection. However, the Board held that it was possible to grant a utility patent (available for inventions since 1790) for the corn variety.[20]
- US Supreme Court ruling in the case of Asgrow Seed Co. *vs.* Winterboer *et al.*, in 1995, which extended the prohibition of farmers selling seeds (of protected varieties) to varieties developed prior to April 4, 1995 (when the 1994 amendment to the PVPA came into force). This amendment forbids farmers to sell seeds of protected varieties without prior consent from the IP title-holder). The Supreme Court interpreted the PVPA to mean that a farmer could save only the amount of seed of a protected variety necessary to replant his own fields.
- US Supreme Court ruling in the case of J.E.M. Ag Supply *v.* Pioneer Hi-Bred International, in 2001, which upheld that both forms of protection (utility patent and plant breeder's right) could be granted to sexually reproduced plants. In doing so, the Court rejected the argument that the 1930 Plant Patent Act and the 1970 PVPA preclude the patenting of plants through utility patents. The US Supreme Court further held

that these three federal statutes can coexist and provide overlapping IP protection for plants.

Even though it is not directly related to plant genetic resources, the ruling in the case of Association for Molecular Pathology *et al. vs.* Myriad Genetics, University of Utah Research Foundation, and the US Patent and Trademark Office, is worth mentioning because of its potential to reshape IP law, especially in relation to patents on genes and on biotechnological inventions. Despite dealing with human genes, the judicial decision (if upheld by higher courts) can end up having an impact on IP rights over plant genetic resources too, since it ruled on the possibility of patenting isolated DNA molecules. On March 29, 2010, Judge Robert W. Sweet, of the US District Court for the Southern District of New York, invalidated seven patents granted to Myriad Genetics (a private company) on human genes which are associated with breast and ovarian cancer (known as breast cancer susceptibility genes 1 and 2, or *BRCA1* and *BRCA2*).[21] The US District Court held that genomic DNA that has merely been isolated from the human body, without further alteration or manipulation, is not patent eligible. According to the ruling, isolation does not transform a product of nature into a man-made invention. Myriad Genetics filed a notice of appeal on June 16, 2010, to the US Court of Appeals for the Federal Circuit, which has not been decided yet (as of January 30, 2011).

No European patents for essentially biological breeding processes: the broccoli and the tomato cases

On December 9, 2010, the European Patent Office (EPO)'s Enlarged Board of Appeal[22] rendered its decision in the so-called "broccoli" and "tomato" patent cases.[23] The Enlarged Board interpreted the term "essentially biological processes for the production of plants (or animals)," used in the European Patent Convention (EPC, Article 53b), to exclude classical plant breeding processes from patentability. In its decision, the Board concluded that a process for the production of plants involving sexually crossing whole plant genomes, and the subsequent selection of plants, is not patentable. The mere inclusion of a technical step that serves to enable or assist the performance of the steps of sexually crossing the whole genomes of plants or of subsequently selecting plants does not override the exclusion from patentability. According to the Board's decision, while technical devices or means, such as genetic markers, may themselves be patentable inventions, their use does not make an essentially biological process patentable. The Board also held that processes for producing plants by inserting or modifying a trait in the genome by using genetic engineering (which produces GMOs) do not rely on sexual crossing of whole genomes and may therefore be patentable.[24]

The legal questions referred to the Enlarged Board of Appeal arose in two plant breeding patent cases:

Case G2/07.[25] Plant Bioscience (a British private company) claimed a patent on a method of producing broccoli (*Brassica oleracea*) with elevated levels of glucosinolates[26] (organic compounds which are thought to prevent cancer). The method involves crossing wild *Brassica oleracea* (broccoli) species with broccoli double-haploid breeding lines. The patent holder (Plant Bioscience) argued that the wild *Brassica* strains grew in remote geographical locations and were not likely to hybridize with broccoli breeding lines unless specifically brought into contact with them by human intervention. According to Plant Bioscience, the crossing involved the use of a non-natural starting material, that is, a double-haploid strain, which was made by technical steps. Additionally, the use of molecular markers in the claimed process was a technical step requiring removal and *in vitro* analysis of plant tissues, argued the patent holder. Plant Bioscience held that there were at least three levels of human invention which brought the claimed invention outside the exclusion from patentability of "essentially biological processes." The patent owned by Plant Bioscience was opposed by Syngenta and by Limagrain.

The Board, however, ruled that "The use of molecular markers such as DNA markers is a well-known step in the selection of plants with desired characteristics. Methods to discover and produce molecular markers that segregate with a desired trait were commonly known in the art and had already been used in the context of *Brassica* species. This feature is therefore not able to contribute anything beyond a trivial level to the claimed invention. Double haploid breeding lines are, as such, well known in plant breeding, and techniques to obtain them in broccoli were publicly available before the priority date. The derivation of such breeding lines can therefore not be regarded as being the essence of the claimed invention or as contributing anything beyond a trivial level to it." Thus, the Board decided that the claimed broccoli breeding process is an "essentially biological process," and is therefore excluded from patentability.

Case G1/08.[27] The Ministry of Agriculture of Israel claimed a patent on a method for breeding tomato plants that produce tomatoes with reduced fruit water content. The method involves crossing at least one *Lycopersicon esculentum* (tomato) plant with another (tomato) species to produce hybrid seed, collecting the first generation of hybrid seeds, growing plants from the first generation of hybrid seeds, pollinating the plants of the most recent hybrid generation, collecting the seeds produced by the most recent hybrid generation, growing plants from the seeds of the most recent hybrid generation, and allowing fruit to remain

> on the vine past the point of normal ripening, as well as screening for reduced fruit water content as indicated by extended preservation of the ripe fruit and wrinkling of the fruit skin.
>
> The patent holder (Israel) argued first that interspecies crossing between *L. esculentum* and a wild tomato species required special intervention in order to achieve reliably fertile offspring, and would not take place in nature as generally individuals belonging to separate species are not capable of interbreeding, and, second, that selection for reduced fruit water content as indicated by extended preservation of the ripe fruit and wrinkling of the fruit skin would not occur in nature. According to the patentee, the exclusion from patentability should apply only if the claimed steps reflected and corresponded to phenomena that could occur in nature without human intervention.
>
> According to the Board's decision, however, "the arguments put forward by the patentee to show that the claimed method requires a high level of human intervention cannot alter the conclusion that the essence of the claimed method is classical plant breeding technology. Neither the necessity of an interspecific cross nor the choice of an unusual selection criterion nor the existence of technical steps such as weighing and drying take the claimed method outside the realm of classical plant breeding technology, which frequently uses corresponding elements of human intervention." Therefore, the Board decided that the claimed tomato breeding process is an "essentially biological process," and is therefore excluded from patentability.

The legal issues referred to the Enlarged Board of Appeal involved mainly the interpretation of Article 53(b) of the EPC, which sets forth that "European patents shall not be granted in respect of: (b) plant or animal varieties or essentially biological processes for the production of plants or animals; this provision shall not apply to microbiological processes or the products thereof." The Board also analyzed Article 26(5) of the Implementing Regulations to the EPC, according to which, "a process for the production of plants or animals is essentially biological if it consists entirely of natural phenomena such as crossing or selection," as well as Article 2(2) of the European Biotechnology Directive[28] (98/44/EC, 1998), which says that "a process for the production of plants or animals is essentially biological if it consists entirely of natural phenomena such as crossing or selection." The Board concluded that the purpose of the exclusion of patentability (established in Article 53b of the EPC) was to avoid the patenting of breeding processes in which technical measures or devices were used only as a means to bring about processes for the production of plants that were otherwise based on biological forces. According to the Board, although plant breeding processes by their nature involve human intervention, the conclusion to be drawn is that a process for the production of plants which is based on the sexual crossing of whole genomes and on the subsequent selection of plants, in which human intervention, including the provision of technical means, serves to enable or assist the performance of the process steps, remains excluded from patentability as being essentially biological within the meaning of Article 53(b) of the EPC. While the decisions of the Enlarged Board of Appeal clarified the legal concept of "essentially biological processes" under the EPC, the application of such concepts to concrete cases will be provided by the Technical Boards of Appeal, which had referred the questions to the Enlarged Board of Appeal, and are now called to decide upon the individual cases in the light of the guidance provided by the Enlarged Board of Appeal.

The 1978 and 1991 Acts of the UPOV Convention: main differences

The UPOV Convention has undergone successive revisions, which have granted increasingly weaker protection to farmers' rights, and brought it closer to the patent system. This is particularly the case of the last revision of the UPOV Convention, which was approved in 1991 and entered into force in 1998, and is referred to as the 1991 Act of the UPOV Convention.

The duration of the breeder's right is extended, from a minimum of 15 years, for most plant species (and 18 years for grapevines and trees), in the 1978 UPOV Act, to at least 20 years (and 25 years for grapevines and trees) in the 1991 UPOV Act. The scope of protection for the breeder's right is also expanded: in the 1978 UPOV Act, it covers production for commercial purposes, offering for sale, and marketing of propagating material of the plant variety. In the 1991 Act, it covers production or reproduction (multiplication) of the propagating material of the protected variety, conditioning, offering for sale, selling, exporting, importing, or stocking for any of the above-mentioned purposes. The object of protection has also expanded: in the 1978 UPOV Act it covers only the reproductive or vegetative propagating material, as such, of the plant variety, whereas in the 1991 UPOV Act it covers not only the reproductive or vegetative propagating material, but also the harvested material (including entire plants and parts of plants) – when obtained through the unauthorized use of propagating material of the protected variety, if the breeder has not had "reasonable opportunity" to exercise his or her right in relation to the propagating material. Furthermore, the 1991 UPOV Act also establishes that countries can extend protection to products made directly from harvested material of the protected variety (e.g., soya oil, soya flour). The concept of plant variety was not defined in the original version of the UPOV Convention, but according to Article 1(vi) of the 1991 Act of the UPOV Convention, "variety" means a plant grouping within a single botanical taxon of the lowest known rank, that, irrespective of whether the conditions for the grant of a breeder's right are fully met, can be defined by the expression of the characteristics resulting from a given genotype or combination of genotypes, distinguished from any other plant grouping by the expression of at least one of the said characteristics and considered as a unit with regard to its suitability for being propagated unchanged.

The 1978 UPOV Act allows breeders to use protected varieties as a source of variation for the development of new varieties, as well as to sell these newly developed varieties, without the breeder's authorization. Authorization by the breeder is required, however, when the repeated use of the variety is necessary for the commercial production of another variety. Under the 1991 Act of the UPOV Convention, the breeder's exemption is maintained, but limited.[29] If a new variety is so close to a protected variety that it can be considered "essentially derived" from it, the breeder's authorization is required, and the same rule applies to varieties that are not clearly distinguishable from the protected variety, and also to varieties whose production requires the use of the protected variety.

The 1991 Act of the UPOV Convention considers a variety to be essentially derived from another variety ("the initial variety") when (1) it is predominantly derived from the initial variety, or from a variety that is itself predominantly derived from the initial variety, while retaining the expression of the essential characteristics that result from the genotype or combination of genotypes of the initial variety; (2) it is clearly distinguishable from the initial variety; and (3) except for the differences which result from the act of derivation, it conforms to the initial variety in the expression of the essential characteristics that result from the genotype or combination of genotypes of the initial variety. Essentially derived varieties may be obtained for example by the selection of a natural or induced mutant, or of a somaclonal variant, the selection of a variant individual from plants of the initial variety, backcrossing, or transformation by genetic engineering (according to Article 14.5b,c of the 1991 Act of the UPOV Convention).

However, the concept of "essentially derived" variety is highly controversial,[30] and there is no consensus among breeders on a definition of a minimum genetic distance required for second-generation varieties to be treated as not essentially derived from an earlier variety and thus outside the first breeder's control. Nevertheless, the overall effect of this provision has been to narrow the breeders' exemption and expand the rights of first-generation breeders (Helfer, 2004, p. 28). "Essentially derived" varieties are, in some cases, the result of genetic engineering in which the breeder's innovation consists only of insertion of a new gene or a new DNA molecule. One of the objectives of the introduction of the notion of "essentially derived" varieties in UPOV 1991 was to limit the possibility of appropriation (through IP rights) of plant varieties developed through classical breeding methods, by GMO breeders, using just a quick genetic manipulation (i.e., the insertion of one gene). In addition, the "essential derivation" concept aims to reward actual breeding as opposed to cosmetic breeding or other forms of copying, such as the use of inbred lines derived from existing lines by repeated crossing (Louwaars, 2008), that is, to separate what is really an important breeding contribution from unimportant derivatives. However, as Cubero (2011) points out, the concept of an "essentially derived" variety is easier to understand than to put in practice, because this concept is not well defined in UPOV 1991 (e.g., Would a transgenic variety with a simple but very valuable transferred gene be an essentially derived variety?).

Another important distinction between the UPOV 1978 and 1991 Acts is related to farmers' rights. Under the 1978 UPOV Act, farmers could save seeds of protected varieties for use in coming harvests, without the breeder's authorization. Under the 1991 UPOV Act, the right of farmers to use saved seeds to replant their fields is subjected to recognition by national laws, which can establish such an exception to the breeder's right only "within reasonable limits and as long as the legitimate interests of the breeder are safeguarded, protected," and as long as farmers use saved seeds only on their own holdings.[31] The exchange of seeds among farmers is not allowed because farmers must multiply seeds on their own lands, to be used also on their own lands only, and not in someone else's holdings. Sales of seeds of protected varieties to other farmers are not permitted either. Under the 1991 UPOV Act, national laws may decide that farmers can no longer re-utilize saved seeds in future harvests, or that only some farmers (e.g., small-scale farmers) have this right, or that royalties must be paid to the breeders for farmers to be allowed to maintain this tradition. National laws can also limit the size of the lands, the quantity of seeds, and the plant species to which the farmers' rights to save seed are applicable. According to the European Council Regulation 2100/94, on Community Plant Variety Rights (Article 14), farmers' rights are governed by the following rules: the farmers' right to save seeds is restricted to approximately 20 species (including fodder plants, cereals, potatoes, oils, and fiber plants) and requires that they pay an "equitable remuneration" to the holder of breeders' rights, which must be "sensibly lower than the amount charged for the licensed production of propagating material of the same variety in the same area." Small-scale farmers (defined as those who grow plants on an area no bigger than the area that would be needed to produce 92 tonnes of cereals) are exempt from the payment of royalties. On the other hand, Directive 98/44 (Article 11.1) on the Legal Protection for Biotechnological Inventions of the European Parliament and Council sets forth that the farmers' right to save and re-use seeds of plant varieties that contain patented components (usually genetically modified) must be maintained. Therefore, farmers can also reproduce seeds with genetically modified and patented components for their own use (within the limits imposed by Article 14 of the European Council Regulation 2100/94, on Community Plant Variety Rights).

Another difference between the 1978 UPOV Act and the 1991 UPOV

Act is that the former explicitly prohibits double protection (by breeder's right and patent) for a single variety, and countries that allow both forms of protection must allow only one form in relation to a single botanical genus or species. The prohibition of double protection was removed from the 1991 UPOV Act. The UPOV system thus ceased to be an alternative to the patent system, in its original form, as the 1991 UPOV Act permits the breeder's right to be used as an additional protection for patents. The UPOV system is actually becoming increasingly similar to the patent system, specially for countries that ratified the 1991 UPOV Act. Currently, new countries wishing to join UPOV must adopt the 1991 UPOV Act, since UPOV membership under the 1978 Act was only possible until 1998. The main differences among the two UPOV Acts and the patent system are summarized in Table 5.2.

Some countries that said no to UPOV

It is possible to be a WTO member and meet obligations under the TRIPS Agreement without necessarily having to adopt one of the UPOV Convention Acts or even having to become a member of UPOV. India, for example, approved its Plant Variety Protection and Farmers' Rights Act in 2001. The Indian law grants rights to both (professional) breeders and farmers. Breeders are granted exclusive rights over new varieties, and farmers have the right to save, use, plant, replant, exchange, and sell seeds. However, when farmers sell seeds of protected varieties, they cannot use the same or even similar packaging to that of the breeder, nor can they use the same or a similar name or denomination of a registered variety. The Indian law combines aspects of the UPOV Convention, regarding breeders' rights, and principles established by the Convention on Biological Diversity regarding access to genetic resources and associated traditional knowledge, and will be analyzed in more depth in Chapter 8 of this book. It is important to note that India is a member of the WTO, and signed the TRIPS Agreement, but adopted a national law that does not correspond to either of the UPOV Convention Acts – it is a *sui generis* system.[33]

The option of not becoming a UPOV member, in spite of being part of the WTO and having to abide by the rules of the TRIPS Agreement, has been exercised by many Asian countries. Some examples are Nepal, Bangladesh, Pakistan, and Sri Lanka, which are not UPOV members.[34] In 2003, Nepal was subjected to strong pressure from the Unites States to join UPOV, as part of its obligations under the TRIPS Agreement. However, as the result of a widespread campaign carried out by the Nepalese civil society, showing the negative impacts of the UPOV system on traditional agricultural systems, the Nepalese government ended up resisting the pressure from the United States, and did not join UPOV (Andersen and Winge, 2008, pp. 53–56).

Thailand also chose not to become a member of UPOV, although its national legislation has an orientation similar to that of the 1978 UPOV Act, which allows more flexibility to developing countries and a broader recognition of farmers' rights. The Thai law, passed in 1999, establishes several levels of protection for different categories of plant varieties: new varieties (developed or discovered by breeders), local domestic varieties (which only exist in one specific location within the country and have never been registered), general domestic varieties (originating from or existing in the country with no previous registration), and wild varieties (which exist in their natural habitat and are not cultivated). It also establishes the participation of members appointed by the farmers in the Committee for Protection of Plant Varieties, responsible for registration of different plant varieties. However, Thailand has signed free trade agreements with Bahrain, China, India, and Australia, and is in the process of negotiating with Japan and the United States, its two largest trading partners. The free trade agreements, especially the proposed Thai–US version, contain detailed provisions on enforcement of IP rights protection which may force Thailand to step up its level of IPR protection (Naboriboon, 2007).

Some African countries, such as Namibia and Uganda, have proposed national laws based on the model adopted by the Organization of African Unity, an international organization founded in 1963, preceding the African Union, which was constituted in 2002 and is currently made up of all countries in the African continent, except for Morocco, with the objective of promoting cooperation among African countries. The African Model Law[35] was passed during the 34th Meeting of the Organization of African Unity, held in Burkina-Faso in 1998, at which it was decided that African countries should adopt that model in their national laws, adapting it to the peculiarities of each country. The African model law aims to implement both the Convention on Biological Diversity and the WTO TRIPS Agreement, particularly regarding protection for plant varieties, and both farmers' and breeders' rights are covered (the African Model Law is further discussed in Chapter 8 of this book). The only African countries that became UPOV members are Tunisia and Morocco (which ratified the 1991 UPOV Act) and Kenya and South Africa (which ratified the 1978 UPOV Act).

Other countries, however, joined UPOV under the 1978 UPOV Act and have no intention of ratifying the 1991 UPOV act, owing to restrictions imposed on farmers' rights. In 2005, the Norwegian government rejected a legal bill which made breeders' rights stronger, aimed at adapting Norwegian legislation to the 1991 UPOV Act. The bill was rejected for two main reasons: too many limitations would be imposed on farmers' rights to save, re-use, and exchange seeds, and Norwegian farmers would be forced to buy seeds each year. Norway is a UPOV member based on the 1978 UPOV Act and upholds its right to remain a UPOV member under the 1978

Table 5.2 Main differences among the 1978 and 1991 UPOV Acts and the patent system

	1978 UPOV Act	*1991 UPOV Act*	*Patent system (TRIPS Agreement)*
Protection coverage	Plant varieties of nationally defined genera or species (member countries may limit botanical genera and species covered by protection[32])	Plant varieties of all genera and species must be protected	Any invention, be it a product or a process, in all technological sectors (exceptions are listed on Article 27) must be protected
Conditions for protection	Novelty, distinctness, homogeneity, stability, variety denomination	Novelty, distinctness, uniformity, stability, variety denomination	Novelty, inventive step (or non-obviousness) activity and industrial application
Protection duration	At least 15 years (18 years for vines and trees)	At least 20 years (25 years for vines and trees)	No less than 20 years
Protection scope	Production for purposes of commercial marketing, offering for sale, and marketing of the propagating material of the plant variety. Protection covers vegetative or reproductive propagating material, as such, of the variety	Production, conditioning, offering for sale, selling, exporting, importing, or stocking for above purposes of propagating materials of the variety. Protection may also cover the harvested product, if obtained through an unauthorized use of propagating material, and if the breeder has had no "reasonable opportunity" to exercise his or her right in relation to the propagating material. Protection may also be extended to products made directly from harvested material of the protected variety	In respect of a product: making, using, offering for sale, selling or importing, stocking for purposes of offering for sale, etc. In respect of a process: using the process, doing any of the above-mentioned acts in respect of a product obtained directly by means of the process

Breeders' exemption	Yes, the breeders' exemption is recognized. However, when the repeated use of the variety is necessary for the commercial production of another variety, the breeder's authorization is required	Yes, the breeders' exemption is recognized. However, "essentially derived" varieties and varieties that are not distinguishable from the protected variety are not included in the breeder's exemption. When the repeated use of the variety is necessary for the commercial production of another variety, the breeder's authorization is also required	Depends on the laws in each country, but exemptions are usually for research/experimental use
Farmers' "privilege"	Not explicitly mentioned, but, since the breeder's authorization is required only for production with commercial purposes, farmers can use saved seeds (of protected varieties) for replanting on their own lands, as well as exchange seeds among themselves. Sales of protected seeds depend on exceptions to the breeder's right established in national laws	Each country may decide whether to allow (or not) "within reasonable limits and as long as the legitimate interests of the breeders are safeguarded" that farmers use, for propagation purposes only, and on their own lands, the product of the harvest of protected varieties. Exchanges among farmers are not allowed	Generally no provision is made, but national laws may establish it
Double protection (under patents and plant breeders' right)	Not allowed. Only one form of protection (patent or breeders' rights) is allowed for the same botanical genus or species	Allowed	Up to national laws

UPOV Act (Andersen, 2008, pp. 53–56). China is also a member of UPOV under the 1978 UPOV Act, as are Brazil, Argentina, Paraguay, Uruguay, Chile, Colombia, Ecuador, and Mexico.[36] Many countries in the Americas have been forced, however, to adopt stricter IP regimes owing to bilateral or regional free trade agreements with the United States and the EU. Such agreements are known as "TRIPS-plus," and they establish obligations such as joining UPOV 1991 and "making reasonable efforts to provide patents on plants" (in the case of the US regional agreement with Central American countries and the US bilateral agreements with Chile, Colombia, and Peru (Rajotte, 2008, pp. 142–145).

Patents and the UPOV system: compulsory cross-licenses

Although WTO member countries can choose to adopt a *sui generis* plant variety protection system (in accordance with Article 27.3b of the TRIPS Agreement), they are forced, by this agreement, to allow patenting of microorganisms and nonbiological and microbiological processes (such as transgenic plants and genetic engineering processes). Therefore, a plant variety can also incorporate an invention, which can be protected by a patent, most likely in the form of a genetically engineered component (a gene or a gene sequence). The question thus arises: can a patent over a genetic component of a plant variety restrict its use in development of new varieties? The reason for this question is that the breeders' exemption, established in the UPOV system, grants access to protected varieties for the development of new varieties, and the patent system in general does not grant this exemption, thus restricting access to the patented component of the variety, even for breeding and research purposes.

To avoid the possibility that a patent blocks access to plant genetic resources needed for the development of new varieties, and to regulate the cases of interdependence in plant biotechnologies, European Community Directive 98/44 on the Legal Protection for Biotechnological Inventions sets forth that member countries must allow breeders to apply for a compulsory license for nonexclusive use of the patented invention, upon payment of royalties to the holder of the patent. When such a license is granted, the holder of the patent will be entitled to a cross-license to use the protected variety. On the other hand, if the holder of a patent over a biotechnological invention cannot exploit it without infringing a prior plant variety right, he may apply for a compulsory license for non-exclusive use of the plant variety protected by that right, subject to payment of an appropriate royalty. When such a license is granted, the holder of the plant variety right will also be entitled to a cross-license to use the protected invention, upon payment of royalties (according to Article 12 of the European Community Directive 98/44). Applicants for such compulsory licenses must, however, demonstrate that (a) they have applied unsuccessfully to the holder of the patent or of the plant variety right to obtain a contractual license; and (b) the plant variety or the invention constitutes significant technical progress of considerable economic interest compared with the invention claimed in the patent or the protected plant variety.

Such compulsory cross-licenses aim to guarantee a "remunerated open access" to genetic resources. Van Wijk and Louwaars (2002) point out, however, that the above-mentioned conditions for compulsory cross-licenses are too vague and costly for breeders. Development of a new plant variety usually takes many years, and it is too difficult to show what economic and technical value a possible end product may have at the start of the research, and that is when the compulsory license (for use of the protected component) must be requested.

When transposing Directive 98/44 into national laws, France[37] and Germany[38] adopted more flexible laws to protect the breeders' exemption and the advancement of scientific research, by expressly allowing breeders to use genetic materials which contain patented components. In the French law, access to genetic diversity is safeguarded, including to genetically modified varieties that integrate one or several patented genes, as the patent covering a gene in a GMO is not extended to the plant variety as a whole (Trommetter, 2008). There is thus free access to the genetic diversity of the GMO minus the patented gene(s). However, if a new variety is developed and it contains the patented genetic component, the breeder's authorization is required for marketing of the new variety. If the patented genetic component is bred out of the new variety, however, the holder of the patent has no right over the new variety. This keeps the genetic background of the variety in the public domain while protecting the rights of the patent holder.

Notes

1 Plant breeding is the science of changing the genetics of plants for different purposes, such as disease resistance, drought tolerance, higher yields, etc. The term "plant breeding" is often used as a synonym for "plant improvement." Even though the phrase "to breed plants" often connotes the involvement of a sexual process in effecting a desired change, modern plant breeding also includes the manipulation of asexually reproducing plants. Breeding is hence about manipulating plant attributes, structure, and composition, to make them more useful to humans. Source: Acquaah (2007, p. 3).

2 The introduction of hybrids (particularly hybrid corn, starting in the 1920s and 1930s), partially overcame this "difficulty" pointed out by plant breeders: the increased yield in hybrids (known as hybrid vigor or heterosis) is not transmitted to new generations, which forces farmers to buy new seeds each year, in order to maintain high yields of their crops. Therefore, a stable market for hybrids is created. However, plant breeders developed IP rights over plant varieties, to legally oblige farmers to buy seeds (of nonhybrid varieties) every year, and to prevent them from multiplying seeds for sale. In the breeders' view, plant varieties should be protected so that their seeds cannot be replanted and sold, in competition with the breeder himself (Dutfield, 2008, p. 30). The IP protection

represents a "legal control," as opposed to "biological" or "natural control," represented by hybrids. Hybrid corn had a major role in the development of a private breeding and seed production sector, especially in the United States. Several of the world's major seed companies – such as Monsanto, Syngenta, and Dupont – have developed successful breeding programs of hybrid corn varieties (Dutfield, 2008, p. 30).

3 The International Association for the Protection of Industrial Property (AIPPI) and the International Association of Plant Breeders for the Protection of Plant Varieties (ASSINSEL) had decisive roles. In May 2002, ASSINSEL merged with the International Seed Trade Federation to become the International Seed Federation. The International Community of Obtainers of Asexual Reproduction Ornamental and Fruit Plants and the International Chamber of Trade also had strong influence on the Convention and its revisions. Source: Dutfield (2008, pp. 27–47).

4 Directive 98/44/EC, of the European Parliament and Council, on the Legal Protection of Biotechnological Inventions, forbids patenting of plant and animal varieties, as well as essentially biological processes for the production of plants and animals. Therefore, individual plant varieties per se are *not* patentable. Nevertheless, the Directive allows patenting of inventions (concerning plants and animals) whose technical feasibility is not confined to a particular plant or animal variety (Article 4.2). The preamble (No. 32) of Directive 98/44/EC also specifies that if an invention consists only in genetically modifying a particular plant variety, and if a new plant variety is bred, it will still be excluded from patentability. Thus, patents cannot be granted for a single plant variety, but they can be granted to an invention concerning a plant grouping which is characterized by a particular gene (and not its whole genome).Transgenic plants are patentable if they are not restricted to a specific plant variety but represent a broader plant grouping. See also the decision of the Enlarged Board of Appeal of the EPO, in case G0001/98 (Transgenic Plants/Novartis II) issued on December 20, 1999 (http://legal.european-patent-office.org/dg3/pdf/g980001ex1.pdf; accessed January 20, 2011).

5 Both processes and the resulting products are eligible for patents.

6 The expression "farmers' privilege" is politically biased. Since the dawn of agriculture, farmers have produced their own seeds, and saved the best ones for future harvests. The IP rights regime reduces such a traditional practice to a "privilege," as if it were a generous concession to farmers made by professional breeders. In reality, it was the breeders who obtained the "privilege" of freely accessing seeds that farmers have cultivated for millennia.

7 In 1972, minor adjustments were made in the UPOV Convention, regarding financial contributions of member countries.

8 UPOV 1978 Act entered into force in 1981, and the 1991 UPOV Act in 1998.

9 Up until the TRIPS Agreement was signed, in 1994, South Africa was the only developing country to be a member of UPOV. South Africa joined UPOV in 1977.

10 The TRIPS Agreement was incorporated as Annex 1C of the Marrakesh Agreement, which established the WTO. The Marrakesh Agreement was concluded on April 15, 1994, and entered into force on January 1, 1995, and the TRIPS Agreement came into effect a year later, on January 1, 1996 (Article 65.1). In this book, we are discussing issues that are more closely related to plant variety protection. For a more comprehensive analysis of the TRIPS Agreement, see Correa (2000) and Carvalho (2002). The WTO succeeded the GATT, created in 1947 as a permanent forum for matters related with international trade and for promoting its liberalization. It was during the negotiations of the Uruguay

Round (which lasted from 1986 to 1994) that GATT, as an organization, was replaced by the WTO. GATT was kept only as an agreement, pertaining to goods. There is also WTO's General Agreement on Trade in Services, and other agreements dealing with specific sectors.

11 In 2000, the WIPO General Assembly established an Intergovernmental Committee on IP and Genetic Resources, Traditional Knowledge and Folklore, to examine, among other things, IP issues arising in the context of access and benefit-sharing and traditional knowledge. However, as it does not deal specifically with plant genetic resources for food and agriculture, this book will not focus on its activities.

12 Brazil's previous patent law (No. 5772, of 1971) did not allow patenting of food products, chemical–pharmaceuticals, and medications of any kind. After the ratification of the TRIPS Agreement by the Brazilian Congress, a new patent law (9279, of 1996) was enacted, allowing the patenting of such products and processes.

13 See www.bilaterals.org (accessed December 10, 2010).

14 In 2006, President Hugo Chávez announced the withdrawal of Venezuela from the Andean Community, arguing that free trade agreements signed by Colombia and Peru with the Unites States had caused irreparable damage to Andean institutions.

15 The transition period (for the implementation of the TRIPS Agreement) for developing countries expired on January 1, 2000. However, an additional period of five years (until 2005) was granted for developing countries that had to extend patent protection to areas of technology which were not patentable in its territory until the application of the TRIPS Agreement. In the case of the least developed countries, the TRIPS Council extended the deadline for implementation of the agreement until July 1, 2013, and until January 1, 2016, for pharmaceutical patents.

16 Article 27.2 (of the TRIPS Agreement) sets forth that member countries may exclude inventions from patentability when it is necessary to protect the public order or morality, including protection of human, animal, or plant life or health or to avoid serious damage to the environment. Diagnostic, therapeutic, and surgical methods for treatment of humans or animals can also be excluded from patentability (Article 27.3).

17 The first Patent Law passed in the USA, in 1790, and its later amendments did not allow patenting of new plant varieties; on the contrary, biological innovations (such as plant varieties) were considered "products of nature," and could therefore not be patented. The US Congress revised the Patent Law in 1793 and then in 1952.

18 Modes of asexual reproduction in plants include grafting, bulbs, apomictic seeds, rhizomes, and tissue culture.

19 Appeal No. 645–91. Source: www.uspto.gov/go/dcom/bpai/index.html (accessed December 12, 2009).

20 A utility patent is often sought for products related to genetic engineering because utility patents can be granted to the process used to genetically engineer a plant, to the genetic sequences that are inserted, and to the plant that results from the process.

21 The *BRCA1* and *BRCA2* genes are located on human chromosomes 17 and 13, respectively, and they express proteins that assist in the repair of damaged DNA and the suppression of tumors. Mutations in these genes are associated with significantly increased risks of breast and ovarian cancer. Source: US Court of Appeals for the Federal Circuit. Case No. 2010-1406.

22 Decisions of the first instances of the EPO can be appealed (challenged) before the Boards of Appeal of the EPO. In addition to the Boards of Appeal, the EPO includes an Enlarged Board of Appeal, which is the highest instance in the EPO's judiciary and only takes decisions when important points of law arise. Its purpose is to ensure uniform application of the law and to clarify and interpret important points of law in relation to the EPC.

23 Source: http://www.ipeg.eu/blog/wp-content/uploads/Plant-Breeding-Method_EPO_G1_08_en.pdf (accessed January 20, 2011).

24 Source: http://www.epo.org/topics/news/2010/20101209a.html (accessed January 20, 2011).

25 Case G2/87 arose from a referral of Technical Board of Appeal 3.3.04 in case T83/05 – Plant Bioscience/Broccoli.

26 The glucosinolates are a class of organic compounds that contain sulfur and nitrogen and are derived from glucose and an amino acid. Glucosinolates are well known for their toxic effects in both humans and animals at high doses. In contrast, at subtoxic doses, they act as chemoprotective agents against chemical carcinogens by blocking the initiation of tumors in a variety of tissues, such as the liver, colon, mammary gland, and pancreas.

27 Case G1/08 arose from a referral by Technical Board of Appeal 3.3.04 in case T1242/06 – State of Israel/Tomatoes.

28 EU directives are a form of EU legislation which contains deadlines for the implementation ("transposition") of the rights and obligations in the directive into the law of the member states.

29 Exceptions to the breeder's right under the 1991 Act of the UPOV Convention also include acts done privately and for noncommercial purposes, acts done for experimental purposes, and acts done for the purpose of breeding other varieties where the varieties so obtained are not essentially derived (Article 15).

30 The International Seed Federation has adopted guidelines for dealing with disputes on essentially derived varieties of perennial ryegrass, maize, oilseed rape, cotton, and lettuce. It has also adopted a protocol for assessing the genetic distance between diploid perennial ryegrass varieties, as well as technical rules for establishing a threshold for essential derivation, which are available on its website (http://www.worldseed.org/isf/edv.html; accessed January 20, 2011).

31 A common mistake regarding the 1991 UPOV Act is to assume that it forbids farmers from using saved seeds for use in coming harvests in any circumstance. The UPOV Convention and any legislation based on it are applicable only to protected varieties (by means of IP rights). Varieties in the public domain do not have these restrictions (although many seed laws also impose restrictions on farm-saved seed)

32 The 1978 UPOV Act establishes that protection may be applied to all botanical genera and species, and member countries agreed to "adopt all measures necessary for the progressive application of protection to the largest possible number of botanical genera and species."

33 India applied for UPOV membership in 2002, but its application has not been decided (as of January 20, 2011), which shows that UPOV will probably not accept a *sui generis* system different from the one established in its own convention.

34 As of January 10, 2011, according to the UPOV website (http://www.upov.int).

35 Its complete name is African Model Law for the Protection of the Rights of Local Communities, Farmers and Breeders and for the Regulation of Access to Biological Resources.

36 As of January 10, 2011.

37 French Law 2004-1338, of December 8, 2004, Articles 5 and 6.

38 The transposition of the European Directive in Germany was made by the Biopatent Act of January 21, 2005, which came into force on February 28, 2005.

Bibliography

Acquaah, G. (2007) *Principles of Plant Genetics and Breeding*, Blackwell Publishing, Malden, MA.

Andersen, R. (2008) *Governing Agrobiodiversity: Plant Genetics and Developing Countries*, Ashgate, Aldershot, UK.

Andersen, R. and Winge, T. (2008) *The Farmers' Rights Project – Background Study 7: Success Stories from the Realization of Farmers' Rights Related to Plant Genetic Resources for Food and Agriculture*. FNI Report 4/200, Fridtjof Nansen Institute, Lysaker, Norway (http://www.farmersrights.org/fr-project/second_phase_5.html; accessed March 20, 2010).

Carvalho, N.P. (2002) *The TRIPS Regime of Patent Rights*, Kluwer Law International, London,

Correa, C. (2000) *Intellectual Property Rights, the WTO and Developing Countries: The TRIPS Agreement and Policy Options*, Zed Books, London, and Third World Network, Penang, Malaysia.

Cubero, J.I. (2011) 'The point of view of a plant breeder on the International Treaty on Plant Genetic Resources for Food and Agriculture', in Frison, C., López, F. and Esquinas-Alcázar, J.T. (eds.) *Plant Genetic Resources and Food Security: Stakeholder Perspectives on the International Treaty on Plant Genetic Resources for Food and Agriculture*, Earthscan, London.

Curci, J. (2010) *The Protection of Biodiversity and Traditional Knowledge in International Law of Intellectual Property*, Cambridge University Press, Cambridge, UK.

Dutfield, G. (2002) *Intellectual Property Rights, Trade and Biodiversity*, Earthscan, London.

Dutfield, G. (2008) 'Turning plant varieties into intellectual property: the UPOV Convention', in Tansey, G and Rajotte, T. (eds.) *The Future Control of Food: A Guide to International Negotiations and Rules on Intellectual Property, Biodiversity and Food*, Earthscan, London, pp. 27–47.

Erbisch, F. and Maredia, K. (eds.) (2003) *Intellectual Property Rights in Agricultural Biotechnology* (Biotechnology in Agriculture Series), Cabi, Wallingford, UK.

Helfer, L.R. (2004) *Intellectual Property Rights in Plant Varieties: International Legal Regimes and Policy Options for National Governments* (FAO Legislative Study, 85), FAO, Rome.

Hermitte, M.A. and Kahn, P. (eds.) (2004) *Les Ressources Génétiques Végétales et le Droit dans les Rapports Nord-Sud* (Travaux du Centre René-Jean Dupuy pour le Droit et le Développement et du Centre de Recherche sur le Droit des Sciences et Techniques, v. II), Bruylant, Brussels.

Leskien, D. and Flitner, M. (1997) *Intellectual Property Rights and Plant Genetic Resources: Options for a Sui Generis System* (Issues in Genetic Resources no. 6), International Plant Genetic Resources Institute, Rome.

Lightbourne, M. (2009) *Food Security, Biological Diversity and Intellectual Property Rights*, Ashgate, Farnham, UK.

Louwaars, N. (2008) 'Seed systems and plant genetic resources for food and agriculture' (www.fao.org/docrep/013/i1500e/i1500e21.pdf; accessed January 20, 2011).

Mahop, M.T. (2010) *Intellectual Property, Community Rights and Human Rights: The Biological and Genetic Resources of Developing Countries*, Routledge, London.

Marin, P.L.C. (2002) *Providing Protection for Plant Genetic Resources: Patents, Sui Generis Systems and Biopartnerships*, Kluwer Law International, London.

Naboriboon, P. (2007) 'Plant variety protection in Thailand' (www.tillekeandgibbins.com/Publications/Articles/IP/plant_variety_protection.pdf; accessed July 10, 2010).

Parvin, M.R (2009) 'Patentability of plants: technical and legal aspects', *Journal of Intellectual Property Rights*, vol. 14, pp. 203–213.

Rajotte, T. (2008) 'The negotiations web: the complex connections', in Tansey, G. and Rajotte, T. (eds.) *The Future Control of Food: A Guide to International Negotiations and Rules on Intellectual Property, Biodiversity and Food*, Earthscan, London, pp. 141–167.

Rimmer, M. (2008) *Intellectual Property and Biotechnology: Biological Inventions*, Edward Elgar Publishing, Villinstone, VT.

Roffe, P. (2008) 'Bringing minimum global intellectual property standards into agriculture: the Agreement on Trade-Related Aspects of Intellectual Property Rights (TRIPS)', in Tansey, G. and Rajotte, T. (eds.) *The Future Control of Food: A Guide to International Negotiations and Rules on Intellectual Property, Biodiversity and Food*, Earthscan, London, pp. 48–68.

Tripp, R, Eaton, D. and Louwaars, N. (2006) Intellectual property rights: designing regimes to support plant breeding in developing countries (Report 35517-GLB), World Bank, Washington, DC.

Trommetter, M. (2008) *Intellectual Property Rights in Agricultural and Agro-food Biotechnologies to 2030*, OECD International Futures Project on 'The Bioeconomy to 2030: Designing a Policy Agenda', OCDE, Paris (http://www.oecd.org/dataoecd/11/56/40926131.pdf; accessed March 15, 2011).

Van Overwalle, G. (1999) 'Plant Protection for plants: a comparison of American and European approaches', *Journal of Law & Technology*, vol. 39, pp. 143–194 (www.ipmall.org/hosted_resources/IDEA/39_IDEA/39–2_IDEA_143_VanOverwalle.pdf; accessed January 10, 2008).

Van Wijk, A. and Louwaars, N. (2008) *Framework for the Introduction of Plant Breeders' Rights in Countries with an Emerging Plant Variety Protection System*, Plant Variety Protection Course. Centre for Genetic Resources, Naktuinbouw, the Netherlands.

Würtenberg, G., Kiewit, B. and Van der Kooij, P.A.C.E. (2006) *European Community Plant Variety Protection*, Oxford University Press, New York.

6

ACCESS AND BENEFIT-SHARING LAWS AND PLANT GENETIC RESOURCES FOR FOOD AND AGRICULTURE

The international legal regime

Historical background: FAO conferences in 1961, 1967, and 1973 – discussions on *ex situ* and *in situ* conservation of plant genetic resources

Plant genetic resources are the basis of any agricultural system and of agrobiodiversity at all levels. Along with water and soil, they are essential for any agricultural model and, therefore, for food security. The set of genes in a plant is fundamental for determining characteristics such as resistance to diseases and insects, droughts or floods, color, flavor, nutritional value, etc. Hereditary characteristics are transmitted from one generation to the next by genes, and both farmers and professional breeders depend upon wide access to diversified genetic materials in order to develop and/or improve agricultural varieties and to adapt them to new environmental and sociocultural conditions. Wild relatives of cultivated plants are also very important, since their capacity for survival in adverse situations is usually very high, and this makes them especially valuable as sources of traits for developing the so-called "climate-ready" crops.

The concept of genetic resources – any genetic material with real or potential value – was developed in the 1960s and 1970s to emphasize the strategic social and economic value of genetic information contained in them, which justifies them being treated as "resources." Genetic diversity should be protected in order to ensure food security for all humankind, both now and in the future. Genetic resources were regarded, essentially, as the indispensible input for plant breeding. The concept of genetic resources – later adopted by many legal instruments[1] – emphasized the economic and utilitarian value of these resources, underestimating the cultural and identity value they have for farmers and local communities. Conservation of agrobiodiversity has far broader and more comprehensive implications than conservation of plant genetic resources. In this book, we shall focus on agrobiodiversity as biological and cultural heritage, and plant genetic resources as one of the components of such heritage, considering that policies and legal instruments must consider both the biological support of agricultural biodiversity and the knowledge and practices associated with it.

Loss of agricultural biodiversity and genetic erosion reached alarming levels in the 1960s and 1970s. Expansion of the Green Revolution provoked – and continues to provoke – widespread substitution of traditional and local varieties, which are genetically heterogeneous by modern cultivars, with high yields and narrow genetic bases. Countless agricultural species and varieties have disappeared, along with agricultural knowledge associated with agrobiodiversity. And it is not only species domesticated by humans that have been disappearing, but also their wild relatives, owing to the rapid devastation of natural ecosystems.

Concern with extinction of agricultural species and varieties motivated a series of conferences and technical meetings to discuss plant genetic resources and the most appropriate strategies for their conservation, held by FAO in 1961, 1967, and 1973. According to Robin Pistorius (1997), the three events laid the foundation for the scientific premises behind current international *ex situ* conservation (out of their natural habitats, in germplasm banks) of plant genetic resources. During the 1967 conference, it was decided that *ex situ* conservation was considered preferable to *in situ* conservation, and during the 1973 conference the scientific criteria for *ex situ* conservation were established.

The decision to give priority to *ex situ* conservation generated diverging opinions in 1967 among scientists gathered at the FAO conference. As different conservation strategies, both *ex situ* and *in situ*, are still at the center of current debate, we will highlight below some of the main points of divergence between two renowned scientists, Otto Frankel (Australian) and Erna Bennett (Irish), who led the debates at the FAO. They were the ones who, in the 1960s, coined the expression "genetic resources" to describe the strategic importance of gene conservation and the risks that genetic erosion would impose on food security.

Otto Frankel and Erna Bennett[2] had fundamental roles in international discussions on plant genetic resources conservation. In 1970, Otto Frankel and Erna Bennett (in association with R. Brock, A. Bunting, J. Harlan, and E. Schreiner) published the book *Genetic Resources in Plants: Their Exploration and Conservation*, which became a reference for debates regarding conservation and use of plant genetic resources in the 1970s and 1980s. The book helped to convince participants of the UN Conference on the Human Environment, held in Stockholm, Sweden, in 1972, of the need to adopt a global plant genetic resource conservation program. In 1975, Otto Frankel and Jack Hawkes edited another reference book, *Crop*

Genetic Resources for Today and Tomorrow, more action oriented, offering specific scientific, technical, and organizational solutions to start programs to collect and conserve threatened genepools. The books resulted from FAO conferences of 1967 and 1973, respectively (Pistorius, 1997).

Otto Frankel argued mainly in favor of the advantages of *ex situ* conservation. In Frankel's opinion, to make genetic material useful for breeding purposes it had to be preserved under controllable conditions and should not be left in the field exposed to continuously changing agricultural practices. As a corn breeder, Frankel saw germplasm banks mainly as genetic material stocks for breeding purposes. Frankel believed that *ex situ* conservation created a safe niche for genetic resources, in which they would be protected against alterations provoked by external factors. Frankel did not rule out *in situ* conservation, but considered it more complex, difficult, and risky (Pistorius, 1997, p. 26).

Erna Bennett and other scientists agreed and held that there was an urgent need to adopt *ex situ* conservation, due to the alarming genetic erosion in the fields. They feared, however, that if *ex situ* conservation became dominant, local varieties would lose their capacity to adapt. Erna Bennett went as far as claiming that "static" conservation of seeds, through storage, meant "adopting museum concepts." To Bennett, "The purpose of conservation is not to capture the present moment of evolutionary time, in which there is no special virtue, but to conserve material so that it will continue to evolve," and, according to her, "such continued evolution could only be possible in *in situ* collections" (Pistorius, 1997, p. 27).

According to Pistorius (1997, p. 27), there were, however, other areas of disagreement between Erna Bennett and Otto Frankel. Frankel thought that locally adapted variability would probably be of little use, as this would only meet local demands, and that genetic materials used on a global scale should have priority (for conservation), as they "feed millions of people." Bennett thought, on the contrary, that the limited use of local varieties was not a problem, as these contribute to maintenance of genetic diversity in the field and feed the local population. At that time, the Green Revolution was at its peak, and the points of view of the two scientists regarding it were also divergent. Frankel was basically in line with major international agricultural research institutes, including FAO, oriented toward the development of high-yielding varieties. Bennett, however, was not convinced of the success of the Green Revolution, and rejected breeding for high-yielding varieties without adaptation to local environmental conditions. She also thought that the Green Revolution was not effectively eliminating hunger worldwide. Furthermore, with the expansion of IP rights over plant varieties, Bennett began to worry about the control of multinationals over genetic materials conserved *ex situ*, holding, strongly, that these should be kept in public domain.

Although *in situ* conservation was discussed at the 1967 FAO conference (and later at the 1973 conference), the position that prevailed was that *ex situ* conservation should be given priority, and national and international policies became mainly aimed at this type of conservation. Thus, expeditions for germplasm[3] collection around the world increased drastically during the 1970s and 1980s. The prevailing idea was that agricultural varieties should be collected and stored before devastation of their natural habitats resulted in them being lost forever. A study performed by the International Board for Plant Genetic Resources (IBPGR[4]) in 1975 found that, at that time, there were only eight long-term genetic resource conservation centers, mainly located in developed countries. Just seven years later, this total had gone up to 33 (Damania, 2008, p. 14). Historically, germplasm collections have been created for different purposes, which are not mutually exclusive, for example ensuring the country's autonomy regarding these resources (e.g., the former Soviet Union), introducing new plants (in countries with many immigrants, such as the United States), and conserving genetic resources for plant breeding programs.

In 1971, the Consultative Group on International Agricultural Research (CGIAR[5]) was established with financial support from the Ford and Rockefeller Foundations, the United Nations Environment Programme (UNEP), FAO, the United Nations Development Programme (UNDP), and the World Bank. CGIAR gathered international agricultural research centers in a network and coordinated discussions regarding the priorities for international agricultural research and financial support. The CGIAR system currently comprises 15 international agricultural research centers, 11 of which have germplasm banks. CGIAR germplasm banks have approximately 650,000 samples[6] of crop, forage, and agroforestry genetic resources in the public domain, which represent around 12 percent of the world's plant genetic resources kept *ex situ*. The International Board for Plant Genetic Resources (IBPGR) was founded in 1974, and became the International Plant Genetic Resources Institute (IPGRI) in 1991, and it is also part of the CGIAR network. In 2006, IPGRI and the International Network for the Improvement of Banana and Plantain became one organization: Bioversity International. Since 1994, these germplasm collections have come under the auspices and jurisdiction of the FAO, and access to these materials is regulated by agreements between FAO and CGIAR centers (called "in trust" agreements), which forbid any claims of IP rights over genetic resources held in their collections (Fowler, 2003). According to such agreements, *ex situ* collections held by the CGIAR centers are not the property of individual nations, nor are they the property of the centers themselves, but are held by the centers "in trust" for the international community (Moore and Frison, 2011). In 2006, the 11 international agricultural research centers (IARCs) of the CGIAR which hold *ex situ* germplasm collections signed agreements with the governing body of the ITPGRFA, placing their collections under the treaty. Under these agreements, the centers recognize the authority of the governing body of the treaty to provide policy guidance relating to their *ex situ* collections.[7]

In the 1980s and 1990s, however, the international *ex situ* conservation system came under heavy criticism from many civil society organizations, led mainly by the Rural Advancement Foundation International (RAFI, which later became the ETC Group[8]). They argued that the system served mainly the interests of developed countries and of the international seed industry, and developing countries were losing control over their own genetic resources. In their view, the CGIAR system focused too much on research to develop genetically uniform high-yielding varieties of commercial crops that are dependent on expensive inputs (chemical fertilizers and pesticides) and to which poor farmers have no access. Locally important crops remained under-researched and little priority was given to locally important breeding goals. Moreover, they believed that the introduction of new homogeneous varieties tended to undermine local varieties and biodiversity. *Ex situ* conservation became increasingly associated with the Green Revolution, and farmers have always had limited access to plant genetic resources kept in *ex situ* germplasm banks.

In the 1980s, it was also realized that genetic erosion was taking place in germplasm banks, owing to the precariousness of many facilities, especially in poor countries. Lack of seed regeneration, and consequent loss of its viability, and the lack of resources for characterization and evaluation are some of the problems.[9] Furthermore, the evolution of plants kept in germplasm banks is frozen in time and space, which does not happen when they are left in their natural environments (*in situ*), where they evolve and adapt to environmental and sociocultural changes. *In situ* conservation maintains not only plants, but also their agricultural ecosystems, conserving agrobiodiversity at all of its levels, and giving farmers control over their plant genetic resources.

Nevertheless, it was only in the 1980s and 1990s that *in situ* conservation became more recognized as a necessary and complementary strategy for genetic diversity conservation. The Keystone International Dialogue Series on Plant Genetic Resources, chaired by Dr. M.S. Swaminathan, played an important role in the recognition that genetic diversity should be maintained *in situ* and on-farm (in local agricultural systems, with the participation of farmers) as well as in germplasm banks. The "Keystone Dialogues" took place in 1988, 1990, and 1991, in Keystone (USA), Madras (now Chennai, India), and Oslo (Norway), respectively, and gathered 92 specialists from 30 different countries to discuss many topics related to conservation and the sustainable use of plant genetic resources. Different stakeholders (private, public, NGOs, and academics) were involved in these informal meetings, and one of their agreed conclusions was that plant genetic resource conservation programs should include both *ex situ* and *in situ* strategies, which are complementary.

Plant genetic resources conservation strategies were central to the international agenda during the 1980s and 1990s. However, another important part of this agenda was dominated by diverging opinions and conflicting interests about access, control, and ownership over plant genetic resources that were regulated by successive international legal instruments, such as the International Undertaking on Plant Genetic Resources for Food and Agriculture (1983) and the ITPGRFA, which was approved in 2001 (and came into force on June 29, 2004).

The International Undertaking on Plant Genetic Resources

The International Undertaking on Plant Genetic Resources was adopted at the 22nd Meeting of the FAO conference, through Resolution 08/83.[10] It was the first comprehensive international agreement dealing with plant genetic resources for food and agriculture, but it was not legally binding. Adopted in 1983, the Undertaking was signed by 113 countries, and it was based on the "universally accepted principle that plant genetic resources are a heritage of humankind and consequently should be made available for use without restriction" (Article 1). It was approved with the reservations of eight countries (Canada, France, Germany, Japan, New Zealand, Switzerland, the UK, and the United States).

According to Article 1 of the International Undertaking, its objective is to "ensure that plant genetic resources of economic and/or social value, particularly in agriculture, are explored, preserved, evaluated and made available for plant breeding and scientific purposes." The International Undertaking established, in fact, two categories of genetic resources, subject to distinct legal regimes: (1) plant genetic resources in the public domain, freely accessible to all (in reality, these are the genetic resources recognized by the International Undertaking as "common heritage of humankind") and (2) plant genetic resources under private control. The International Undertaking is signed by states, not by private companies: plant genetic resources under private control are not covered by the free access principle. Private companies have no obligation to make their collections accessible to others. The International Undertaking addressed mainly the interests of breeders from private institutions, interested in guaranteeing access to public collections and in freely collecting genetic materials located in their centers of origin and diversity, in tropical and subtropical countries (Hermitte and Kahn, 2004, p. 31). The IU makes no reference to farmers' rights, but applies both to varieties developed by farmers, called "obsolete" or "primitive," and to "elite" varieties.

During all negotiations of the International Undertaking, developing countries defended the principle of free and unrestricted access to plant genetic resources, which is more coherent with farmers' traditional practices.

The concept of "common heritage of humankind" was incorporated into the International Undertaking, but many developed countries (led mainly by the United States) did not sign the International Undertaking because they considered that it did not recognize and offer sufficient protection to breeders' rights. To increase the number of signatories of the International Undertaking, the FAO Commission on Plant Genetic Resources started to negotiate its controversial aspects with different countries, and this process resulted in the adoption of "agreed interpretations" of the International Undertaking, set out by Resolutions 04/89, 05/89, and 03/91 of the FAO conference.

Resolutions 04/89, 05/89, and 03/91 of the FAO conference

The three resolutions approved by the FAO conference (04/89, 05/89, and 03/91), with "agreed interpretations" of the International Undertaking, meant so many concessions for developed countries that the International Undertaking lost its original purpose (to guarantee free access to plant genetic resources), and became an empty and incoherent instrument, with very little practical applicability. In the words of Regine Andersen (2008), these three resolutions marked the "beginning of the end of the International Undertaking."

On November 29, 1989, the FAO conference approved two resolutions that recognized breeders' and farmers' rights, in an attempt to balance relations between industrialized countries (biotechnology rich) and developing countries (agrobiodiversity rich). These were the resolutions adopted (annexed to the International Undertaking):

1 Resolution 04/89 reaffirms that plant genetic resources are a "common heritage of humankind" and that they "must be freely accessible for use, in benefit of present and future generations." Furthermore, it recognized that plant breeders' rights, as provided for under the UPOV, were not inconsistent with the International Undertaking, and simultaneously recognized farmers' rights defined in Resolution 5/89. Resolution 04/89 also established that countries should impose restrictions to free interchange of genetic materials only if they are strictly necessary for meeting national and international obligations.
2 Resolution 04/89 also states that, by adhering to the International Undertaking, countries "recognize the enormous contribution that farmers of all regions have made to the conservation and development of plant genetic resources, which constitute the basis of plant production throughout the world, and which form the basis for the concept of farmers' rights." According to Resolution 04/89, they "consider that the best way to implement the concept of farmers' rights is to ensure the conservation, management and use of plant genetic resources, for

the benefit of present and future generations of farmers." According to Resolution 04/89, these objectives could be reached through the appropriate means, including, in particular, the International Fund for Plant Genetic Resources (which is part of Resolution 03/91, discussed below).

3 Resolution 05/89 recognized farmers' rights as "rights arising from the past, present and future contributions of farmers in conserving, improving, and making available plant genetic resources, particularly those in the centres of origin/diversity." These rights were "vested in the international Community, as trustee for present and future generations of farmers, for the purpose of ensuring full benefits to farmers, and supporting the continuation of their contributions, as well as the attainment of the overall purposes of the International Undertaking." It was the first international instrument to recognize farmers' rights, but it was not legally binding and had very little concrete efficacy (for more information on the history of farmers' rights, see Chapter 8).

According to Regine Andersen (2008, p. 95), these two resolutions (04/89 and 05/89) were the results of tense negotiations. There had been fierce resistance to plant breeders' rights among developing countries, and the agreed interpretations providing for the acceptance of such rights could be adopted only with the simultaneous recognition of farmers' rights. Access to and control of plant genetic resources had become a highly political issue, dividing northern and southern countries.

Two years later, the FAO conference adopted a new resolution (Resolution 03/91), which made access to plant genetic resources and its interactions with IP rights even more ambiguous and confusing. Resolution 03/91 states that "the concept of common heritage of humankind, as set forth by the International Undertaking on Plant Genetic Resources, is subject to the sovereignty of countries over their resources." In other words, access to plant genetic resources is no longer free, but subject to approval by the country of origin. These countries may agree or not to allow access to their resources, just as they can establish conditions and/or restrictions in light of IP rights. The resolution itself recognizes that "conditions for access to plant genetic resources need further clarification." Resolution 03/91 also created an international fund for plant genetic resources, aimed at implementing farmers' rights, especially in developing countries. This fund received very few voluntary contributions, and never became a reality.

Thus, while farmers' rights recognition did not go beyond formal statements, IP rights over plant varieties became increasingly stronger in the 1980s and 1990s, which exacerbated developing countries' fears of losing control over their own genetic resources. This situation contaminated, once and for all, north–south relations. The UPOV Convention, approved in 1961, was revised in 1972, 1978, and 1991, always with the purpose of granting more effective protection for plant breeders' rights, and with greater restrictions for farmers' traditional practices. The legal space for traditional agricultural practices, such as seed saving and exchange among farmers, was increasingly reduced, making inequalities more marked in the international legal system. The question was: if the rights of holders of biotechnology were recognized in the form of breeders' rights, patents, etc., then why not the rights of germplasm holders (Diaz-Silveira, 2011)?

According to Marie Angèle Hermitte (and Kahn 2004, p. 29), IP rights became a "political obstacle" to free access to genetic resources, leading developing countries to defend the principle of sovereignty over genetic resources located in their territories, which was later incorporated into the Convention on Biological Diversity (CBD). Ironically, access and benefit-sharing (ABS) laws, based on the CBD, have produced effects that are very similar to those of IP rights: they further restrict access to genetic resources, keeping a larger part of resources and knowledge out of the public domain. Access and benefit-sharing laws, on the other hand, have generated very few (if any) benefits for local communities, and contributed very little to biodiversity conservation.

The Convention on Biological Diversity and agriculture

The International Undertaking adopted in 1983 was formally in force until the adoption of the ITPGRFA, in 2001, but received very little attention after the CBD. The CBD was opened for signature on June 5, 1992, at the United Nations Conference on Environment and Development (UNCED, the "Earth Summit"), held in Rio de Janeiro. The CBD was the first international instrument to address biological diversity in a comprehensive way, and was signed by 157 countries during UNCED. Currently, 193 countries are parties to the CBD. The United States signed but did not ratify it. The Nagoya Protocol on Access to Genetic Resources and the Fair and Equitable Sharing of Benefits Arising from their Utilization is a new international agreement, adopted under the auspices of the CBD,[11] in Nagoya, Japan, on October 29, 2010, and it will enter into force 90 days after 50 countries have ratified it.

The CBD represented a paradigm shift, and it broke with the concept that genetic resources are a "common heritage of humankind." It recognizes the sovereign rights of states over their natural resources and establishes that the authority to determine access to genetic resources rests with the national governments and is subject to national laws. It also sets forth that access must be granted in "mutually agreed upon terms," and is subject to prior informed consent from the country providing the resources, as well as to fair and equitable sharing of benefits deriving from their use (Article 15).[12] The CBD entered into force on December 29, 1993 (90 days after the 30th ratification), and it has three main objectives: conservation of biological diversity, sustainable use of the components of biological diversity, and fair and equitable sharing of the benefits arising out of the utilization of genetic resources.

In the Nairobi (Kenya) conference, which approved the agreed text of the CBD, in May 1992, Resolution 3 was also adopted, which deals with the interactions between biodiversity and promotion of sustainable agriculture. Resolution 3 of the Nairobi Final Act recognizes the need to seek solutions for outstanding matters concerning plant genetic resources, in particular (a) access to *ex situ* collections not addressed by the Convention and (b) farmers' rights. Resolution 3 acknowledges the importance of CBD's principles for plant genetic resources for food and agriculture, and the need for measures to promote complementary and cooperative relations between the CBD and the FAO Global System of Conservation and Use of Genetic Resources for Food and Agriculture (of which the International Undertaking on Plant Genetic Resources was a key component). FAO became responsible for implementation of the CBD regarding plant genetic resources for food and agriculture.

In 1993,[13] the FAO conference approved a new resolution (Resolution 07/93), requesting FAO's general director to start negotiations with member countries to bring the International Undertaking on Plant Genetic Resources into harmony with the CBD. This was the starting point for long negotiations that resulted in the adoption, in 2001, of the ITPGRFA. In 1995, the Second Conference of the Parties of CBD, held in Jakarta (Indonesia), also recognized the "special nature of agricultural biodiversity, its distinct features and problems, needing distinct solutions" (Decision II/15).

Adapting the International Undertaking on Plant Genetic Resources to CBD's principles was not an easy task. These instruments have very different approaches: the International Undertaking was aimed mainly at promoting sustainable agriculture and food security, as well as facilitating access to plant genetic resources for food and agriculture. The International Undertaking was negotiated by specialists from the agricultural area, generally representatives of the ministries of agriculture, through FAO. The CBD was negotiated mainly by environmentalists and representatives of ministries of environment, primarily concerned with wild fauna and flora conservation, through the UNEP, according to Regine Andersen (2008).

The CBD focuses mainly on conservation and sustainable use of wild biodiversity, and establishes a bilateral system of access benefit-sharing. According to CBD, member countries (called 'contracting parties') must "respect, preserve and maintain knowledge, innovations and practices of Indigenous and local communities." They must also encourage the wide application of such knowledge, innovations, and practices (with the approval of their holders), as well as encourage equitable sharing of the benefits arising from the utilization of such knowledge, innovations, and practices (Article 8j).

The CBD sets forth that conditions for access and sharing of benefits must be established between countries providing and using genetic resources and associated traditional knowledge, through bilateral contracts and on a case-by-case basis. Each contract is negotiated with the country of origin and with local communities which hold traditional knowledge, including benefit-sharing for each concrete case. This system is difficult to apply to plant genetic resources (used for food and agriculture) and agricultural knowledge and practices, as it was conceived to regulate access to genetic resources of wild fauna and flora species used for chemical and pharmaceutical purposes.

According to the CBD, the country of origin of genetic resources means the country that possesses those genetic resources in *in situ* conditions. "*In situ* conditions" are defined as "conditions where genetic resources exist within ecosystems and natural habitats, and, in the case of domesticated or cultivated species,[14] in the surroundings where they have developed their distinctive properties." The CBD does not have a definition of "center of origin" or of "center of crop diversity," but such concepts are defined by the FAO ITPGRFA, which is a specific agrobiodiversity treaty.[15] Therefore, CBD requires that, in the case of cultivated species, the "surroundings where they have developed their distinctive properties" be identified.

Identification of the country of origin of many agricultural species and varieties can be a complex task because of all the migrations and exchanges that have occurred throughout history. Identifying the "surroundings where they have developed their distinctive properties" is even more complex. The country of origin is not always the same as the country in which the species developed its distinctive properties. The same crop species can develop new traits in geographical locations different from its place of origin. To make this point clearer, the studies of some scientists who have carried out research on the origin of crops are presented below.

In the famous 1859 book by Charles Darwin, *On the Origin of Species*, the origin of cultivated and domesticated plants was not the main object of study. However, in a later, less well-known book, *The Variation of Animals and Plants Under Domestication*, published in 1868,[16] Charles Darwin came to some conclusions on crop domestication: there is a wild species in the origin of domesticated species; modifications among wild and cultivated species are so important that most cultivated species lose their ability to survive in nature, without assistance from humans; the morphological diversity (in a broad sense) found in cultivated species is far greater than that found in wild species (David, 2008).

Alphonse de Candolle was a Swiss botanist who in 1882 published *Origin of Cultivated Plants*, republished in English in 1959, in which he tried to identify the geographical regions of origin of cultivated plants through botanical, archeological, historical, and linguistic criteria. He investigated the distribution of wild relatives of cultivated plants and the variation patterns. Alphonse de Candolle was one of the collaborators of *Flora Brasiliensis*, a classical study about Brazilian flora, produced in Germany between 1840 and 1860 by Carl Friedrich Philipp von Martius, August Wilhelm Eichler, and Ignatz Urban.[17]

Nikolai Vavilov was a Russian agronomist and geneticist who went on more than 100 expeditions to collect genetic materials worldwide, for plant breeding programs developed by the National Institute of Plant Industry,[18] in the former Soviet Union. He visited over 50 countries, in Asia, the Americas, Africa, and Europe, collecting approximately 50,000 plant samples, and is considered a pioneer in germplasm collection according to scientific and systematic bases. He defended the existence of eight "centers of origin" for the main plants cultivated around the world, which would be the geographical locations from which species originated. They are (1) China, (2) India and Indo-Malaysia, (3) Central Asia, (4) the Near East, (5) the Mediterranean, (6) Ethiopia (Abyssinia), (7) South Mexico and Central America, and (8) South America (Peru, Ecuador and Bolivia, the Chiloe Archipelago, in southern Chile, and the southern region of Brazil–Paraguay).

Vavilov became interested in the origins of cultivated plants because he was interested in genetic diversity, and believed they were linked. In 1926, Vavilov wrote the essay "On the origin of cultivated plants," dedicated to Alphonse de Candolle (who had a major influence on his work), in which he argued that the center of origin of a cultivated plant was located in the region with the greatest genetic diversity and, in particular, where their wild relatives could be found (Harlan, 1975, pp. 52–53). Subsequent research showed that this is not always the case, and that center of diversity and center of origin are not necessarily the same. Plants migrated with humans, and were taken to places different from those where they originated, developing new traits in other geographic regions, which may also become centers of diversity. Many scientists pointed out gaps in Vavilov's studies, since he did not investigate, in greater depth, sub-Saharan Africa and the lowlands of South America, where many important agricultural crops were domesticated.[19]

However, many of the concepts and theories developed by Vavilov are still widely accepted and used by geneticists.[20] Vavilov's studies were important to show that plant genetic resources are not uniformly distributed across the globe. They are concentrated, in large part, in centers of origin and diversity of cultivated plants and their wild relatives, located mainly in tropical and subtropical regions (Africa, Asia, and the Americas).

In 1975, Jack Harlan, a US American agronomist and geneticist wrote the classical book *Crops and Man*, based on the previous works of Alphonse de Candolle and Nikolai Vavilov. Harlan considered that many cultivated plants did not originate in the centers proposed by Vavilov, and that some plants do not have a center of diversity, whereas others may have more than one. Harlan (1975, pp. 149–167) argued that centers of origin are not the same as centers of genetic diversity, and that cultivated plants have different patterns of variation and evolution in time and space. He held, however, that centers of diversity do exist for many cultivated plants, and that this concept is useful for the study of genetic diversity.

Many scientists have contested, reformulated, and added new elements to the theories of Candolle, Vavilov, and Harlan,[21] proposing other centers of origin and diversification, and such issues are beyond the scope of this book.[22] The point is that it is not always easy to define a precise geographical location where a given agricultural species originated or diversified, in order to determine who is the legitimate party to authorize access to the plant genetic resource and receive benefits, as set forth in the bilateral CBD system.

Furthermore, from the standpoint of farmers, the bilateral CBD regime creates another problem: to whom do the knowledge and practices associated to cultivated plants belong? Who can authorize and is entitled to receive benefits which may arise out of the use of plant genetic resources and associated knowledge? Ethnobotanist Laure Emperaire (see Emperaire and Santilli, 2006) explains that, in local agricultural systems "there have been, and there continues to be, selection, genetic improvement, exchange of seeds, knowledge and experiences and wide plant circulation." And that this is a dynamic process: "Plants circulate among families, communities or ethnic groups; new varieties originating from other regions or locally produced are evaluated and incorporated into the farmer's stock of varieties; there is an interest in diversity for itself." Plant diversity is a value in itself, promoted by many traditional communities.

How can farmers, used to sharing and promoting exchange of genetic materials, knowledge, and agricultural experiences through social networks and regulated by local norms, define to whom these resources belong? If local/traditional farming communities are to determine who the "owners" of these resources are, access and benefit-sharing laws may encourage disputes and rivalries that can end up restricting circulation and exchange of genetic material, which is fundamental for conservation of agricultural biodiversity.

Although the CBD does not exactly establish an "owner" of genetic resources and associated traditional knowledge, its principles – prior informed consent and benefit-sharing with countries of origin and local communities – are based on the assumption that there are "providers" and "recipients" (or users) of genetic resources, and that they must establish, contractually, the conditions for access and benefit-sharing (ABS). Genetic resources and traditional knowledge end up becoming commodities or merchandise, negotiated at "market prices," which goes against the collective logic of how these resources and knowledge are generated and shared by local communities. In addition, the mercantile approach adopted by the CBD does not take into consideration agricultural species of great local and regional importance for food security, which are not commodities, and, therefore, of little commercial interest.

The complexity of processes to obtain access authorizations (and ambiguities in national ABS laws) has, in many countries, discouraged research on genetic resources and traditional knowledge,[23] and, at the same time, brought very few concrete benefits for local communities and biodiversity. There is no word, to date, of a benefit-sharing contract with local farming communities resulting from the implementation of ABS laws based on the CBD. The public domain is becoming increasingly restricted – be it by means of private appropriation by IP rights or through the principle of sovereignty of countries of origin over their genetic resources.

The CBD treats genetic resources as goods with an economic and utilitarian value, without considering the biological and sociocultural processes that lead to the construction of agrobiodiversity and associated knowledge. It ignores local perceptions and values associated with plant genetic resources, how local norms determine ownership over genetic resources, the intrinsic link between resource and knowledge, the circulation and exchange of plant genetic materials, and sharing of these materials by several communities, etc. It does not take into consideration the complexity of social and cultural processes that enrich agrobiodiversity. It tends to undermine the free circulation of plant genetic material, encourage monopolies, and restrict public domain, and potentially can have a negative impact on local agricultural systems (Emperaire and Santilli, 2006).

CBD raised unrealistic expectations in many biodiversity-rich countries that resources from benefit-sharing contracts, signed between providers and users of genetic resources, would be large and sufficient to finance biodiversity conservation, which did not happen in the vast majority of cases. Nor did the CBD solve political imbalances among countries that are biodiversity rich (southern developing countries) and those that are biotechnology rich (northern developed countries). The principle of sovereignty of countries of origin over their own genetic resources, set forth by the CBD, was an attempt to balance out historical inequalities in north–south relations, related to the appropriation of these resources by northern countries through IP rights. However, CBD did not provide a solution to the negative impacts of IP rights on biodiversity, and, at the same time, legitimized them (indirectly). Access to genetic resources and associated knowledge became more restricted, and the special nature of plant genetic resources for food and agriculture led to the adoption of the ITPGRFA, which will be discussed below.

The International Treaty on Plant Genetic Resources for Food and Agriculture

The ITPGRFA was adopted (by consensus, with only two abstentions: Japan and the United States) during the 31st meeting of the FAO conference, held in Rome on November 3, 2001, and came into effect on June 29, 2004.[24] It is the first legally binding international instrument to deal exclusively with plant genetic resources for food and agriculture. The objectives of the ITPGRFA are the conservation and sustainable use of plant genetic resources for food and agriculture and the fair and equitable sharing of the benefits arising out of their use, in harmony with the CBD, for sustainable agriculture and food security. The preamble to the Treaty is important because it facilitates understanding of the Treaty's main premises, and therefore it is analyzed below.

According to the preamble to the Treaty, contracting parties are "convinced of the special nature of plant genetic resources for food and agriculture, their distinctive features and problems needing distinctive solutions." The "special" nature of plant genetic resources (for food and agriculture) is highlighted by many studies (Cooper, 2002; Stannard *et al.*, 2004; Moore and Tymowski, 2005; Halewood and Nnadozie, 2008; Frison *et al.*, 2011; among others) to justify a differentiated legal regime for these resources, distinct from the legal regime established for genetic resources in general. Some of the distinctive features of plant genetic resources are:

- Human intervention had (and still has) a fundamental role in the domestication of crop species and conservation of agrobiodiversity. Throughout history, farmers have been domesticating wild plants and, through selection and improvement, adapting them to agriculture and to human needs. Useful traits, such as resistance to diseases and extreme weather conditions, more nutritious grains, rapid germination, and uniform maturation were bred in by farmers, whereas others, such as seed dormancy, bitter-tasting grains, or toxic components, were bred out by farmers. Any cultivated plant variety is the result of selection and improvement activities developed by many generations of farmers. Agrobiodiversity is the result of complex and dynamic management carried out by farmers. Conservation and sustainable use of agrobiodiversity are intrinsically related, and cannot be dealt with separately. This difference between wild and cultivated biodiversity is, however, relative, as biodiversity cannot, under any circumstances, be reduced to a mere natural phenomenon; it is also a cultural phenomenon (Balick and Cox, 1996; Diegues *et al.*, 2001; Santilli, 2005). Nevertheless, cultivated plants depend more heavily on humans, as many domesticated species lose their capacity to survive in wild environments.
- Exchanges among different countries and farmers led to the development of varieties whose composition is based on genetic materials from various geographical origins, which makes it difficult, in many situations, to attribute a single geographical origin to genetic materials used in the development and/or breeding of the variety. Several varieties are generally used in selection and breeding processes, by both farmers and professional breeders. Local agricultural systems are neither closed nor static, and farmers constantly incorporate new varieties, brought by other farmers or by agricultural research institutions.

Plant varieties developed by professional breeders also have a complex pedigree, which may make it difficult to identify all varieties that contributed to the development of the new variety, which is the end product of plant breeding. Evaluating the importance of each variety used in the breeding process, giving it a specific value, is a complex task. After a long process of selection, breeding and cross-breeding of several varieties, it is not simple to establish which genetic component determined that specific trait, present in the end variety. The amount of resources and time spent in doing such an identification would likely surpass the resulting economic benefits. In addition, many genetic resource collections are located outside their regions of origin, in locations very distant from where they were collected.

The spring bread wheat variety known as Veery, developed by the International Maize and Wheat Improvement Center (CIMMYT), was the product of 3,170 different crosses involving some 51 parental varieties, from at least 26 different countries. The IR 36 rice variety has 15 landraces and one wild species in its heritage (Moore and Tymowski, 2005, p. 24). The Orofen wheat variety is included in the pedigree of 245 released varieties in China (Louwaars, 2007, p. 82). Estimates indicate that for the development of each new variety of wheat, the average number of varieties used increased from 12 to 64, in 1992 (Visser, 2008).

The preamble to the ITPGRFA also sets forth that contracting parties recognize that "plant genetic resources for food and agriculture are a common concern of all countries, in that all countries depend very largely on plant genetic resources for food and agriculture that originated elsewhere." All countries are dependent, to a greater or lesser extent, on plant genetic resources originating from other parts of the world. Currently, no country is self-sufficient in terms of plant genetic resources, all are interdependent, to an average extent of 50 percent (Palacios, 1997). This interdependence among countries is greater in regard to plant genetic resources (for food and agriculture) than regarding other genetic resources. Thus, countries frequently need to access and use plant genetic resources originating from other countries, for scientific research and breeding, as well as for direct use in their agricultural and food systems. Maintaining the flow and exchange of plant genetic resources is essential to breeders, farmers, and consumers.

Climate change will also have a great impact on countries' interdependence on genetic resources for food and agriculture, according to a more recent study, prepared at the request of the Commission on Genetic Resources for Food and Agriculture (CGRFA) (Fujisaka *et al.*, 2009). According to the study,

> germplasm interdependence will be the greatest for crops, augmenting the already high (and well-documented) international movement of plant genetic resources for food and agriculture (PGRFA) that

> has been taking place for a long time. Interdependence on PGRFA will likely increase in association with adaptive crop improvement achieved through both conventional plant breeding and biotechnological methods. Interdependence will also increase as climate change creates the need to adopt new crops in particularly stressed areas – millets and sorghum in the place of maize, for example.
> (Fujisaka *et al.*, 2009)

The ITPGRFA addresses adaptation to climate change through facilitated access to plant genetic resources (through a multilateral system) and through its benefit-sharing fund, which supports on-farm conservation of plant genetic resources in developing countries, as well as adaptation to climate change. During the fourth session of its governing body, the ITPGRFA's benefit-sharing fund was officially recognized as an international adaptation-funding mechanism under the Adaptation Fund of the UN Framework Convention on Climate Change (UNFCCC). The interfaces between agricultural biodiversity and climate change, and the need to promote closer collaboration between the ITPGRFA and the UNFCC, have been highlighted in several international fora. The Cordoba Declaration on Agricultural Biodiversity in Addressing Hunger and Climate Change, adopted in September 2010,[25] has encouraged the adoption, by UNFCCC, of a program of work on agriculture and the development of synergies between UNFCCC mechanisms and agricultural biodiversity fora. The Ministerial Conference on Biodiversity, Food Security and Climate Change, held on March 11, 2011, in Bali (Indonesia), has also adopted the Bali Ministerial Declaration on the Role of the ITPGRFA on Biodiversity, Climate Change and Food Security.[26] During its fourth session, the governing body of the ITPGRFA asked its secretary to explore possible areas of cooperation with the UNFCC, given the connection between climate change and adaptation in agriculture and genetic resources, such as the participation of the Treaty as a partner in the initiative on Reducing Emissions from Deforestation and Forest Degradation in Developing Countries (REDD). In relation to mitigation of climate change, the ITPGRFA has the following limitation: its multilateral system applies only to Annex I crops used for food and feed purposes, and not for greenhouse gas emission reduction and CO_2 fixation or for bioenergy production (Grugel, 2009).

The Treaty preamble also sets forth that contracting parties

> acknowledge that the conservation, exploration, collection, characterization, evaluation and documentation of plant genetic resources for food and agriculture are essential in meeting the goals of the Rome Declaration on World Food Security and the World Food Summit Plan of Action and for sustainable agricultural development for this and future generations.

The Rome Declaration was adopted in 1996, during the World Food Summit, and countries committed to "achieving food security for all and to an ongoing effort to eradicate hunger in all countries, with an immediate view to reducing the number of undernourished people to half their present level no later than 2015." In order to meet these objectives, the World Food Summit Plan of Action sets forth that governments, in collaboration with civil society, must "promote access, by farmers and farming communities, to genetic resources for food and agriculture." According to the Plan of Action, countries must also

> promote an integrated approach to conservation and sustainable utilization of plant genetic resources for food and agriculture, through *inter alia* appropriate *in situ* and *ex situ* approaches, systematic surveying and inventorying, approaches to plant breeding which broaden the genetic base of crops, and fair and equitable sharing of benefits arising from the use of such resources.[27]

The Treaty recognizes the close relationship between conservation of plant genetic resources, and, when referring to the Rome Declaration on World Food Security, it contributes to create awareness that hunger and undernourishment can be eradicated only if access to plant genetic resources is ensured for farming communities, free of restrictions, and if genetic diversity is conserved not only *ex situ*, in collections, but also *in situ* and on-farm, in the agroecosystems, with the participation of local farmers.

Another part of the Treaty's preamble says that contracting parties

> recognize that, in the exercise of their sovereign rights over their plant genetic resources for food and agriculture, states may mutually benefit from the creation of an effective multilateral system for facilitated access to a negotiated selection of these resources and for the fair and equitable sharing of the benefits arising from their use.

The Treaty recognizes sovereign rights of states over their plant genetic resources, and that authority to determine access to these resources rests with national governments and is subject to national laws. However, in the exercise of their sovereign rights over their plant genetic resources, the Treaty's contracting parties agree to establish a multilateral system of ABS, both to facilitate access to plant genetic resources for food and agriculture, and to share, in a fair and equitable way, the benefits arising from the utilization of these resources. Such a multilateral system is restricted, nonetheless, to 35 food crops and 29 forage crops (legumes, grasses, and others) that are listed in Annex I of the Treaty and which are under the management and control of the contracting parties and in the public domain. Forage crops were included because they are mainly destined for feeding animals, which are then consumed by humans.

Access to the plant genetic resources included in Annex I of the Treaty, through the multilateral system, is provided solely for the purposes of utilization and conservation for research, breeding, and training for food and agriculture, and as long as it does not include chemical, pharmaceutical, and/or other non-food/feed industrial uses. Maize (corn) genetic resources, for instance, can only be accessed through multilateral systems for food and agriculture purposes, not for production of agrofuels, for example. If access is aimed at non-food/feed industrial uses, the multilateral system is not applicable, and the interested party must follow CBD's bilateral regime of access and benefit-sharing, which depends on "mutually agreed terms" between providers and users of genetic resources.[28]

Inclusion of agricultural crops in Annex I of the Treaty met, in theory, the criteria of interdependence and food security, but political criteria were also decisive. The choice of crops to be included in the multilateral system led to many controversies during Treaty negotiations.[29] The extensive list of crops originally presented ended up substantially reduced, and important crops were excluded, such as soybean (removed by China, after a conflict over the occupation of China airspace by a US aircraft), peanuts, tomatoes, sugarcane, many wild relatives of cultivated plants, onions, garlic, cucumber, pumpkins and squashes, pepper, tropical forages, tea, coffee, and cacao, among others. On the other hand, minor crops, with questionable importance for food security, were included, such as strawberries and asparagus. Crops of great local or regional importance were excluded, such as wild relatives of manioc/cassava (needed for genetic improvement of the species, and a staple food in sub-Saharan Africa and many Latin American countries), many species of millet used in human and animal diets in mid-Asia and the Near East, and tropical forages which are widely used by pastoralist communities. The 64 agricultural crops included in Annex I account for approximately 80 percent of the world's supply of food, but represent the main commercial crops. Inclusion of a new crop in Annex I depends on consensus between all contracting parties to the Treaty.[30]

Many developing countries tried to limit the scope of the multilateral system because they believed that the CBD bilateral regime would be more advantageous,[31] as access and benefit-sharing are negotiated directly with the provider country, and benefits go to that country, not to the multilateral system. (See Table 6.1 for the main differences between the CBD bilateral regime and the FAO Treaty's multilateral system.) Developed countries, on the other hand, strongly resisted any restriction to IP rights over plant genetic resources, and this made developing countries even less willing to make concessions, and to allow the inclusion of further crops in the multilateral system. Retaliation among countries also took place. Brazil and

Table 6.1 Main differences between the CBD bilateral regime and the multilateral system (FAO Treaty)

	CBD	*FAO Treaty (multilateral system)*
Scope	All forms of biodiversity, including both wild and domesticated species	The Treaty covers all plant genetic resources for food and agriculture, but the multilateral system includes only those that are listed in Annex I *and* which are under management and control of the contracting parties and in public domain
Objectives	Conservation of biological diversity, sustainable use of the components of biological diversity and fair and equitable sharing of the benefits arising out of the utilization of genetic resources	Conservation and sustainable use of plant genetic resources *for food and agriculture* and the fair and equitable sharing of the benefits arising out of their use, in harmony with CBD, *for sustainable agriculture and food security*
Access purposes	Conservation and sustainable use for any purposes, in principle, but CBD was conceived mainly for chemical, pharmaceutical, and/or other non-food/feed uses	Utilization and conservation for research, breeding, and training for *food and agriculture*, and as long as it does *not* include chemical, pharmaceutical and/or other non-food/feed industrial uses
Access and benefit-sharing	CBD recognizes the sovereign rights of states over their natural resources, and the authority to determine access to genetic resources rests with national governments and is subject to national access and benefit-sharing laws. Access depends on "mutually agreed upon terms," established through bilateral contracts between providers and users on a case-by-case basis. Access is subject to prior informed consent of the country of origin of resources, and to fair and equitable sharing of benefits deriving from their use. Access to associated traditional knowledge depends also on prior informed consent of Indigenous peoples and local communities	Contracting parties (of the FAO Treaty) agree to establish a multilateral system of access and benefit-sharing (which applies only to the 35 food crops and 29 forages listed in Annex I of the Treaty, under the management and control of the contracting parties and in the public domain). The SMTA establishes ABS conditions, and access is facilitated and expeditious. Benefit-sharing is mandatory only when commercialized products (that incorporate material accessed from the multilateral system) are *not* available without restriction to others for further research and breeding (e.g., patented genetic materials) The equitable share corresponds to 1.1% of gross product sales minus 30%, which represents 0.77% *or* 0.5% of all product sales resulting from the same crop. Other benefit-sharing mechanisms are exchange of information, access to and transfer of technology, and capacity-building

Forms of conservation	*Ex situ* conservation is considered as complementary to *in situ* conservation, and must preferably take place in the country of origin of genetic resources	Articles 5 and 6 of the FAO treaty establish principles and guidelines for *in situ*, on-farm, and *ex situ* conservation of *all* plant genetic resources for food and agriculture. The multilateral system of ABS (Articles 12 and 13) applies only to plant genetic resources listed in Annex I and under management and control of the contracting parties and in the public domain. Access to plant genetic resources for food and agriculture found in *in situ* conditions must be provided according to national laws

Source: CBD, Articles 1, 8, 9, and 15 (www.cbd.int/convention/text/; accessed March 10, 2011) and FAO Treaty, Articles 1, 5, 6, 10, 11, 12, and 13 (/www.planttreaty.org/texts_en.htm)

Bolivia excluded peanuts, Andean countries excluded tomato, and African countries excluded tropical forages.

In reality, the Treaty establishes a double legal regime for plant genetic resources that are under the management and control of the contracting parties and in the public domain:

1 When included in Annex I and access is aimed at research, breeding, and training for food and agriculture, plant genetic resources are considered global commons (Hermitte and Kahn, 2004, p. 81), and crops can be accessed expeditiously, free of charge (or when a fee is charged, it may not exceed the minimal cost involved), through the multilateral system.
2 When included (or not) in Annex I, but access is aimed at chemical, pharmaceutical, and/or other industrial uses, crops are subject to the sovereignty of their countries of origin, and to the bilateral regime established by the CBD (if the plant genetic resource is located on the territory of a country that has ratified the CBD and approved a national ABS law making this instrument operational in its territory).

Very important components of the Treaty are the general provisions regarding conservation and sustainable use of plant genetic resources for food and agriculture (Articles 5 and 6), that apply to *all* crops (the multilateral system applies only to crops listed in Annex I).[32] Articles 5 and 6 set forth the fundamental principles and guidelines that must orient policies and actions aimed at conservation and sustainable use of plant genetic resources. These principles and guidelines are mainly based on the rolling Global Plan of Action for Conservation and Sustainable Use of Plant Genetic Resources for Food and Agriculture (GPA) – a voluntary instrument adopted by 150 countries during the Fourth International Technical Conference on Plant Genetic Resources, held in Leipzig (Germany) from June 17 to 23, 1996, which is one of the supporting components of the Treaty. An updated version of the Global Plan of Action was approved by the Commission on Genetic Resources for Food and Agriculture in July 2011, based primarily on the gaps and needs identified in the Second Report, and in light of new challenges such as climate change.[33]

According to Article 5.1c of the Treaty, contracting parties must promote and support farmers and local communities' efforts to manage and conserve on-farm their plant genetic resources for food and agriculture. It is the first time that a legally binding international treaty acknowledges the role of local farmers and communities in the conservation of agrobiodiversity, obliging countries to adopt actions, policies, and programs to support on-farm conservation, although the Treaty recognizes that it is up to individual countries to decide which specific actions to take. On-farm conservation plays several important functions besides conservation, such as empowerment of local communities, strengthening of traditional and local agricultural systems, and keeping farmers farming.

Some authors see on-farm conservation as a type of *in situ* conservation, and the expression "on-farm *in situ* conservation" is very commonly used. Others believe that the term *in situ* refers mainly to conservation of wild species in their natural habitats, and that, when domesticated species are managed by farmers, in local agroecosystems, it is more appropriate to use the expression "on-farm conservation." Walter de Boef prefers to use the expression "community biodiversity management," pointing out that the term "management" is more appropriate than "conservation," because it better reflects the dynamic nature of human and ecological processes, which cannot be controlled or "conserved." "Conservation" is a concept developed by conservationists, rather than a goal of farmers (Boef, 2000).

In situ conservation is in Article 5.1d of the Treaty, which establishes the obligation of contracting parties to promote *in situ* conservation of wild crop relatives and wild plants for food production, including in protected areas, by supporting, *inter alia*, the efforts of Indigenous and local communities. *In situ* conservation of cultivated plants may take place both in protected areas and beyond their borders, in the environments where they developed their distinctive properties. When conserved *in situ*, cultivated plants maintain their capacity to evolve and adapt. Furthermore, *in situ* conservation maintains not only the plants but the whole agricultural ecosystem, and it does not isolate the plants from the social and cultural contexts in which they evolved.

Ex situ conservation is covered in Article 5.1e, which sets forth that countries must cooperate to promote the development of an efficient and sustainable system of *ex situ* conservation, giving due attention to the need for adequate documentation, characterization, regeneration, and evaluation of genetic resources. "Documentation" refers to the totality of the documentation that should be available for genebank accessions, including that related to the characterization, regeneration, and evaluation of individual accessions. "Characterization" refers to the categorization of data on highly heritable characteristics of genebank accessions, such as color of flowers, that are constant in any environment. "Regeneration" refers to the need to grow out stored seeds periodically to ensure that they remain viable. "Evaluation," on the other hand, relates to the assessment of the agronomic characteristics of the material, including disease or drought resistance, including molecular techniques (Moore and Tymowski, 2005, p. 47). To a large extent, the accessibility of germplasm, and its usefulness for farmers and breeders, will depend on the adequacy of the documentation, characterization, regeneration, and evaluation of that germplasm, as Moore and Tymowski (2005) point out. Hawtin and Fowler (2011) point out that

> conserving genetic diversity e*x situ* is vital if plant breeders are to have ready access to the traits and genes they need to do their work. It would be impossibly complicated and expensive if new

> materials had to be freshly collected from the wild or from farmers' fields, often in far away countries, every time a plant breeder needed new genetic diversity.

Furthermore,

> *ex situ* collections provide a "safety net" – a last resort – that enables locally adapted varieties and/or unique traits to be reintroduced back into farming systems after they have been lost due to natural or human-induced disasters, changing production systems, or as a result of their replacement by new varieties.
>
> Hawtin and Fowler (2011)

Two significant developments (aimed at *ex situ* conservation) have been the creation, in 2004, of the Global Crop Diversity Trust (an endowment fund to support *ex situ* conservation of plant genetic resources and an essential element of the funding strategy of the ITPGRFA), and the opening, in 2008, of the Svalbard Global Seed Vault (see Chapter 3 for more information on this vault, whose primary operational costs are covered by the Global Crop Diversity Trust).

It is very difficult, and perhaps impossible, to determine how representative *ex situ* collections are of the total agricultural diversity found *in situ*/on-farm. All over the world, about 7.4 million plant accessions are currently conserved in germplasm banks, that is, in *ex situ* collections, and there are more than 1,750 genebanks worldwide, about 130 of which hold more than 10,000 accessions each. Of the 7.4 million accessions worldwide, national government genebanks conserve about 6.6 million, 45 percent of which are held in only seven countries. These accessions represent a limited number of major crops, and about 45 percent of all the accessions in the world's genebanks are cereals. Food legumes are the next largest group, accounting for about 15 percent of all accessions, while vegetables, fruits, and forage crops each account for 6–9 percent of the total number of accessions maintained *ex situ*. Roots and tubers, as well as oil and fiber crops, each account for 2–3 percent of the total (according to the *Second Report on the State of the World's Plant Genetic Resources for Food and Agriculture*). Landraces and local varieties, as well as underutilized and neglected plant species and crop wild relatives, are under-represented in *ex situ* collections. According to Laure Emperaire (2008, p. 420), the genetic diversity of manioc/cassava found in a traditional agroecosystem in the Negro River Basin (in the Brazilian Amazon) is wider than the genetic diversity found in manioc/cassava *ex situ* collections of the International Center for Tropical Agriculture (CIAT), in Cali (Colombia). In addition, various analyses suggest that between 25 and 30 percent of the total holdings (or 1.9–2.2 million accessions) are distinct, with the remainder being duplicates held either in the same or, more frequently, in a different collection. Much of this duplication is unintended and the situation is most serious for vegetatively propagated species and species with recalcitrant seeds (according to the Second Report). It is widely recognized that there is a need for greater rationalization within and among collections.

Article 6 of the treaty establishes that the contracting parties must develop and maintain appropriate policy and legal measures that promote the sustainable use[34] of plant genetic resources for food and agriculture, such as:

a pursuing fair agricultural policies that promote the development and maintenance of diverse farming systems that enhance the sustainable use of agricultural biological diversity and other natural resources;
b strengthening research that enhances and conserves biological diversity by maximizing intra and interspecific variation for the benefit of farmers, especially those who generate and use their own varieties and apply ecological principles in maintaining soil fertility and in combating diseases, weeds, and pests;
c promoting plant breeding efforts that, with the participation of farmers, particularly in developing countries, strengthen the capacity to develop varieties particularly adapted to social, economic, and ecological conditions, including in marginal areas;
d broadening the genetic base of crops and increasing the range of genetic diversity available to farmers;
e promoting the expanded use of local and locally adapted crops, varieties, and underutilized species;
f supporting the wider use of diversity of varieties and species in on-farm management, conservation, and sustainable use of crops and creating strong links to plant breeding and agricultural development in order to reduce crop vulnerability and genetic erosion, and promote increased world food production compatible with sustainable development; and
g reviewing and adjusting breeding strategies and regulations concerning variety release and seed distribution.

This list of measures to be adopted by countries is merely illustrative, and the measures listed above are based on Decision III/11[35] of the Third Conference of the CBD parties, held in Buenos Aires, in 1996, and on the main elements of the GPA. They constitute, once more, recognition that agrobiodiversity is the result of complex and dynamic management of agricultural crops by farmers, and that public policies and legal instruments must promote an integrated approach to agrobiodiversity, which takes both biological and cultural diversity into consideration.

The Treaty (Article 4) also sets forth that member countries must ensure the conformity of their laws, regulations, and procedures with obligations established by this international instrument. Therefore, national legal systems of countries that ratified the treaty must be in accordance with the international obligations related not only to the exchange of plant genetic resources (through the multilateral system), but also regarding *in situ* and on-farm conservation of agrobiodiversity, with participation by local farmers. According to Article 7 of the Treaty, each contracting party must integrate into its agriculture and rural development programs and policies, activities aimed at conservation and sustainable use of plant genetic resources (set forth in Articles 5 and 6, mentioned above), and cooperate with other countries, directly or through FAO, and with other international organizations.

The multilateral system of access and benefit-sharing

Facilitated access

Next, we will present the main rules of the multilateral system of ABS. We shall start with the rules related with facilitated access:

> The multilateral system applies only to plant genetic resources in Annex I of the Treaty that are under the management and control of the contracting parties and in public domain, and facilitated access is provided to natural or legal persons in the jurisdiction of any treaty contracting party (or member country).

The multilateral system is a global common pool of PGRFA (plant genetic resources for food and agriculture) for a specific group of crop species (listed in Annex I of the treaty), to which access will be facilitated for research and breeding for food and agriculture. (See Table 6.2 for the list of food crops and Table 6.3 for the list of forages included in Annex I.) It is a geographically distributed "virtual" genebank, and materials in the multilateral system are distributed over a large number of individual collections located in different countries (Chiarolla and Jungcurt, 2011). In the multilateral system, sovereignty of countries of origin over their plant genetic resources is meant not to establish, on a case-by-case basis, the conditions that access, but rather to create a system which provides access to all contracting parties, under standard conditions, to all 64 plant genetic resources included in the multilateral system (and listed in Annex I of the treaty). Regardless of how many plant genetic resources each country made available to the multilateral system, all contracting parties have facilitated

Table 6.2 List of crops covered under the multilateral system (Annex I of the Treaty): food crops

Crop	*Genus*	*Observations*
Breadfruit	*Artocarpus*	Breadfruit only
Asparagus	*Asparagus*	
Oat	Avena	
Beet	*Beta*	
Brassica complex	*Brassica et al.*	Genera included are *Brassica*, *Armoracia*, *Barbarea*, *Camelina*, *Crambe*, *Diplotaxis*, *Eruca*, *Isatis*, *Lepidium*, *Raphanobrassica*, *Raphanus*, *Rorippa*, and *Sinapis*. This comprises oilseed and vegetable crops such as cabbage, rapeseed, mustard, cress, rocket, radish, and turnip. The species *Lepidium meyenii* (maca) is excluded
Pigeon pea	*Cajanus*	
Chickpea	*Cicer*	
Citrus	*Citrus*	Genera *Poncirus* and *Fortunella* are included as root stock
Coconut	*Cocos*	
Major aroids	*Colocasia*, *Xanthosoma*	Major aroids include taro, cocoyam, dasheen, and tannia
Carrot	*Daucus*	
Yams	*Dioscorea*	
Finger Millet	*Eleusine*	
Strawberry	*Fragaria*	
Sunflower	*Helianthus*	
Barley	*Hordeum*	
Sweet potato	*Ipomoea*	
Grass pea	*Lathyrus*	
Lentil	*Lens*	
Apple	*Malus*	
Cassava	*Manihot*	*Manihot esculenta* only
Banana/ plantain	*Musa*	Except *Musa textilis*
Rice	*Oryza*	
Pearl millet	*Pennisetum*	
Beans	*Phaseolus*	Except *Phaseolus polyanthus*
Pea	*Pisum*	
Rye	*Secale*	

Crop	*Genus*	*Observations*
Potato	*Solanum*	Section tuberose included, except *Solanum phureja*
Eggplant	*Solanum*	Section melongena included
Sorghum	*Sorghum*	
Triticale	*Triticosecale*	
Wheat	*Triticum et al.*	Including *Agropyron*, *Elymus*, and *Secale*
Fava bean/ vetch	*Vicia*	
Cowpea *et al.*	*Vigna*	
Maize	*Zea*	Except *Zea perennis*, *Zea diploperennis*, and *Zea luxurians*

Table 6.3 List of crops covered under the multilateral system (Annex I of the Treaty): forages

Genus	*Species*
Legume forages	
Astragalus	*chinensis, cicer, arenarius*
Canavalia	*ensiformis*
Coronilla	*varia*
Hedysarum	*coronarium*
Lathyrus	*cicera, ciliolatus, hirsutus, ochrus, odoratus, sativus*
Lespedeza	*cuneata, striata, stipulacea*
Lotus	*corniculatus, subbiflorus, uliginosus*
Lupinus	*albus, angustifolius, luteus*
Medicago	*arborea, falcata, sativa, scutellata, rigidula, truncatula*
Melilotus	*albus, officinalis*
Onobrychis	*viciifolia*
Ornithopus	*sativus*
Prosopis	*affinis, alba, chilensis, nigra, pallid*
Pueraria	*phaseoloides*
Trifolium	*alexandrinum, alpestre, ambiguum, angustifolium, arvense, agrocicerum, hybridum, incarnatum, pratense, repens, resupinatum, rueppellianum, semipilosum, subterraneum, vesiculosum*

Genus	*Species*
Grass forages	
Andropogon	*gayanus*
Agropyron	*cristatum, desertorum*
Agrostis	*stolonifera, tenuis*
Alopecurus	*pratensis*
Arrhenatherum	*elatius*
Dactylis	*glomerata*
Festuca	*arundinacea, gigantea, heterophylla, ovina, pratensis, rubra*
Lolium	*hybridum, multiflorum, perenne, rigidum, temulentum*
Phalaris	*aquatica, arundinacea*
Phleum	*pratense*
Poa	*alpina, annua, pratensis*
Tripsacum	*laxum*
Other forages	
Atriplex	*halimus, nummularia*
Salsola	*vermiculata*

access to all 64 plant genetic resources included in the multilateral system. According to the Treaty (Article 10.2), contracting parties agree to "establish a multilateral system, which is efficient, effective, and transparent, both to facilitate access to plant genetic resources, and to share, in a fair and equitable way, the benefits arising from the utilization of these resources, on a complementary and mutually reinforcing basis."

Under Article 12.2 of the Treaty, contracting parties have agreed to provide facilitated access to plant genetic resources for food and agriculture (within the multilateral system) *to other contracting parties and to legal and natural persons under the jurisdiction of any contracting party* (emphasis added). A literal interpretation would lead to the conclusion that only researchers and institutions from contracting parties can access resources included in the multilateral system. However, this interpretation is not correct: individual researchers and research institutions located in countries that did not ratify the treaty can also access plant genetic resources through the multilateral system, as long as they sign the standard material transfer agreement (SMTA). The SMTA will be binding on its parties, which are the provider and recipient institutions, and not the countries themselves (these are contracting parties to the treaty, not to the SMTA). The SMTA is a bilateral contract, and a legally enforceable agreement. The SMTA can be enforced within the jurisdictions of the countries where providers and recipients are based, regardless of whether or not the Treaty has been ratified by their country. This is especially true in relation to materials held in trust by CGIAR centers worldwide, which must continue to make material available to recipients from noncontracting parties. In 2006, when CGIAR centers signed agreements with the governing body of the ITPGRFA, they clarified their understanding that, although the agreements mention only making samples of plant genetic resources available to contracting parties, this would not prevent the centers from also making germplasm available to noncontracting parties, using the SMTA, as well as to farmers for direct cultivation and use (Moore and Frison, 2011). Restricting access to genetic resources would go against the spirit of the Treaty, which is intended to facilitate access.

Thus, everyone (including private companies and scientific research institutions, either public or private) has facilitated access to plant genetic resources made available through the multilateral system. In contrast, there is no obligation for private companies to make their own *ex situ* collections available to the multilateral system. In other words, private companies benefit from facilitated access to resources kept by the multilateral system without having to commit themselves to share their own germplasm collections. For this reason, civil society organizations claim that, in its original conception, the objective of the Treaty was to strengthen farmers' rights, but it ended up becoming an instrument which "grants new privileges to industry." In reality, there is a serious inequality in the system: access to resources maintained by public and international agricultural research institutions is provided to private institutions through the multilateral system and in a facilitated way, but there is no obligation for private institutions to make their own germplasm collections available to public and international institutions. Despite this, many organizations recognize that it is a viable alternative to the bilateral CBD regime, which restricts access and circulation of genetic resources by imposing complex and costly bilateral negotiations that are incompatible with the nature of plant genetic resources (ETC Group, 2001).

The Treaty sets forth that all natural and legal persons holding plant genetic resources listed in Annex I are "invited" to include their resources in the multilateral system, and that contracting parties must take appropriate measures to "encourage" them to do so (Article 11.2).[36] It also establishes that within two years of the entry into force of the Treaty (which occurred on June 29, 2004), the governing body must assess the progress that such natural and legal persons have made in the inclusion of their own resources in the multilateral system. The governing body must then decide whether access will continue to be facilitated to those natural and legal persons that have not included their resources in the multilateral system (Article 11.4). According to a recent study (Chiarolla and Jungcurt, 2011), as of

March 2011, no natural and legal persons that are not part of national PGRFA systems, such as private plant breeding companies, had decided to voluntarily place their collections of Annex I materials directly in the multilateral system. Two legal persons are listed on the Treaty website – the Association pour l'Etude et l'Amélioration du Maïs (Pro-Maïs) and the Association Française des Semences de Céréales à Paille et Autres Espèces Autogames (AFSA) – but these are both part of the French National Institute for Agricultural Research (INRA) and thus their collections have to be considered to be part of the materials under the management and control of a party (France). This means that to date (March 2011), no collections of truly separate natural and legal persons (i.e., collections from the private sector) have been included in the multilateral system, and thus there are currently no natural and legal persons outside national PGRFA systems that have included their collections in the multilateral system, according to the study (Chiarolla and Jungcurt, 2011).

At its second session, in Rome (2007), the governing body of the Treaty decided to postpone the assessment of progress in the inclusion of PGRFA in the multilateral system until its third session. At its third session in Tunis (2009), the governing body conducted an assessment of progress of inclusion of materials, but decided to postpone the assessment of whether facilitated access should continue to be granted to legal and natural persons who have not included their PGRFA in the multilateral system until its fourth session in March 14–18, 2011, in Bali (Indonesia), when, once again, this decision was postponed until its fifth session, expected to be held in 2013. As such a decision has not yet been made, there is currently no legal basis to refuse access to natural and legal persons from contracting parties who have not made their collections available to the multilateral system. The above-mentioned study (Chiarolla and Jungcurt, 2011) recommends that the governing body of the Treaty consider options for restricting access to natural and legal persons (from contracting and noncontracting parties) that have not made their genetic materials available to the multilateral system. It considers two options. The first option would be to simply follow the principle that only those entities that share their own materials are entitled to benefit from facilitated access through the multilateral system, but, according to the study, such an option must be considered with caution, as a direct restriction could lead to adverse effects and deter private sector participation and undermine the multilateral system in the long run. The second option discussed by the study would be to devise a payment scheme for access for those natural and legal persons who have decided not to make their materials available. The scheme would make access subject to additional contributions to the benefit-sharing fund for those entities that benefit from facilitated access under the multilateral system but which refuse to grant the same access to their own materials by not putting their collections in the system. There would be two options for such entities:

1 *Pay per accession.* At the time of accession, the recipient would pay a fixed fee per accession to the benefit-sharing fund. Benefit-sharing payments would be made once a product incorporating that material is commercialized, that is, based on a percentage of the sales of the individual product.
2 *Pay per crop.* Recipients that opt to make payments based on a percentage of all sales of a given Annex I crop would have to contribute a higher percentage of the sales of products of that crop as long as their own collections are not available to the multilateral system. If they make their materials available, the payment is reduced to the normal rate (Chiarolla and Jungcurt, 2011).

Access to plant genetic resource must be provided "expeditiously," without the need to track individual accessions and free of charge, or, when a fee is charged, it must not exceed the minimal cost involved (that is why it is called facilitated access). Facilitated access to plant genetic resource must be provided along with all available passport data and other associated available non-confidential descriptive information.

According to a recent study (Chiarolla and Jungcurt, 2011), seven years after the Treaty's coming into force, progress in the implementation of the multilateral system has been slow: fewer than one-sixth of the contracting parties have notified which collections they are placing in the multilateral system and provided the documentation necessary to facilitate access. According to the study (Chiarolla and Jungcurt, 2011), although all materials under the management and control of contracting parties are legally part of the multilateral system, their effective inclusion requires that parties identify which collections are under their management and control, and inform the Treaty secretariat where information on how to access these materials is publicly available. As of January 2011, the total number of accessions for which such complete information is available was estimated at around 1 million. Roughly two-thirds are accessions made by the international agricultural research centers (IARCs) of the CGIAR and one-third by parties and institutions of the European region. Accessions by other regions make up approximately 2.6 percent, with African countries contributing 2.1 percent, those from the Near East 0.3 percent, and those from Latin America and the Caribbean 0.2 percent. Only 22 of the 127 contracting parties have provided notification of their collections and access to the relevant information. Of these, 13 have made all necessary information directly available to the secretariat, and six parties have made partial information available, but information about their collections can be accessed through the Eurisco[37] catalogue or the website of ICARDA (International Center for Agricultural Research in Dry Areas).

Facilitated access to plant genetic resources must be provided pursuant to a standard material transfer agreement (SMTA), which was adopted by the treaty's governing body during its first session, held between June 12 and 16, 2006, in Madrid. The treaty's governing body adopted the SMTA through Resolution 01, of October 16, 2006.

The SMTA[38] is a contract between the provider and the recipient of a plant genetic resource, which establishes the terms and conditions for the genetic material transfer. The contracting parties of the Treaty are the (member) countries, but the parties in the SMTA are natural or legal persons providing (the "provider") and receiving (the "recipient") plant genetic resources through the multilateral system.[39] Once a country ratifies the Treaty, the adoption of the SMTA becomes mandatory for crops listed in Annex I. Other types of material transfer agreements can be used only for plant genetic resources not included in the multilateral system.

The SMTA contains a clause (pursuant to Article 12.4 of the Treaty) stating that the recipient of plant genetic resources must require that the conditions of the SMTA be applied to all subsequent transfers of plant genetic resources. According to Article 6.4 of the SMTA, when the recipient transfers the genetic material supplied (through the multilateral system) to another person or entity ("the subsequent recipient"), the recipient must do so under the terms and conditions of the SMTA, through a new material transfer agreement, and notify the governing body.[40] However, the recipient has no further obligations regarding the actions of any subsequent recipient.

The recipient must also make available to the multilateral system all nonconfidential information resulting from research and development performed on the material received. It is not clear, however, who decides what information is confidential and what information is not: the recipient or the Treaty's governing body? The objective of this clause is to force those who access resources included in the multilateral system to provide information about them, in order to share with other users of the system. Nevertheless, criteria for definition of what is and what is not confidential information are not established. This could create space for users to simply withhold information about accessed resources, under the pretext of "confidentiality."

The multilateral system also includes plant genetic resources listed in Annex I and held in *ex situ* collections of the IARCs of the CGIAR, and access to such resources will be provided according to the same rules of the multilateral system, and through the SMTA.

The IARCs started using the SMTA for genetic materials of Annex I crops on January 1, 2007 (see Tables 6.2 and 6.3). At the second session of the Treaty's governing body, held from October 29 to November 2, 2007, in Rome, contracting parties decided that plant genetic resources *not* listed in Annex I, held by IARCs, should also be provided according to the SMTA. The governing body agreed to the addition of an explanatory footnote to the SMTA clarifying its application to Annex I as well as non-Annex I materials. Therefore, the SMTA is used for all material collected prior to the entry into force, in 2004, of the ITPGRFA (Annex I and non Annex I materials), and for Annex I material collected after the entry into force of the ITPGRFA (which are held in trust by IARCs). However, non-Annex I material received and conserved by IARCs after the coming into force of the ITPGRFA will be available for access on terms consistent with those mutually agreed between the IARCs that receive the material and the country of origin of such resources (or the country that has acquired those resources in accordance with the CBD or other applicable laws), as Article 15.3 of the Treaty sets out. In other words, non-Annex I material collected after the entry into force of the Treaty fall outside the scope of the multilateral system in the IARCs, and are subject to the CBD and to the Nagoya Protocol. On the other hand, the contracting parties in whose territory the PGRFA were collected from *in situ* conditions (and are conserved in *ex situ* collections of IARCs) will be provided with samples of such plant genetic resources on demand, *without any MTA* (Article 15.1b.ii of the treaty). The international legal personality of the IARCs holding *ex situ* collections is recognized and forms the basis of the agreements signed by the centers with the governing body of the treaty.

According to the study by Chiarolla and Jungcurt (2011), up to December 31, 2009, a total of 1.15 million samples of Annex I crops were distributed using the SMTA, of which approximately 84 percent were sent to recipients in developing countries or countries with economies in transition, 9.5 percent to developed countries, and 6.5 percent to other CGIAR centers. According to the *Second Report on the State of the World's Plant Genetic Resources for Food and Agriculture*, germplasm of crops listed under Annex 1 of the ITPGRFA is conserved in more than 1,240 genebanks worldwide and adds up to a total of about 4.6 million samples. Of these, 13 percent are stored in the collections of the CGIAR centers and about 51 percent are conserved in more than 800 genebanks of the contracting parties of the ITPGRFA. However, a study conducted by Edward Hammond, and released in March 2011 (Hammond, 2011), indicates that approximately half of sorghum genebank collections maintained by the International Crops Research Institute for the Semi-Arid Tropics (ICRISAT)[41] are being distributed outside of the multilateral system, through the US Department of Agriculture (USDA) genebanks. USDA and ICRISAT hold some sorghum germplasm collections in common, and, according to the study, corporate and commercially oriented breeders are avoiding the benefit-sharing requirements of the multilateral system by accessing the USDA sorghum genebanks instead of ICRISAT's collections. The United States has not ratified the ITPGRFA, and the USDA distributes its sorghum collections freely, whereas ICRISAT's collections are available only under the multilateral system, and access can be granted only through the SMTA, which imposes mandatory benefit-sharing requirements, according to the Treaty's system. Thus, recipients of the USDA distribution of sorghum are not obligated to share benefits and do not have to comply with the restrictions of the SMTA on patenting parts of the genetic material received (Hammond, 2011).

Access to plant genetic resources under development, including materials being developed by farmers, is at the discretion of their developer, during the period of their development.

That is, there is no obligation to provide access to materials under development, and if the developer opts to grant access, he or she may establish additional conditions (apart from the conditions established by the SMTA). According to the SMTA, plant genetic resources "under development" means "material derived from the genetic material, and hence distinct from it, that is not yet ready for commercialization and which the developer intends to further develop or to transfer to another person or entity for further development." The period of development is deemed to have ceased "when those resources are commercialized as a product" (Article 2 of the SMTA). Article 12.3d of the Treaty sets out the following rule (emphasis added):

Recipients shall not claim any intellectual property or other rights that limit the facilitated access to the plant genetic resources for food and agriculture, *or their genetic parts or components, in the form received from the multilateral system.*

Natural or legal persons who receive plant genetic resources through the multilateral system cannot prevent third parties from receiving the same resources through the system, through IP rights over them. This is one of the most controversial issues of the Treaty, and this norm resulted from tense negotiations between developed countries, which opposed any restriction or limitation on IP rights established by the Treaty, and developing countries, which intended to expressly prevent IP rights from being granted over genetic materials provided through the multilateral system, which would limit access to them.

Most developed countries understand that IP rights can be requested regarding plant genetic resources or their parts or components if some innovation or modification has taken place, that is, as long as the material is no longer in "the form received from the multilateral system." When the EU joined the Treaty, on March 31, 2004, it declared that it interprets Article 12.3d (above-mentioned) of the Treaty as recognizing that PGRFA or their genetic parts or components which have undergone innovation may be the subject of IP rights provided that the criteria relating to such rights are met. Austria, Belgium, Denmark, Finland, Germany, Greece, Ireland, Italy, Luxembourg, Poland, Spain, Sweden, and the United Kingdom made similar declarations.[42] In other words, any innovation would be enough to allow IP rights' claims. It is controversial, however, if the mere isolation of genes (contained in genetic material accessed through the multilateral system) would be enough to allow patenting because, if an unaltered gene (simply isolated from a plant in the multilateral system) can be patented, access to these materials would be restricted, which goes against the spirit of the Treaty which seeks to facilitate access to plant genetic resources. According to Visser and Borring (2011), it remains a matter of legal interpretation and jurisprudence "what really constitutes a 'product' in biotechnological use, and in particular whether the mere isolation and independent multiplication and use of a DNA sequence in its original form but in a different genetic environment is sufficient to define that as a 'product'." However, on the view of the seed industry (represented mainly by the International Seed Federation –ISF), cells, organelles (specialized subunits within cells that have specific functions), genes, or molecular constructs isolated from the material may be protected by the recipient through patents, if the criteria for patentability are met (Den Hurk, 2011).

Access to plant genetic resources found in *in situ* conditions must be provided according to national legislation or, in the absence of such legislation, in accordance with standards set by the governing body of the Treaty (Article 12.3h).

In the case of countries that ratified the CBD and approved national ABS laws (based on CBD principles), access to *in situ* plant genetic resources (including both wild relatives of crop species and local/traditional varieties and landraces) is subject to prior informed consent and equitable benefit-sharing with the countries of origin and local communities (CBD, Article 8j). Not all countries that have ratified the CBD already have national ABS laws, which creates operational difficulties for the implementation of CBD principles. Article 12.3h of the Treaty establishes that, in the absence of such national legislation, contracting parties may adopt standards set by the governing body of the Treaty. However, such (international) standards have not yet been adopted by the governing body, and national laws must regulate access to plant genetic resources in *in situ* conditions (for a more detailed discussion on national laws on access to *in situ* and on-farm plant genetic resources, see Chapter 7).

In emergency disaster situations, contracting parties agree to provide facilitated access to plant genetic resources for food and agriculture in the multilateral system for the purpose of contributing to the re-establishment of agricultural systems, in cooperation with disaster relief coordinators (Article 12.6).

Restoring agriculture and food security is an essential step in helping any community or nation recover from natural disasters or conflicts.[43] Germplasm banks can be useful for recomposition of areas devastated by wars or natural catastrophes, which has already occurred on several occasions. When a tsunami killed thousands and devastated 12 Asian countries in 2004, local agricultural systems were also affected. In Sri Lanka and Malaysia, one of the effects of advancing sea waves over the sand was excessive salt contents on coastal regions, and the germplasm bank in the Philippines sent six varieties of salt-tolerant rice to the affected regions in order to rebuild their farming systems. Something similar happened in Rwanda, which, in 1994, was devastated by genocide, war, and hunger. Approximately 800,000 people died in a few months and 3 million more became refugees in neighboring countries. An emergency program was established by the International Tropical Agriculture Center, with support from other international centers, agricultural institutions in neighboring countries, and nongovernmental organizations, in order to ensure that farmers received seeds and technical assistance to restore their agricultural systems, and restart planting of local varieties. Seeds kept in germplasm banks were used, as well as bean and corn seeds stored by farmers themselves.[44] There are also examples of threats facing genebanks when political instability and civil unrest occurs: in Egypt, recent political unrest led to the looting of the Egyptian Desert Genebank at Sheikh Zowaid Station in North Sinai. At the Desert Genebank, home to a prized collection of fruit and medicinal plants, equipment was stolen, the facility's cooling system was destroyed, and data that represented more than a decade worth of research were ruined (in February 2011). The Egyptian bank is specialized in desert plants.[45]

Access and benefit-sharing

We will now present the main rules related with benefit-sharing in the multilateral system (Article 13.1).

> The contracting parties shall recognize that facilitated access to PGRFA constitutes a major benefit of the multilateral system. The following benefit-sharing mechanisms are established: (1) exchange of information: contracting parties agree to make available information which encompasses, *inter alia*, catalogues and inventories, information on technologies, the results of technical, scientific, and socioeconomic research, including characterization, evaluation, and utilization, regarding those PGRFA under the multilateral system (such information shall be made available, when nonconfidential, through the information system);[46] (2) access and transfer of technology for the conservation, characterization, evaluation, and use of PGRFA which are under the multilateral system; (3) capacity-building;[47] sharing of monetary and other benefits of commercialization, taking into account the priority activity areas in the rolling GPA, and under the guidance of the governing body of the Treaty.

There are two different benefit-sharing frameworks. The first (which includes exchange of information, access to and transfer of technology, and capacity-building) is not connected with any specific genetic material access or transfer, as it comprehends general mechanisms, which are independent of specific transactions. The second benefit-sharing framework (sharing of monetary and other benefits of commercialization) is associated with a specific access or transfer, and depends on further commercialization of resources and other conditions (Article 16d.ii and Article 6.7 of the SMTA). According to the Treaty, a recipient (of plant genetic resources) who commercializes a product that is a PGRFA and which incorporates material accessed from the multilateral system must pay to the Treaty's benefit-sharing fund an equitable share of the benefits arising from the commercialization of that product, except whenever such a product is *available without restriction to others for further research and breeding*, in which case the recipient who commercializes will be "encouraged" to make such payment (that is, the payment is not mandatory, but voluntary). In other words, if recipients (who accessed the resources maintained by the multilateral system) commercialize an end product (which is also a plant genetic resource[48]) and do not allow others to use this product for research or breeding, they are obliged to pay a fixed percentage of the sales of the commercialized product to the benefit-sharing fund of the Treaty. If the developed product is available for further research and breeding, benefit-sharing is voluntary.

When monetary benefit-sharing is mandatory, the recipient may choose one of following forms of payment: (1) a fixed percentage (1.1 percent) of the sales of each commercialized[49] product (that incorporates material received from the multilateral system), minus 30 percent, which represents 0.77 percent, *or* (2) a discounted rate of 0.5 percent on the sales of each product made from the same crop (that is, the payment is made per crop, and not per accession or per product). This second option is valid for a period of 10 years, and is renewed for additional periods of five years unless the recipient notifies his or her intention to opt out, and the choice must be communicated to the governing body. In the second option, the payment obligation is applicable not only to the sales of the product that incorporated material received from the multilateral system, but to *any* products that are PGRFA belonging to the same crop to which the material received from the multilateral system belongs. This means that, by selecting this option, the recipient would pay a royalty on all products of a certain crop regardless of whether these products incorporate the material received from the multilateral system or whether the further use of the material by third parties for research and breeding is limited. A clear advantage of this option from the perspective of contracting parties is that the payment obligation is triggered as soon as the recipient sells any product of the crop. Another feature of this option is that, once the choice is made, it becomes the mandatory form of payment applicable to the recipient. This means the recipient is free to choose but, after selecting the preferred option, is bound by the appropriate terms and conditions of the SMTA, as Carlos Correa (2011) explains.

In any case, no payment will be mandatory if the product is available without restriction to others for further research and breeding; has been purchased or otherwise obtained from another person or entity who either has already made payment on the product or is exempt from the obligation to make payment; or is sold or traded as a commodity (and not as a plant genetic resource). In any case, such monetary benefits do not return to the country of origin of the resources or to the institution that provided the plant genetic resources, but are deposited in a benefit-sharing fund, aimed at implementation of the Treaty. According to the Treaty (Article 13.3), such benefits must flow primarily, directly and indirectly (through the benefit-sharing fund of the Treaty), to farmers in all countries, especially in developing countries and countries with economies in transition, who conserve and sustainably use PGRFA. That is, benefits must be shared not only with farmers who developed plant varieties used in breeding programs, but with all farmers involved in conservation and sustainable use of agrobiodiversity.

The expression "available without restriction to others for further research and breeding," raises many controversies, however, and its interpretation is essential to determine who is and who is not obliged to share monetary benefits arising from the commercialization of a product that incorporates genetic material received from the multilateral system. According to the SMTA (Article 2), a product is considered to be available without restriction to others for further research and breeding when it is available for research and breeding without any legal or contractual obligations, or technological restrictions, that would preclude using it in the manner specified in the Treaty. Currently, when recipients claim breeders' rights over plant genetic resources (that incorporate material accessed from the multilateral system), and independently of adopting the 1978 or 1991 UPOV Acts, they are not obliged to share benefits, as breeders' rights protect the so-called "breeder's privilege,"[50] that is, the possibility of using protected resources as a source of genetic variation in research and breeding. If, however, new products are protected through patents, benefit-sharing is mandatory. However, as some patent laws permit use of patented materials for research purposes (called "research exemption"), some companies claim that, under these circumstances, patent protection should not trigger compulsory benefit-sharing either (Meienberg, 2007). However, the only countries that allow patenting of plant varieties are the United States, Japan, Australia, and New Zealand. The United States signed but did not ratify the Treaty. Japan and New Zealand neither signed nor ratified the Treaty.[51] Only Australia signed and ratified the treaty (as of March 2011), which shows how limited this benefit-sharing mechanism really is.

Another controversy arises if benefit-sharing is also mandatory when hybrids are developed (using genetic material accessed from the multilateral system). In hybrids, parental lineages are kept secret, and new generations lose the so-called "hybrid vigor," which discourages farmers from resowing saved seeds because of loss of yield and productivity in the plant's next generations. Third parties are not legally prevented from using hybrids, but they are discouraged from doing so because of the "biological protection" that hybrids have. The development and commercialization of hybrids (that incorporate genetic material accessed through the multilateral system) should trigger compulsory benefit-sharing, because they represent a serious restriction on the availability of these resources to other parties. Other situations of compulsory benefit-sharing are genetic use restriction technologies (GURTs) and restrictions imposed by contracts and licenses.

It is essential, therefore, that the governing body of the Treaty clarifies all these issues involving mandatory monetary benefit-sharing and, more specifically, the exact meaning of the expression "available without restriction," and who has to pay mandatory benefit-sharing and who is exempt, considering the needs not only of professional breeders, but also of farmers-breeders and on-farm breeding. An excellent study conducted by Chiarolla and Jungcurt (2011) makes the following recommendations:

1 Patents that cover PGRFA products under current IP laws must be presumed to restrict access for research and breeding and to fulfill the mandatory benefit-sharing requirement (independently of eventual research exemptions in national patent laws).
2 UPOV 1991-type plant variety protection impedes informal exchange and sale of seeds, and it reduces opportunities for on-farm breeding, varietal improvement, and selection by farmers. By doing so, UPOV 1991 also imposes restrictions on research and breeding which takes place outside the formal seed system, and it must be considered to fulfill the mandatory benefit-sharing requirement.
3 There are already technical means in widespread use that are restricting access to PGRFA for research and breeding (such as cytoplasmic male sterile varieties and hybrids), and they must be considered to fulfill the mandatory benefit-sharing requirement. Chiarolla and Jungcurt (2011) explain that conventional hybrids are generally deemed not to restrict access for research and breeding, because, in theory, their genetic composition is freely available. However, if a breeder does not have access to the parental lines, it is very complicated to use hybrids as the basis for further development. Thus, hybrids limit on-farm breeding and effectively prevent farmers from using the material for selection and breeding and also – to a large extent – from replanting farm-saved seeds. In crops where this is not feasible (e.g., sunflower, cabbage, etc.) other expedients may be needed, such as cytoplasmic male sterility breeding or incompatibility. Although cytoplasmic male sterility may occur naturally in some species (e.g., radish), proprietary techniques have been used to transfer cytoplasmic male sterility to species in which it does not occur naturally and cannot be hybridized through conventional techniques. Cytoplasmic male sterile varieties prevents the female parent from being selfed because it is male sterile (i.e., it does not produce functioning pollen).
4 Varieties that incorporate genetic use restriction technologies (already covered under the current SMTA) must also be considered to fulfill the mandatory benefit-sharing requirement.

Chiarolla and Jungcurt (2011) recommend, therefore, a broader interpretation of what may constitute an "access restriction," to build a more economically viable, equitable, and ethically sound product base for the calculation of benefit-sharing under the Treaty, and I totally agree with their wise recommendations. After all, the implementation of a funding strategy for the Treaty aims to enhance availability, transparency, efficiency, and effectiveness of financial resources to implement activities under the Treaty. The financial strategy must cover all objectives and activities contained in the Treaty, not only the multilateral system, and give priority to implementation of agreed plans and programs for farmers in developing countries who conserve and sustainably utilize PGRFA (Article 18.5). During its first session, held in Madrid (2006), the governing body adopted the funding strategy for the implementation of the ITPGRFA (through Resolution 1/2006, Appendix F), and decided that benefits arising from the use of PGRFA that are shared under the multilateral system should be used in accordance with priority activity areas of the rolling GPA.[52] Such benefits include both monetary benefits arising from the commercialization of PGRFA (explained above), as well as donations and voluntary contributions from countries, international organizations, private companies, NGOs, etc. According to the secretariat of the Treaty, on the first call for proposals for funding (under the benefit-sharing fund of the Treaty), for the cycle 2009–2011, US$550,000 was granted for 11 projects (submitted by public and private institutions from developing countries from Africa, Asia, Latin America and the Caribbean, and the Near East that are contracting parties to the Treaty[53]). These funds derived entirely from voluntary contributions, and so far (as of March 2011), the secretariat of the Treaty has not received any mandatory monetary payment (arising from the commercialization of plant genetic resources). The governing body also agreed on a strategic plan for the implementation of the Treaty's benefit-sharing fund, and established a target for benefit-sharing over five years (US$116 million between July 2009 and December 2014, in accordance with Resolution 3/2009). So far (as of March 2011), US$14.37 million has been committed to the second call for proposals (2010–2014) under the benefit-sharing fund, mainly from Spain, Italy, Australia, Switzerland, the UNDP, and Norway (0.1 percent of seed sales in perpetuity[54]). The projects to be funded under the second project cycle were announced in July 2011.[55] During the fourth session of the governing body, in March 2011, Indonesia (the second richest country in biodiversity, after Brazil) announced that it will be the first developing country to make a financial contribution to the benefit-sharing fund of the ITPGRFA.[56] The governing body also approved (during its fourth session) a resolution stating that voluntary contributions to the core administrative budget of the Treaty will be based on an indicative scale of contributions, aimed at serving as guidance regarding the possible level of contributions from contracting parties.[57] It clearly appears that mandatory benefit-sharing payments are expected to play a very marginal role in achieving the US$116 million target for benefit-sharing under the funding strategy.[58] According to Visser and Borring (2011), if the breeding cycle is taken as a reference, substantial income can only be expected 7–15 years after distribution of germplasm for breeding has taken place.

Despite the limited scope of mandatory benefit-sharing under the Treaty, some European seed industries have criticized it. When the Treaty's benefit-sharing mechanisms were being discussed, some of their representatives argued that benefit-sharing should take place only if a substantial part of the genetic resources could be found back in the final product; it was proposed that a minimum of 25 percent should be incorporated. In addition, they claimed that benefit-sharing should be triggered only in the case that an identifiable trait of value or essential characteristic of the genetic resource is incorporated into the final product. They further argued that the duration of benefit-sharing mechanisms should be established, but this was left out of the final Treaty text. Similarly, there was no agreement on what part of DNA of the genetic resources should be incorporated to the final product (to trigger benefit-sharing), and the final text of the Treaty established that any incorporation of a genetic resource should trigger benefit-sharing (Den Hurk, 2011).

Article 13.2c.ii of the Treaty sets forth, however, that the governing body of the Treaty may decide to establish distinct levels of payment for various categories of recipients who commercialize such products. It may also decide on the need to exempt from such payments small farmers in developing countries and in countries with economies in transition. The governing body may, from time to time, review the levels of payment with a view to achieving fair and equitable sharing of benefits. It may also assess, within a period of five years from the entry into force of this Treaty (which occurred on June 29, 2004), whether the mandatory payment requirement in the material transfer agreement will apply also in cases where such commercialized products are available without restriction to others for further research and breeding. At its fourth session, held in Bali (Indonesia), from March 14 to 18, 2011, the governing body decided to postpone this decision (on whether mandatory payment would apply for products made available without restriction) until its fifth session, expected to be held in 2013.

Given the scarce resources for programs aimed at conservation and sustainable use of agrobiodiversity, the governing body can – and should – decide that payments must be calculated on a fixed percentage of all sales of products resulting from genetic materials accessed through the multilateral system, regardless of whether or not these products are protected by IP rights, or whether they are available (or not) without restriction for further research and breeding. After all, it would be fair that all users/recipients of the multilateral system channel part of their profits obtained from sales of their products to the conservation of plant genetic resources, and this would contribute to a more solid and sustainable funding strategy in the long term. It would also be important that countries create national benefit-sharing funds, to be used for *in situ* and on-farm conservation of PGRFA (see Chapter 7 for a more detailed discussion on national benefit-sharing funds). If this option is not exercised, the governing body should, at least, clarify all questions concerning the exact meaning of the expression "available without restriction," so that it becomes perfectly clear that all commercialized products that are patented (independently of research exemptions), protected through the 1991 UPOV Act, or that have their access restricted for breeding (such as hybrids and cytoplasmic male sterile varieties) must fulfill the mandatory benefit-sharing requirement. See Table 6.2 for the list of food crops and Table 6.3 for the list of forages which are covered under the multilateral system.

The Nagoya Protocol and its interfaces with the FAO Treaty and other specialized access and benefit-sharing agreements

The Nagoya Protocol on Access to Genetic Resources and the Fair and Equitable Sharing of Benefits Arising from their Utilization is a new international agreement, adopted by the Conference of the Parties to the Convention on Biological Diversity (CBD) at its 10th meeting, on 29 October 2010, in Nagoya (Japan).[59] The Nagoya Protocol will enter into force 90 days after the deposit of the 50th instrument of ratification, acceptance, approval, or accession.[60] It supports the further implementation of one of the three objectives of the CBD: the fair and equitable sharing of benefits arising from the utilization of genetic resources and traditional knowledge.

As the Nagoya Protocol is an instrument for the implementation of ABS provisions of CBD, it applies only to genetic resources and traditional knowledge (associated to genetic resources) that are covered by CBD. It does not apply to such resources when they are covered by other specialized international ABS instruments. To date, ITPGRFA is the only international agreement to establish specialized (multilateral) ABS provisions.

The Nagoya Protocol aims to establish clearer conditions for ABS and to create greater transparency in relationships between providers and users of genetic resources and traditional knowledge. However, the full implementation of the Nagoya Protocol depends on national ABS laws. The core elements of the Nagoya Protocol are:

1 *Access to genetic resources* (Articles 6 and 8)
   a In the exercise of sovereign rights over natural resources, and subject to national laws, access to genetic resources is subject to the prior informed consent of the provider party (country of origin or that has acquired the genetic resources in accordance with CBD).

- b Contracting parties must take measures to ensure that the prior informed consent or approval and involvement of Indigenous and local communities is obtained for access to genetic resources, where they have the established right to grant access to such resources (in accordance with national laws).
- c Contracting parties have the obligation to establish legislative, administrative or policy measures regarding ABS, that create legal certainty, clarity, and transparency; provide fair and nonarbitrary rules and procedures; establish clear rules and procedures for prior informed consent and mutually agreed terms; and provide for issuance of a permit or equivalent when access is granted.
- d In the development and implementation of their national ABS laws, contracting parties must create conditions to promote and encourage research that contributes to the conservation and sustainable use of biological diversity, particularly in developing countries, including through simplified measures on access for noncommercial research purposes; pay due regard to cases of present or imminent emergencies that threaten or damage human, animal, or plant health, considering the need for expeditious ABS; and consider the importance of genetic resources for food and agriculture and their special role for food security.

2 *Access to traditional knowledge associated with genetic resources* (Articles 7 and 12)

- a In accordance with its national law, each contracting party must take measures to ensure that traditional knowledge associated with genetic resources that is held by Indigenous and local communities is accessed with the prior and informed consent or approval and involvement of these Indigenous and local communities, and that mutually agreed terms are established.
- b Due consideration must be given to Indigenous and local communities' customary laws, community protocols, and procedures, with respect to traditional knowledge associated with genetic resources.
- c Contracting parties must support the development, by Indigenous and local communities, of community protocols, minimum requirements for mutually agreed terms, and model contractual clauses.
- d Contracting parties must not restrict the customary use and exchange of genetic resources and associated traditional knowledge within and amongst Indigenous and local communities.

3 *Fair and equitable benefit-sharing (arising from the utilization of genetic resources and associated traditional knowledge)* (Articles 5, 9 and 10)

- a Benefits arising from the utilization of genetic resources, as well as subsequent applications and commercialization, must be shared in a fair and equitable way with the provider party (country of origin or that has acquired the genetic resources in accordance with CBD), based on mutually agreed terms.

- b In accordance with national laws regarding the rights of Indigenous and local communities over genetic resources, contracting parties must take legislative, administrative, or policy measures to ensure that benefits arising from the utilization of genetic resources that are held by Indigenous and local communities are shared in a fair and equitable way with the communities concerned.
- c Benefits may include monetary[61] and nonmonetary benefits,[62] including but not limited to those listed in the Annex of the Protocol.
- d Contracting parties must take legislative, administrative, or policy measures to ensure that benefits arising from the utilization of traditional knowledge associated with genetic resources are shared in a fair and equitable way with Indigenous and local communities holding such knowledge.
- e Contracting parties must encourage users and providers to direct benefits arising from the utilization of genetic resources toward the conservation of biological diversity and the sustainable use of its components.
- f Contracting parties must consider the need for a global multilateral benefit-sharing mechanism to address the fair and equitable sharing of benefits derived from the utilization of genetic resources and traditional knowledge associated with genetic resources that occur in transboundary situations or for which it is not possible to grant or obtain prior informed consent.
- g The benefits shared by users of genetic resources and traditional knowledge associated with genetic resources through this mechanism must be used to support the conservation of biological diversity and the sustainable use of its components globally.

4 *Compliance with national ABS legislation* (Articles 15, 16, 17, 18, and 19)

- a Contracting parties must take appropriate, effective, and proportionate legislative, administrative, or policy measures to ensure that genetic resources utilized within their jurisdiction have been accessed in accordance with prior informed consent and that mutually agreed terms have been established, as required by domestic ABS legislation. Such measures must also ensure that traditional knowledge associated with genetic resources was accessed in accordance with prior informed consent or approval and involvement of Indigenous and local communities.
- b Contracting parties must take appropriate, effective, and proportionate measures to address situations of noncompliance.
- c Contracting parties must cooperate in cases of alleged violation of domestic ABS legislation.
- d In order to support compliance, contracting parties must designate one or more checkpoints, encourage users and providers of genetic resources to include provisions, in mutually agreed terms, to share

information on the implementation of such terms, including through reporting requirements, and encourage the use of cost-effective communication tools and systems.

e Contracting parties must encourage providers and users of genetic resources and/or traditional knowledge associated with genetic resources to include provisions in mutually agreed terms to cover dispute resolution, including the jurisdiction to which they will subject any dispute resolution processes, the applicable law, and/or options for alternative dispute resolution, such as mediation or arbitration. They must also ensure that an opportunity to seek recourse is available under their legal systems.

f Contracting parties must encourage the development, update, and use of sectoral and cross-sectoral model contractual clauses for mutually agreed terms, voluntary codes of conduct, guidelines, and best practices and/or standards in relation to ABS.

g An internationally recognized certificate of compliance will serve as evidence that the genetic resource which it covers has been accessed in accordance with prior informed consent and that mutually agreed terms have been established, as required by national ABS legislations (of the provider party). Such a certificate of compliance must be made available to the Access and Benefit-Sharing Clearing-House, a mechanism established in Article 14 of the Nagoya Protocol as a means for sharing of information related to ABS.

The Nagoya Protocol sets out that its provisions do not "affect the rights and obligations of any party deriving from any existing international agreement" and do not "prevent parties from developing and implementing other international agreements, including other specialized ABS agreements," if "they are supportive of and do not run counter to the objectives of CBD and the Nagoya Protocol" (Article 4). It makes several explicit mentions to agricultural biodiversity in its preamble, and it recognizes:

- "the special nature of agricultural biodiversity, its distinctive features and problems needing distinctive solutions";
- "the interdependence of all countries with regard to genetic resources for food and agriculture as well as their importance for achieving food security worldwide and for sustainable development of agriculture in the context of poverty alleviation and climate change";
- "the fundamental role of the International Treaty on Plant Genetic Resources for Food and Agriculture and of the FAO Commission on Genetic Resources for Food and Agriculture."

The Nagoya Protocol also recalls "the multilateral system of ABS established under the International Treaty on Plant Genetic Resources for Food and Agriculture, developed in harmony with the Convention."

Therefore, there is no doubt that the Nagoya Protocol does *not* apply to PGRFA that are subject to the multilateral system of ABS that is established by the ITPGRFA. As we have seen before, the multilateral system applies to plant genetic resources that are listed in Annex I of the Treaty (35 food crops and 29 forages), and that are under the management and control of the contracting parties and in the public domain, and as long as they are used for research, breeding, and training for *food and agriculture*, and not for chemical, pharmaceutical, and/or other nonfood/feed industrial purposes. This means that plant genetic resources that are not included in Annex I (of the Treaty), or which are used for any chemical, pharmaceutical, and/ or other non-food/feed purposes, fall under the scope of the CBD, and of the Nagoya Protocol, as they are not subject to the multilateral system of the ITPGRFA. Furthermore, only genetic materials that are in the public domain and under the management and control of contracting parties are excluded from the scope of the CBD and the Nagoya Protocol, which means that genetic materials held by private companies, NGOs, and farmers are not part of the multilateral system (unless they decide to put their collections in the multilateral system, but this is voluntary and not mandatory) and, therefore, such genetic materials are not excluded from the scope of the Nagoya Protocol. While they remain outside the multilateral system, they are subject to the rules of the CBD and of the Nagoya Protocol.

It must also be taken into account that the ITPGRFA explicitly mentions that contracting parties may widen the scope of the multilateral system, to include other crops and forages, as long as such decision is taken by consensus of contracting parties to the Treaty (Articles 23 and 24). Therefore, if contracting parties to the Treaty decide to widen the scope of the multilateral system, other crops and forages that are included in the multilateral system will also be left out of the CBD and Nagoya Protocol bilateral system of ABS. The Nagoya Protocol does not "prevent parties from developing and implementing other international agreements, including other specialized ABS agreements" (Article 4.3). This means that contracting parties to the Treaty may decide to expand the scope of an already existing specialized ABS multilateral system, such as the one established by the ITPGRFA.

Andersen *et al.* (2010) raise some interesting questions regarding the exclusion, from the scope of the Nagoya Protocol, of the multilateral system established by the ITPGRFA. They call attention to the fact that the IARCs of the CGIAR have not only included genetic materials that are listed in the Annex I of the International Treaty in the multilateral system, but are also using the SMTA for transfers of plant genetic resources for food and agriculture of non-Annex I materials, collected prior to the entry into force of the Treaty, and that some countries are also doing so, to avoid different systems for different genetic resources within genebanks. However, Andersen *et al.* (2010) point out that it is still uncertain whether non-Annex I materials, transferred with the SMTA, can also be regarded as included in the multilateral system, and out of the scope of the Nagoya Protocol. Thus, they suggest that not only Annex I materials be excluded from the scope of the Nagoya Protocol, but all "material in the multilateral system and covered by the SMTA" (Andersen *et al.*, 2010). The Nagoya Protocol is not so specific in relation to the exclusion of non-Annex I crops that are transferred with the SMTA, but as the Nagoya Protocol recognizes "the special nature of agricultural biodiversity, its distinctive features and problems needing distinctive solutions" as well as "the fundamental role of the International Treaty on Plant Genetic Resources for Food and Agriculture and the FAO Commission on Genetic Resources for Food and Agriculture" and the multilateral system of ABS established under the Treaty, all countries (including noncontracting parties to the Treaty[63]) can use the SMTA also for the transfer of non-Annex I materials collected prior to the entry into force of the Treaty and held in their *ex situ* collections (and as long as they are used for research, breeding, and training for *food and agriculture*). This is not explicit in the Nagoya Protocol, but it is a viable interpretation. However, non-Annex I crops collected after the entry into force of the ITPGRFA fall outside the scope of the multilateral system, and are subject to the CBD and to the Nagoya Protocol (for countries that are parties to the CBD and the Nagoya Protocol). Nevertheless, the ITPGRFA, in its Article 15.3, opens the possibility that IARCs of CGIAR use the SMTA for transfers of non-Annex I crops, received and conserved by IARCs after the coming into force of the Treaty, as long as the provider country agrees. It is important, however, to take into consideration the fact that this provision of the Treaty deals specifically with transfers of genetic resources, and not with access to *in situ* material, which is subject to national legislation, according to Article 12.3.h of the Treaty (which will be discussed in Chapter 7).

It is also worth mentioning that the IARCs have proposed that the SMTA functions as a certificate of source, with the source or origin of the PGRFA being the multilateral system itself (SGRP, 2007). The SMTA would be a certificate of compliance with the ITPGRFA. As mentioned above, the Nagoya Protocol sets out that an internationally recognized certificate of compliance will serve as evidence that the genetic resource which it covers has been accessed in accordance with prior informed consent and that mutually agreed terms have been established. Contracting parties to the Nagoya Protocol must establish checkpoints and decide on what type of information should be requested at such checkpoints. Chiarolla and Jungcurt (2011) propose that a possible way to enhance transparency and the mutual supportiveness between the Nagoya Protocol and the ITPGRFA would be to amend the SMTA in order to request recipients to disclose, at plant variety protection and patent offices, that the materials for which protection is sought have been obtained from the multilateral system and to inform the governing body (of the Treaty) accordingly. According to them, contracting parties who endeavor to implement the Treaty and the Nagoya

Protocol in a mutually supportive manner may envision using the SMTA as an internationally recognized certificate of compliance to be presented by resource users at all relevant checkpoints. During the fourth session of the governing body, held in Bali (Indonesia), on March 14–18, 2011, the European region suggested that the governing body request the Conference of Parties of the CBD to formally recognize the use of the SMTA as being in harmony with the ABS protocol and an equivalent to the international certificate of compliance (under the Nagoya Protocol), and a resolution was adopted on the relationship of the SMTA with the CBD.[64]

As mentioned above, the Nagoya Protocol to the CBD leaves room for the development and implementation of specialized ABS agreements (Article 4), other than the ITPGRFA. There are ongoing discussions about the potential usefulness of the development of specialized international ABS regimes for animal genetic resources (see Chapter 9), agricultural microbial genetic resources (SGRP, 2010), human pathogens (for public health preparedness and response purposes), and for specific regions and subregions, which could establish a multilateral system of ABS for non-Annex I crops that are particularly important for the regions concerned (SGRP, 2006).

Notes

1 The International Treaty on Plant Genetic Resources for Food and Agriculture defines plant genetic resources for food and agriculture as "any genetic material of plant origin of actual or potential value for food and agriculture." Genetic material is defined as "any material of plant origin, including reproductive and vegetative propagating material, containing functional units of heredity."

2 Erna Bennett directed the FAO Genetic Resource and Plant Ecology Unit in the 1960s and 1970s. She was responsible for several expeditions for collection of genetic materials and genetic resource conservation programs.

3 Historically, expeditions for collection of genetic resources, whether systematic or not, have always taken place. Perhaps the most ancient record of an expedition organized for collection of plants is from 1495 BC, when queen Hatshepsut, of Egypt, sent ships to eastern Africa (currently Somalia, Eritrea, Ethiopia, and Djibouti) to collect plants. Her interest was mainly the incense tree, which she intended to place in her mortuary temple, and myrrh seeds (also used as incense in funerals and cremations). There are also records of an Egyptian pharaoh, Sankhkara, sending ships to the Gulf of Aden (in Yemen) to collect cinnamon and cassia, plants used for embalming the dead. Source: Damania (2008). To learn more about the history of germplasm collection around the world, and about scientific expeditions for collection of botanical material in Brazil, see Walter *et al.* (2005a). These authors report famous expeditions of the Royal Botanic Gardens, Kew (England), Alexander Von Humboldt (German, 1769–1859), Alphonse de Candolle (Swiss, 1808–1893), Nikolai Vavilov (Russian, 1916 until the 1930s), and many other plant collectors.

4 For more information: www.bioversityinternational.org.

5 This group became internationally known by its acronym: CGIAR. Its centers are Africa Rice Center (former West Africa Rice Development Association), in Benin, International Center for Tropical Agriculture (CIAT), in Colombia, International Maize and Wheat Improvement Center (CIMMYT), in Mexico,

the International Potato Center, in Peru, International Center for Agricultural Research in the Dry Areas (ICARDA), in Syria, International Rice Research Institute (IRRI), in the Philippines, International Institute of Tropical Agriculture (IITA), in Nigeria, Center for International Forestry Research (CIFOR), in Indonesia, International Crops Research Institute for the Semi-Arid Tropics (ICRISAT), in India, International Livestock Research Institute (ILRI), in Kenya, International Water Management Institute (IWMI), in Sri Lanka, World Agroforestry Centre, in Kenya, WorldFish Center, in Malaysia, International Food Policy Research Institute (IFPRI), in Washington (United States), and Bioversity International, in Rome (Italy). For more information: www.cgiar.org,

6 Source: http://www.cgiar.org/impact/genebanksdatabases.html (accessed January 24, 2011).

7 Source: http://www.cgiar.org/pdf/model_agreement_centers_2007.pdf (accessed January 24, 2011).

8 The Action Group on Erosion, Technology and Concentration or ETC Group. See http://www.etcgroup.org/en/. The International Coalition for Development Action (ICDA, which later became GRAIN – Genetic Resources Action International) also played an important role in the second half of the 1980s. Two reference books were published during this period: *Seeds of the Earth: A Private or Public Resource*, by Pat Mooney, of RAFI, published in 1979, and *New Hope or False Promise: Biotechnology and Third World Agriculture*, by Henk Hobbelink, of ICDA/GRAIN, published in 1987.

9 Other limitations of *ex situ* collections include the fact that some seeds cannot withstand very low humidity and/or are not resistant to temperatures below 0°C, and therefore cannot be stored in cold chambers (called recalcitrant seeds). Vegetative propagation species (such as potatoes, yams, manioc/cassava, etc.) must also be conserved in field or *in vitro* collections, making *ex situ* conservation more expensive, complex and difficult to implement in poor countries. Walter de Boef *et al.* (2007, p. 46) point out that "passport data" (information about genetic resources) seldom include the characteristics described by farmers. Plant collectors generally spend no more than a few minutes with each plant sample. There is not enough time to talk with farmers and record their knowledge, which undermines the connection between farmers and their biological material.

10 The intergovernmental Commission on Plant Genetic Resources was also established during the 22nd meeting of the FAO Conference, through Resolution 09/83. In 1995, the Commission broadened its mandate to include all components of biodiversity for food and agriculture (not only plant but also livestock, forests, fish, etc.), and became the Commission on Genetic Resources for Food and Agriculture (CGFRA). The commission is responsible for the implementation of all FAO agreements on such issues, and is part of the FAO Global System for the Conservation and Sustainable Use of Plant Genetic Resources for Food and Agriculture.

11 As of March 10, 2011. Source: http://www.cbd.int/convention/parties/list/.

12 In this book, aspects of the CBD that are relevant to agrobiodiversity and PGFRA will be more specifically discussed. For more information about the CBD, see its website (http:www.cbd.int) and the IUCN guidebook (Glowka *et al.*, 2004).

13 In 1993, the FAO Conference also adopted the International Code of Conduct for Plant Germplasm Collection and Transfer. It is a voluntary code, which establishes general principles for the collection, conservation, exchange and use of plant germplasm. Source: http://www.fao.org/docrep/x5586E/x5586e0k.htm#xiv (accessed March 10, 2011).

14 The CBD uses the terms "domesticated" and "cultivated" species as synonyms. A "domesticated or cultivated" species is defined as a "species in which the evolutionary process has been influenced by humans to meet their needs." However, domesticating a plant is not the same as cultivating. For the difference between domesticated and cultivated plants, see Chapter 1, in which these concepts are explored.

15 According to the FAO International Treaty on Plant Genetic Resources for Food and Agriculture, "center of origin" means a geographical area where a plant species, either domesticated or wild, first developed its distinctive properties. "Centre of crop diversity" means a geographic area containing a high level of genetic diversity for crop species in *in situ* conditions.

16 The second edition of this book, which is better known, was published in 1883.

17 *Flora Brasiliensis* was organized with the participation of 65 specialists from different countries. It contains taxonomical characterizations of 22,767 species of the Brazilian flora, gathered in 15 volumes and divided into 40 parts, with a total of 10,367 pages. See http://florabrasiliensis.cria.org.br (accessed March 11, 2011).

18 Since 1965, it has been called Vavilov Research Institute of Plant Industry, and is located in Saint Petersburg, in Russia. For more information, see Loskutov (1997) and www.vir.nw.ru/.

19 More recent research in South America has shown that the Amazon may also be recognized as an independent crop domestication center, along with Meso-America and the Andes. The Upper Madeira Basin and its tributaries (currently the state of Rondonia, in Brazil) are supposedly domestication centers for two of the most important plants cultivated in the Amazon: manioc/cassava and *pupunha*. For more information, see Chapter 11 (section "GIAHS, Amazonian dark earths, and agrobiodiversity").

20 Vavilov died in 1943, in Saratov prison, in Russia, victim of the persecution of the Stalinist regime, because he did not agree with theories developed by Trofim Lysenko, who was the head of the Institute of Genetics in the former Soviet Union. Lysenko did not accept the laws of heredity developed by Gregor Mendel, on which all modern genetics is based.

21 Zeven and De Wet (1982), for example, propose three birthplaces for agriculture – eastern Asia (China and Myanmar), Near East (Fertile Crescent), and Central America, as well as 12 other centers of diversity.

22 For further information, see: Walter *et al.* (2005b).

23 See also Clement (2007).

24 For more information on the negotiations of the Treaty, see Coupe and Lewins (2007). To learn more about the Treaty and its member countries, see http://www.planttreaty.org. The United States signed the Treaty on November 1, 2002, but did not ratify it (as of March 11, 2011). To ratify the Treaty, the state must be recognized by the United Nations, but it does not need to be a member of FAO or of its Commission on Genetic Resources for Food and Agriculture.

25 http://www.itpgrfa.net/International/sites/default/files/cordoba_declaration_en.pdf (accessed March 10, 2011).

26 http://biodiversity-l.iisd.org/news/ministers-adopt-bali-declaration-on-the-itpgr/ (accessed March 10, 2011).

27 Source: www.fao.org/docrep/003/w3613p/w3613p00.htm (accessed March 10, 2011).

28 According to Article 12.3b, in the case of multiple-use crops (food and non-food), their importance for food security must be the determinant for their inclusion in the multilateral system and availability for facilitated access.

29 For a more detailed description of the negotiations of crops to be included in the multilateral system, see Engsiand and Halewood (2008) and Visser and Borring (2011).
30 All governing body decisions are made by consensus, except if consensually agreed otherwise. The Treaty's amendments, however, must necessarily be approved by consensus, including amendments to the Annexes. Institutional provisions in the Treaty are in Articles 19 to 35, and they cover compliance, settlement of disputes, amendments of the treaty, ratification, acceptance or approval, etc.
31 Lightbourne (2009) makes an interesting analysis of two cash crops, soybean and coffee, that were maintained outside the scope of the multilateral system, and whether they can be traded under profitable terms by their countries of origin (China and Ethiopia), in the framework of bilateral agreements complying with the CBD.
32 All components of the Treaty (farmers' rights, supporting components, financial, and institutional provisions) apply to *all* plant genetic resources for food and agriculture, except the multilateral system of ABS, which is restricted to Annex I crops. Annex II contains arbitration and conciliation rules.
33 *Ex situ* collections of plant genetic resources held by the IARCs of the CGIAR, international plant genetic resources networks and the Global Plant Genetic Resource Information System are also supporting components of the Treaty.
34 During its fourth session (held in Bali), the governing body approved a resolution that establishes an *Ad Hoc* Technical Committee on Sustainable Use of PGRFA, subject to availability of financial resources, and requests the secretary to further explore development of a toolbox (to assist countries in designing appropriate measures for sustainable use), develop steps for implementation of the Global Plan of Action, organize stakeholders' consultations, and invite submissions from parties and others on sustainable use, and work with regional networks and partners to promote locally adapted and underutilized crops. Source: *Earth Negotiations Bulletin*, vol. 9, no. 550, March 12, 2011 (http://www.iisd.ca/vol09/enb09550e.html; accessed March 25, 2011).
35 Source: http://www.cbd.int/decision/cop/?id=7107 (accessed February 20, 2011).
36 Evidently, public and private institutions can include plant genetic resources that are not in Annex I list in the multilateral system, but it is a voluntary initiative.
37 EURISCO is a web-based catalogue that provides information about *ex situ* plant collections maintained in Europe (http://eurisco.ecpgr.org/static/about_eurisco.html; accessed March 10, 2011).
38 The SMTA can be consulted on ftp://ftp.fao.org/ag/agp/planttreaty/agreements/smta/SMTAe.pdf (accessed March 11, 2011).
39 The SMTA is said to be "shrink-wrapped" when a copy of the SMTA is included in the packaging of the material, and the recipient's acceptance of the material constitutes acceptance of the terms and conditions of the SMTA. The SMTA is "click-wrapped" when the agreement is concluded on the internet and the recipient accepts the terms and conditions of the SMTA by clicking on the appropriate icon on the website or in the electronic version of the STMA, as appropriate.
40 According to Article 5e of the SMTA, the provider of genetic resources must periodically inform the governing body about the material transfer agreements entered into, and this information must be made available by the governing body to the third party beneficiary (FAO, which is in charge of monitoring compliance with the SMTA).

41 ICRISAT is an international agricultural research center supported by CGIAR, headquartered in India, with two regional hubs and four country offices in sub-Saharan Africa.

42 Source: http://www.fao.org/Legal/treaties/033s-e.htm (accessed March 10, 2011).

43 See the report "The healing wounds: how the International Agricultural Research Centers of the CGIAR help rebuild agriculture in countries affected by conflicts and natural disasters," published by CGIAR (2005) at http://www.cgiar.org/pdf/healingwounds.pdf (accessed March 10, 2011).

44 Source: IRRI (2005) "Restore agriculture, aid can help developing countries recover faster from natural disasters" (http://irri.org/news-events/media-releases/restore-agriculture-aid-can-help-developing-countries-recover-faster-from-natural-disasters (accessed March 10, 2011).

45 Source: "Egyptian genebank looted," February 1, 2011 (http://agro.biodiver.se/2011/02/egyptian-genebank-looted/; accessed March 10, 2011).

46 According to Article 17, contracting parties must cooperate to develop and strengthen a global information system to facilitate the exchange of information on scientific, technical, and environmental matters related to PGRFA (called the Global Information System on Plant Genetic Resources for Food and Agriculture).

47 According to Article 13b,c, contracting parties agree to give priority to (i) establishing and/or strengthening programs for scientific and technical education and training in conservation and sustainable use of PGRFA; (ii) developing and strengthening facilities for conservation and sustainable use of plant genetic resources for food and agriculture, in particular in developing countries, and countries with economies in transition; and (iii) carrying out scientific research preferably, and where possible, in developing countries and countries with economies in transition, in cooperation with institutions of such countries, and developing capacity for such research in fields where they are needed.

48 According to the SMTA (Article 2), a "product" means plant genetic resources for food and agriculture that incorporate (for example, by pedigree or gene insertion) the material or any of its genetic parts or components that are ready for commercialization, excluding commodities and other products used for food, feed, and processing.

49 According to the SMTA (Article 2), "sales" means the gross income resulting from the commercialization of a product or products, by the recipient, its affiliates, contractors, licensees, and lessees. "To commercialize" means to sell a product or products for monetary consideration on the open market, and "commercialization" has a corresponding meaning. Commercialization does not include any form of transfer of plant genetic resources under development.

50 For more detailed information on the "breeder´s privilege," see Chapter 5.

51 Source: http://www.fao.org/Legal/treaties/033s-e.htm (accessed March 11, 2011).

52 The governing body established the following initial priorities, included in the GPA upon indication by the *ad hoc* Advisory Committee on the Funding Strategy: exchange of information, transfer of technology and capacity-building, management and conservation of on-farm plant genetic resources, and sustainable use of plant genetic resources.

53 The following projects were funded under the benefit-sharing fund, within the first project cycle (2009–2011): (1) Kenya: characterization, genetic enhancement, and revitalization of finger millet in Western Kenya (Maseno University); (2) Morocco: on-farm conservation and mining of local durum and bread wheat landraces of Morocco for biotic stresses and incorporating UG99 resistance

(National Agricultural Research Institute); (3) Senegal: conservation of local cultivars of millet, maize, and sorghum, promotion of the major use of local varieties adapted to agroclimatic conditions, and increasing the diversity of the germplasm of these crops available to farmers; (4) Tanzania: strengthening on-farm conservation and use of local varieties of sorghum, finger millet, lablab beans, and yams for food security and adaptation to climate change (National Plant Genetic Resources Centre); (5) India: conservation, dissemination, and popularization of location-specific farmer-developed varieties by establishing village-level enterprises (Peermade Development Society); (6) Costa Rica: identification of useful potato germplasm adapted to biotic and abiotic stresses caused by global climate change (Universidad de Costa Rica, Centro de Investigaciones Agronómicas); (7) Cuba: contribution of traditional methods for *in situ* conservation and management of maize and bean for food security of farming communities (Fundamental Research Institute on Tropical Agriculture); (8) Nicaragua: rescue, conservation, and sustainable management of teosinte in the Apacunca Genetic Reserve (Universidad Nacional Agraria); (9) Peru: conservation and sustainable use of native potato diversity in the Potato Park, Cusco (Association for Nature and Sustainable Development; ANDES); (10) Uruguay: broadening of potato genetic basis through introgression of local wild species (Instituto Nacional de Investigaciones Agropecuarias); (11) Egypt: on-farm conservation and *in vitro* preservation of citrus local varieties and sustainable utilization in Egypt (National Genebank and Genetic Resources). Source: ftp://ftp.fao.org/ag/agp/planttreaty/funding/pro_list09_01_en.pdf (accessed March 11, 2011). In July 2011, the Treaty's secretariat announced the projects to be funded within the second project cycle. See http://www.planttreaty.org/funding-en.htm, accessed August 10, 2011.

54 Norway decided to make a permanent annual contribution to the benefit-sharing fund of the Treaty that amounts to 0.1 percent of the value of all seeds that are sold in the country. The value of this contribution was US$101,368 and it was received on June 15, 2010. The reference to 0.1 percent of seed sales refers only to the method that is used to calculate the amount of donations to the benefit-sharing fund; in practice, the contribution is paid with government funds and not directly by the private seed sector.

55 According to the requirements established by the governing body of the ITPGRFA, all plant genetic resources for food and agriculture listed in Annex 1 of the Treaty resulting from projects funded by the benefit-sharing fund must be made available according to the terms and conditions of the multilateral system. Furthermore, information generated by projects funded through the benefit-sharing fund must be made publicly available within one year of the completion of the project. Grant conditions of the Global Crop Diversity Trust also request that germplasm regenerated or characterized with support from the Trust will be available under the conditions of the SMTA. The Trust currently supports the regeneration and characterization of both international collections of the CGIAR and national collections in a large number of countries.

56 Source: http://www.itpgrfa.net/International/content/indonesia-first-developing-country-contribute-crop-benefit-sharing-fund (accessed March 14, 2011).

57 This voluntary indicative scale of contributions must be based on the scale of contributions adopted from time to time by the UN, and adjusted so as to ensure that no party contributes less than 0.01 percent of the total, that no contribution exceeds 22 percent of the total, and that no contribution from a least developed country party exceeds 0.01 percent of the total. The scale shall be maintained and updated by the secretariat, in accordance with the work program and budget for each biennium, as approved by the governing body.

Source: *Earth Negotiations Bulletin*, vol. 9, no. 550, March 12, 2011 (http://www.iisd.ca/vol09/enb09550e.html; accessed March 25, 2011).

58 As monetary benefits are not destined to countries and/or institutions of origin of the plant genetic resources, and to address the gap in enforcement incentives, the SMTA (Article 8.2) establishes that an "entity designated by the governing body" has the right, as a "third party beneficiary," to initiate dispute settlement procedures regarding rights and obligations of the provider and recipient of resources. FAO was designated the "third party beneficiary" and therefore, is responsible for monitoring compliance with the conditions set by the SMTA and the Treaty. As a "third party beneficiary," FAO acts on behalf of the governing body of the Treaty and its multilateral system.

The SMTA establishes, in Article 8, a dispute settlement procedure, for cases of noncompliance. A dispute settlement may be initiated by the provider, the recipient or the third-party beneficiary (FAO). Any dispute arising from the SMTA will be resolved in the following manner: (a) amicable dispute settlement: the parties will attempt in good faith to resolve the dispute by negotiation; (b) mediation: if the dispute is not resolved by negotiation, the parties may choose mediation through a neutral third-party mediator, to be mutually agreed; (c) arbitration: if the dispute is not settled by negotiation or mediation, any party may submit the dispute for arbitration under the arbitration rules of an international body, if agreed by the parties to the dispute. Failing such agreement, the dispute will be finally settled under the Rules of Arbitration of the International Chamber of Commerce, by one or more arbitrators appointed in accordance with the said rules. The result of such arbitration is binding.

59 During its fourth session (held in Bali), the governing body approved a resolution containing the procedures and operational mechanisms to promote compliance and address issues of noncompliance. It also approved a resolution containing the rules for mediation of a dispute in relation to the SMTA, but no decision was made on applying the third-party beneficiary procedures to non-Annex I material transferred with the SMTA.

According to the *Earth Negotiations Bulletin* (vol. 9, no. 550; http://www.iisd.ca/vol09/enb09550e.html; accessed March 25, 2011), one of the main issues hampering negotiations on compliance was related to recognition of the special needs of developing countries and countries with economies in transition. This is a standard provision in existing compliance-related instruments. Participants from North America took a principled stance that compliance obligations apply equally to all parties, whereas a distinction could be drawn when it comes to remedies. On the other hand, developing countries insisted that the principle of common but differentiated responsibilities applies to compliance. Consensus was achieved only on the last day after an all-night session, using compromise text based on the Basel Convention Compliance Mechanism.

The 10th meeting of the Conference of the Parties to CBD also adopted a revised and updated Strategic Plan for Biodiversity, for the 2011–2020 period, which can be accessed at http://www.cbd.int/decision/cop/?id=12268 (last accessed March 26, 2011).

60 As of March 23, 2011, the following countries had signed the Nagoya Protocol: Algeria, Brazil, Colombia, Mexico, Rwanda, and Yemen. Indonesia, the country which is the second richest in biodiversity, after Brazil, also announced, on March 17, 2011, that it would ratify the Nagoya Protocol. The 11th meeting of the Conference of the Parties to the CBD will take place in India, on October 8–19, 2012, and it is expected to convene the first meeting of the parties to the Nagoya Protocol. For this target to be met, the Nagoya Protocol must enter into force no later than October 8, 2012, with the 50th instrument of

ratification deposited no later than July 10, 2012. Sources: http://www.cbd.int/abs/nagoya-protocol/signatories/ and http://embassyofindonesia.it/indonesia-to-ratify-nagoya-protocol-on-access-to-genetic-resources (accessed March 23, 2011).

61 The Annex of the Nagoya Protocol lists the following monetary benefits: access fees/fee per sample collected or otherwise acquired, up-front payments, milestone payments, payment of royalties, license fees in case of commercialization, special fees to be paid to trust funds supporting conservation and sustainable use of biodiversity, salaries and preferential terms, research funding, joint ventures, and joint ownership of IP rights.

62 The Annex of the Nagoya Protocol lists the following nonmonetary benefits: sharing of research and development results; collaboration, cooperation, and contribution in scientific research and development programs, particularly biotechnological research activities, where possible involving the party providing genetic resources; participation in product development, collaboration, cooperation, and contribution in education and training; admittance to *ex situ* facilities of genetic resources and to databases; transfer to the provider of the genetic resources of knowledge and technology under fair and most favorable terms, including on concessional and preferential terms, in particular, knowledge and technology that make use of genetic resources, including biotechnology, or which are relevant to the conservation and sustainable utilization of biological diversity; strengthening capacities for technology transfer and institutional capacity-building; human and material resources to strengthen the capacities for the administration and enforcement of access regulations; training related to genetic resources with the full participation of countries providing genetic resources and, where possible, in such countries, access to scientific information relevant to conservation and sustainable use of biological diversity, including biological inventories and taxonomic studies; contributions to the local economy; research directed toward priority needs, such as health and food security, taking into account domestic uses of genetic resources by the party providing genetic resources; institutional and professional relationships that can arise from an ABS agreement and subsequent collaborative activities; food and livelihood security benefits; social recognition; and joint ownership of IP rights.

63 All countries that are parties to the ITPGRFA are also parties to the CBD. However, not all countries that are parties to the CBD are also parties to the ITPGRFA. As of March 24, 2011, 127 countries are parties to the Treaty, and 193 countries are parties to the CBD, of which 66 countries are not parties to the ITPGRFA.

64 Source: *Earth Negotiations Bulletin*, vol. 9, no. 550, March 12, 2011 (http://www.iisd.ca/vol09/enb09550e.html; accessed March 25, 2011).

Bibliography

Andersen, R. (2008) *Governing Agrobiodiversity: Plant Genetics and Developing Countries*, Ashgate, Aldershot, UK.

Andersen, R., Tvedt, M.W., Fauchald, O.K., Winge, T., Rosendal, K. and Schei, P.J. (2010) 'International agreements and processes affecting an international regime on access and benefit sharing under the Convention on Biological Diversity: implications for its scope and possibilities of a sectoral approach', Fridtjof Nansen Institute, Oslo (http://www.fni.no/doc&pdf/FNI-R0310.pdf; accessed March 15, 2011).

Aubertin, C. and Filoche, G. (2011) 'The Nagoya Protocol on the use of genetic resources: one embodiment of an endless discussion', *Sustentabilidade em Debate*, vol. 2, no. 1, pp. 51–64.

Balick, M.J. and Cox, P.A. (1996) *Plants, People and Culture: The Science of Ethnobotany*, Scientific American Library, New York.

Boef, W.S. de (2000) *Tales of the Unpredictable: Learning about Institutional Frameworks that Support Farmer Management of Agro-Biodiversity*, Wageningen University, the Netherlands.

Boef, W.S. de, Sthapit, B., Upadhaya, M.P. and Shrestha, P. (2007) 'Estratégias de conservação em unidades de produção familiares', in Boef, W.S. de, Thijssen, M.H., Ogliari, J.B., and Sthapit, B. (orgs.). *Biodiversidade e Agricultores: Fortalecendo o Manejo Comunitário*, L & PM, Porto Alegre, Brazil.

Bulmer, J. (2009) 'Study on the relationship between an international regime on access and benefit-sharing and other international instruments and forums that govern genetic resources. The International Treaty on Plant Genetic Resources for Food and Agriculture and the Food and Agriculture Organisation's Commission on Genetic Resources' (http://www.cbd.int/doc/meetings/abs/abswg-09–3rd/information/abswg-09–3rd-abswg-07-inf-03-part1-en.pdf; accessed March 15, 2011).

Chiarolla, C. and Jungcurt, S. (2011) 'Outstanding issues on access and benefit sharing under the multilateral system of the International Treaty on Plant Genetic Resources for Food and Agriculture', a background study paper by the Berne Declaration and the Development Fund (www.evb.ch/en/p25019093.html and www.utviklingsfondet.no/filestore/ITPGRABSStudy.pdf; accessed March 11, 2011).

Clement, C.R. (2007) 'Um pote de ouro no fim do arco-íris? O valor da biodiversidade e do conhecimento tradicional associado, e as mazelas da lei de acesso – uma visão e proposta a partir da Amazônia', *Amazônia: Ciência & Desenvolvimento*, vol. 3, no. 5, pp. 7–28.

Cooper, D.H. (2002) 'The international treaty on plant genetic resources for food and agriculture', *Review of European Community & International Environmental Law*, vol. 11, no. 1, pp. 1–16.

Correa, C. (2011) 'An innovative and transparent option for royalty payment under the ITPGRFA: implementing the Article 6.11 crop-related modality of the SMTA', in Frison, C., López, F. and Esquinas-Alcázar, J.T. (eds.) *Plant Genetic Resources and Food Security: Stakeholder Perspectives on the International Treaty on Plant Genetic Resources for Food and Agriculture*, Earthscan, London.

Coupe, S. and Lewins, R. (2007) *Negotiating the Seed Treaty*, Practical Action Publishing, Warwickshire.

Damania, A.B. (2008) 'History, achievements and current status of genetic resources conservation', *Agronomy Journal*, vol. 100, no. 1, pp. 9–21.

David, J. (2008) 'L'histoire des plantes cultivées et son role sur leur diversité', presentation given on November 11, 2008, in Montpellier, France, at the École Thématique Internationale Agrobiodiversité: des hommes et des plantes, organized by CIRAD and IRD, with support from the Agropolis Foundation.

Dedeurwaerdere, T. (2010) 'Institutionalizing global genetic-resource commons: towards alternative models for facilitating access in the global biodiversity regime' (http://www.bioecon.ucl.ac.uk/12th_2010/Dedeurwaerdere.pdf; accessed April 12, 2011).

Den Hurk, A.V. (2011) 'The seed industry: plant breeding and the International Treaty on Plant Genetic Resources for Food and Agriculture', in Frison, C., López, F. and Esquinas-Alcázar, J.T. (eds.) *Plant Genetic Resources and Food Security: Stakeholder Perspectives on the International Treaty on Plant Genetic Resources for Food and Agriculture*, Earthscan, London.

Diaz-Silveira, M.F. (2011) 'The Latin American and Caribbean Regional Group: a long and successful process for the protection, conservation and enhancing of PGRFA Treaty', in Frison, C., López, F. and Esquinas-Alcázar, J.T. (eds.) *Plant Genetic Resources and Food Security: Stakeholder Perspectives on the International Treaty on Plant Genetic Resources for Food and Agriculture*, Earthscan, London.

Diegues, A.C., Andrello, G. and Nunes, M. (2001) 'Populações tradicionais e biodiversidade na Amazônia: levantamento bibliográfico georreferenciado', in Capobianco, J.P., Veríssimo, A., Moreira, A., Sawyer, D., Santos, I. and Pinto, L.P. (orgs.) *Biodiversidade na Amazônia Brasileira: Avaliação e Ações Prioritárias para a Conservação, Uso Sustentável e Repartição de Benefícios*, Estação Liberdade, Instituto Socioambiental (ISA), São Paulo, pp. 205–224.

Dinh, T.T.V. (2010) 'The international treaty on plant genetic resources for food and agriculture: an innovative tool for the management of agri-phytodiversity', PhD thesis, University of Limoges, France.

Emperaire, L. (2008) 'Mandioca, raiz do Brasil', in *Almanaque Brasil Socioambiental*, Instituto Socioambiental, São Paulo, p. 420.

Emperaire, L. and Santilli, J. (2006) 'A agrobiodiversidade e os direitos dos agricultores indígenas e tradicionais', in Ricardo, B. and Ricardo, F. (eds.) *Povos Indígenas no Brasil: 2001–2005*, Instituto Socioambiental, São Paulo.

Engsiang, L. and Halewood, M. (2008) 'A short history of the Annex I List', in Tansey, G. and Rajotte, T. (eds.) *The Future Control of Food: A Guide to International Negotiations and Rules on Intellectual Property, Biodiversity and Food Security*, Earthscan, London, pp. 249–250.

ETC Group (2001) 'The law of the seed', *Translator*, vol. 3, no. 1, pp. 2–16.

Frison, C., López, F. and Esquinas-Alcázar, J.T. (eds) (2011) *Plant Genetic Resources and Food Security: Stakeholder Perspectives on the International Treaty on Plant Genetic Resources for Food and Agriculture*, Earthscan, London.

Fowler, C. (2003) 'The status of public and proprietary germplasm and information: an assessment of recent developments at FAO', *IP Strategy Today*, no. 7, pp. 1–19.

Fujisaka, S., Williams, D. and Halewood, M. (eds.) (2009) 'The impact of climate change on countries' interdependence on genetic resources for food and agriculture' (Background Study Paper no. 48), Commission on Genetic Resources for Food and Agriculture, FAO, Rome (ftp://ftp.fao.org/docrep/fao/meeting/017/ak532e.pdf; accessed March 5, 2011).

Genetic Resources Action International (GRAIN) (2005) 'The FAO seed treaty: from farmers' rights to breeders' privileges', *Seedling*, pp. 21–24 (http://www.grain.org/seedling/?id=411; accessed April 15, 2009).

Glowka, L., Burhenne-Guilmin, F., Synge, H., McNeely, J.A. and Gündling, L. (2004) *A Guide to the Convention on Biological Diversity*, IUCN, Gland, Switzerland.

Goedert, C. (2007) 'Histórico e avanços em recursos genéticos no Brasil', in Nass, L. (ed.) *Recursos Genéticos Vegetais*, EMBRAPA Recursos Genéticos e Biotecnologia, Brasília, Brazil.

Grugel, C. (2009) 'The role of the International Treaty in facing the global challenge of mitigation and adaptation to climate change', policy seminar on the International Treaty on Plant Genetic Resources for Food and Agriculture: Global Challenges and Future Direction, Bari, Italy, December 15–18, 2009 (ftp://ftp.fao.org/ag/agp/planttreaty/gb4/bari/Background_paper_day3.pdf; accessed March 11, 2011).

Halewood, M. and Nnadozie, K. (2008) 'Giving priority to the commons: the International Treaty on Plant Genetic Resources for Food and Agriculture', in Tansey, G. and Rajotte, T. (eds.) *The Future Control of Food: A Guide to International Negotiations and Rules on Intellectual Property, Biodiversity and Food Security*, Earthscan, London, pp. 115–140.

Halewood, M., Noriega, I.L. and Louafi, S. (2012) *Crop Genetic Resources as a Global Commons*, Earthscan, London.

Hammond, E. (2011) 'How US sorghum seed distributions undermine the FAO Plant Treaty's multilateral system', a briefing paper by the Berne Declaration, the Development Fund, and the African Center for Biosafety (www.evb.ch/en/p25019094.html and http://www.utviklingsfondet.no/filestore/ITGRFASorghumpaper.pdf; accessed March 11, 2011).

Harlan, J. (1975) *Crops and Man*, American Society of Agronomy and Crop Science Society of America, Madison, WI.

Hawtin, G. and Fowler, C. (2011) 'The Global Crop Diversity Trust: an essential element of the Treaty's Funding Strategy', in Frison, C., López, F. and Esquinas-Alcázar, J.T. (eds.) *Plant Genetic Resources and Food Security: Stakeholder Perspectives on the International Treaty on Plant Genetic Resources for Food and Agriculture*, Earthscan, London.

Henson-Apollonio, V. (2009) 'The International Treaty on Plant Genetic Resources for Food and Agriculture (ITPGRFA): the standard material transfer agreement as implementation of a limited compensatory liability regime', in Van Overwalle, G. (ed.) *Gene Patents and Collaborative Licensing Models*, Cambridge University Press, Cambridge, UK, pp. 289–293.

Hermitte, M.A. and Kahn, P. (eds.) (2004) *Les Ressources Génétiques Végétales et le Droit dans les Rapports Nord-Sud* (Travaux du Centre René-Jean Dupuy pour le Droit et le Développement et du Centre de Recherche sur le Droit des Sciences et Techniques, v. II), Bruylant, Brussels.

Kamau, E.C., Fedder, B. and Winter, G. (2010) 'The Nagoya Protocol on Access to Genetic Resources and Benefit-Sharing: What is New and What are the Implications for Provider and User Countries and the Scientific Community', *LEAD Journal* (*Law, Environment and Development Journal*), vol. 6/3 (http://www.lead-journal.org/content/10246.pdf; accessed March 10, 2011).

Lightbourne, M. (2009) *Food Security, Biological Diversity and Intellectual Property Rights*, Ashgate, Farnham, UK.

Loskutov, I. (1997) *Vavilov and his Institute*, International Plant Genetic Resources Institute, Rome.

Louwaars, N. (2007) 'Seeds of confusion: the impact of policies on seed systems', PhD dissertation, Wageningen, the Netherlands.

Meienberg, F. (2007) 'Access and benefit-sharing under the FAO Seed Treaty', in Informal International Consultation on Farmers' Rights, September 18–10, 2007, Lusaka, Zambia. *Report*. Norwegian Ministry of Agriculture and Food, Oslo, pp. 129–132.

Moore, G. and Frison, E. (2011) 'International Research Centres: The Consultative Group on International Agricultural Research and the International Treaty', in Frison, C., López, F. and Esquinas-Alcázar, J.T. (eds.) *Plant Genetic Resources and Food Security: Stakeholder Perspectives on the International Treaty on Plant Genetic Resources for Food and Agriculture*, Earthscan, London.

Moore, G. and Tymowski, W. (2005) *Explanatory Guide to the International Treaty on Plant Genetic Resources for Food and Agriculture* (IUCN Environmental Policy and Law Paper, 57), IUCN, Gland, Switzerland.

Palacios, X.F. (1997) 'Contribution to the estimation of the interdependence of countries in the field of plant genetic resources', (Background Study Paper no. 7, Rev. 1), Commission on Genetic Resources for Food and Agriculture, FAO, Rome (ftp://ftp.fao.org/docrep/fao/meeting/015/j0747e.pdf; accessed April 15, 2009).

Pistorius, R. (1997) *Scientists, Plants and Politics: A History of the Plant Genetic Resources Movement*, International Plant Genetic Resources Institute, Rome.

Robinson, D.F. (2010) *Confronting Biopiracy: Challenges, Cases and International Debates*, Earthscan, London.

Santilli, J. (2005) *Socioambientalismo e Novos Direitos: Proteção Jurídica à Diversidade Biológica e Cultural*, Peirópolis, Instituto Socioambiental (ISA) and Instituto Internacional de Educação do Brasil (IIEB), São Paulo.

Santilli, J. (2010) *Agrobiodiversidade e Direitos dos Agricultores*, Peirópolis, São Paulo.

System-wide Genetic Resources Programme (SGRP) (2006) with the International Plant Genetic Resources and International Rice Research Institute, 'Developing Access and Benefit-Sharing Regimes: Plant Genetic Resources for Food and Agriculture', Rome.

System-wide Genetic Resources Programme (SGRP) (2007) 'A *de facto* Certificate of Source: the Standard Material Transfer Agreement under the International Treaty', Rome.

System-wide Genetic Resources Programme (SGRP) (2010) and Bioversity International, 'Leaving room in the CBD's ABS protocol for the future development of specialized access and benefit-sharing arrangements – the example of agricultural microbial genetic resources', Rome.

Stannard, C., Vander Graaf, N., Randell, A., Lallas, P. and Kenmore, P. (2004) 'Agricultural biological diversity for food security: shaping international initiatives to help agriculture and the environment', *Howard Law Journal*, vol. 48, no. 1, pp. 397–430.

Varaprasad, K.S. and Sivaraj, N. (2010) 'Plant genetic resources conservation and use in light of recent policy developments', *Electronic Journal of Plant Breeding*, vol. 1, no. 4, pp. 1276–1293.

Visser, B. (2008) Genebank management: what to conserve? Oral presentation made at the international course "Contemporary Approaches in Plant Genetic Resources Conservation and Use", June 18, 2008, Wageningen, the Netherlands.

Visser, B. and Borring, J. (2011) 'The European Regional Group: Europe's role and positions during the negotiations and early implementation of the International Treaty', in Frison, C., López, F. and Esquinas-Alcázar, J.T. (eds.) *Plant Genetic Resources and Food Security: Stakeholder Perspectives on the International Treaty on Plant Genetic Resources for Food and Agriculture*, Earthscan, London.

Wale, E., Drucker, A.G. and Zander, K.K. (eds.) (2010) *The Economics of Managing Crop Diversity On-Farm*, Earthscan, London.

Walter, B.M.T., Cavalcanti, T.B. and Valls, J.F.M. (2005a) 'História da coleta de germoplasma e os coletores', in Walter, B.M.T. and Cavalcanti, T.B. (eds.) *Fundamentos para a Coleta de Germoplasma Vegetal*, EMBRAPA Recursos Genéticos e Biotecnologia, Brasília, pp. 90–136.

Walter, B.M.T., Cavalcanti, T.B., Bianchetti, L. de B. and Valls, J.F.M. (2005b) 'Origens da agricultura, centros de origem e diversificação das plantas cultivadas', in Walter, B. and Cavalcanti, T.B. (eds.) *Fundamentos para a Coleta de Germoplasma Vegetal*, EMBRAPA Recursos Genéticos e Biotecnologia, Brasília, pp. 58–87.

Zeven, A.C. and De Wet, J.M. (1982) *Dictionary of Cultivated Plants and their Regions of Diversity*, Centre for Agricultural Publishing and Documentation, Wageningen, the Netherlands.

7

OPTIONS FOR THE IMPLEMENTATION OF THE INTERNATIONAL TREATY ON PLANT GENETIC RESOURCES FOR FOOD AND AGRICULTURE AT THE NATIONAL LEVEL

Access and benefit-sharing: *in situ* plant genetic resources for food and agriculture

The multilateral system established by the ITPGRFA is very clear when it comes to ABS related to *ex situ* collections of Annex I crops. However, some issues have been raised in relation to access to PGRFA found in *in situ* conditions. According to the Treaty (Article 12.3h), access to PGRFA found in *in situ* conditions must be provided "according to national legislation or, in the absence of such legislation, in accordance with such standards as may be set by the governing body" of the Treaty. This means that international standards apply only in the absence of national laws, and any international standards developed by the governing body of the Treaty, regarding access to *in situ* plant genetic resources, will be subsidiary to national legislation.

No international standards have been set by the governing body (as of April 2011), and it is up to national laws to regulate access to plant genetic resources found in *in situ* conditions. In any case, any international standards for access to *in situ* material would apply only to PGRFA of Annex I crops; PGRFA that are in the public domain and under the management and control of contracting parties; and PGRFA for research, breeding, and training for food and agriculture (Chiarolla and Jungcurt, 2011).

For countries that are parties to the CBD, access to *in situ* plant genetic resources (including both wild relatives of crop species and local/traditional varieties and landraces) is subject to mutually agreed terms between providers and recipients of these resources. Where national laws recognize that Indigenous and local communities have rights to grant access to such resources, their prior informed consent is also necessary, and the same applies to traditional knowledge associated with plant genetic resources.

It is important to stress that such ABS rules apply also to national and international germplasm banks[1] that wish to collect/access *in situ* genetic material: they also must obtain the prior authorization (or permission) from the country of origin (as well as the prior and informed consent of local and Indigenous communities) of the resources that they intend to collect *in situ*. The same applies to any researcher or private company that wants to collect/access genetic material found *in situ*, as crop varieties conserved *in situ* and on-farm were left out of the multilateral system, as were *ex situ* collections kept by NGOs, farmers, cooperatives, private institutions, etc.

However, not all countries that have ratified the CBD have national ABS laws, and others prefer to adopt administrative measures. As of April 18, 2011, the CBD Database on ABS Measures[2] listed 54 countries that had some legislation or regulations on ABS. Such ABS laws and regulations have been developed in response to the CBD, rather than the ITPGRFA, and most countries that have national ABS laws do not establish a special ABS regime for plant genetic resources for food and agriculture. There is an expectation that the Nagoya Protocol will be a strong incentive for countries to adopt and implement national ABS laws, and that these laws will clarify to which ABS regime plant genetic resources for food and agriculture (listed or not listed in Annex I) that are found *in situ* are subject. They must also clarify if they will place any Annex I crops existing in *in situ* conditions in the multilateral system. Some countries, such as the Netherlands and Germany, are choosing to implement the multilateral system of the ITPGRFA through administrative measures rather than through the adoption of new national legislation. Whether through administrative or legislative acts, countries have to decide:

1 If they will apply the same norms and conditions that regulate ABS for genetic resources in general also to PGRFA, or if they will create a differentiated ABS regime for PGRFA (whether included or not in Annex I and in the multilateral system), taking into consideration their "special nature, distinct features and problems, needing distinct solutions." They may approve a specific ABS law for PGRFA or they may include specific provisions on PGRFA in their general ABS laws. In any case, coordination between the ministries of agriculture and environment is essential.
2 For countries that are party to the ITPGRFA, their national ABS laws must leave Annex I crops (included in the multilateral system of the treaty) out of the bilateral system established by the CBD.
3 As not all countries that are parties to the CBD are also parties to the ITPGRFA (as of March 24, 2011, there are 66 countries that are parties to the CBD but not to the ITPGRFA), Annex I crops in countries that are non-parties (to the Treaty) are subject not to the multilateral system, but to the CBD bilateral system (Andersen *et al.*, 2010). However, some non-parties (to the Treaty) have been accessing genetic materials

placed in the multilateral system through the SMTA and, therefore, such countries should make it clear that genetic materials exchanged through the SMTA fall outside the ABS bilateral system, and will be exchanged according to the conditions of the SMTA.

Access and benefit-sharing regimes for plant genetic resources for food and agriculture not included in the multilateral system and national benefit-sharing funds

Countries that are parties to the ITPGRFA cannot create ABS rules for crops included in Annex I, held in *ex situ* public collections, different from those of the multilateral system. However, they can establish, in their national laws, differentiated ABS regimes for PGRFA not included in the multilateral system of the ITPGRFA found *in situ* and *ex situ*. They can voluntarily include their *ex situ* public collections of non-Annex I crops in the multilateral system. Some European countries, such as the Netherlands and Germany, are making germplasm of all PGRFA available under the terms and conditions of the multilateral system (Visser and Borring, 2011). However, developing countries in Latin America, Africa, and Asia (where most centers of origin and diversity of crops are located) are not likely to do this until they are convinced that benefit-sharing mechanisms under the multilateral system are effective. Nevertheless, they can create, in their national laws, specific ABS systems for PGRFA that are different from the general ABS bilateral systems used for wild biodiversity, considering the specific characteristics of PGRFA. One such option is to establish national benefit-sharing funds, covering all PGRFA (that are not included in the international multilateral system), administered by representatives not only of governments, but also of small-scale and traditional farmers, who conserve agrobiodiversity *in situ* and on-farm, and must be the main beneficiaries of the resources destined to benefit-sharing funds. A certain percentage of the value of all seeds that are sold in the country could be destined for a national benefit-sharing fund. These monies would be paid directly by the (national and international) private seed sectors, as a special tax/contribution imposed upon users of plant genetic resources. In Norway, the government pays the equivalent of 0.1 percent of the value of sales of all seeds (and other plant-propagating materials) in the country to the benefit-sharing fund of the ITPGRFA. It is not a tax/contribution paid directly by the private seed sector, and the 0.1 percent value of all seed sales corresponds to the amount of funds that the Norwegian government donates to the benefit-sharing fund of the ITPGRFA. However, in developing countries, it is very unlikely that governments would be willing to do this, because of a lack of public funds, and such a tax/contribution would need to be be paid directly by the (national and international) private seed companies. The payment of a fixed percentage over all seed sales should be mandatory regardless of the availability (or not) of resources to third parties for further research and breeding. Benefit-sharing should not have any connection to IP rights.

A national benefit-sharing fund based on a fixed percentage over the values of all seed sales, to be paid by users of plant genetic resources, would be a mechanism to implement the "user pays" principle, one of the most important principles of environmental law. The user pays principle was recognized by the Rio Declaration on Environment and Development, approved by the 1992 UN Conference on Environment and Development (the Earth Summit). The Rio Declaration consisted of 27 principles intended to guide future sustainable development around the world. According to the user pays principle, national authorities must promote the internalization of environmental costs and the use of economic instruments, taking into account the fact that polluters must bear the costs of pollution, and users of environmental resources must also bear the economic costs of their conservation, with due regard to the public interest. This principle is widely applied in European environmental laws, especially in France and Germany. In Brazil, the National Environment Policy Law (Law 6938/1981, Article 4, VII) explicitly recognizes the user pays principle, which obliges polluters to rehabilitate degraded areas, and determines that users of environmental resources (with economic purposes) must contribute to their conservation. This principle is adopted also in the Brazilian National Water Resources Policy (Law 9433/1997), which obliges users of water resources with economic purposes to pay economic contributions to the agencies in charge of managing water resources. The user pays principle aims at internalizing environmental costs of economic activities, and users of plant genetic resources should also contribute to their conservation. Thus, a percentage of the value of seed sales should go to a national benefit-sharing fund, managed with the participation of representatives of local, family, and traditional farmers, and aimed at supporting plans and programs for *in situ* and on-farm conservation and sustainable use of agrobiodiversity, as well as implementation of farmers' rights. This is a form of benefit-sharing that is more coherent with the nature of plant genetic resources than trying to identify, on a case-by-case basis, the "providers" of resources. Small-scale and traditional farmers could present their proposals of projects, to be supported by the national benefit-sharing fund.

Countries could also establish, in their national ABS laws for PGRFA, some important reciprocity rules that the international multilateral system is unable to, owing to lack of consensus. For instance, countries could establish that access to *ex situ* collections held by public institutions is granted only to private institutions that also make their *ex situ* collections available to public institutions for plant breeding and research. In addition, national ABS laws could establish that private institutions can access plant genetic resources found *in situ* on public domain lands only if they commit themselves (through a legally binding instrument) to make such resources available to public institutions. After all, plant genetic resources are essential to national food security, and must be regarded as "public interest goods," regardless of whether they are in the public or private domain.

Plant genetic resources for food and agriculture held by state and provincial institutions

There is also the question of the inclusion of Annex I crops held by state and provincial institutions, in countries that have adopted federalism in their constitutions. In federal states, the power to govern is shared between national and provincial/state governments; examples include Canada and Brazil, which ratified the Treaty in 2002 and in 2006, respectively.[3] In Brazil, some genebanks run by states initially argued that the Treaty (and the multilateral system) was not legally binding on their collections, as it was signed by the federal government, without any participation of Brazilian states and, therefore, only federal genebanks were bound by it. However, they eventually came to the conclusion that it is in their own interest to include their collections in the multilateral system, and most of them are starting to do so now. The Brazilian states of Acre[4] and Amapa,[5] in the Amazon region, have enacted state laws on access to genetic resources and associated traditional knowledge. While in some countries *ex situ* collections are held by national genebanks, in others they are held by state/provincial institutions, and in still others they may be held by formally separate legal persons but are under the control and management of a national policy framework. In Canada, several provincial collections are held by universities but, although these collections are legally outside the management and control of the federal government, their status with regard to access is similar to that of national collections (Chiarolla and Jungcurt, 2011). Italy, despite being a unitary state, gives its regions and provinces considerable autonomy, and that is why it is commonly referred to as a "regional" state (an intermediary state, between unitary and federal). Italy ratified the ITPGRFA in 2004, through Law No. 101, which states that implementation of the Treaty is the responsibility of the regions and autonomous provinces. Thus, such arrangements must be taken into consideration when contracting parties implement the Treaty at the national level.

The special legal regime of plant genetic resources found in the territories of Indigenous peoples and other ethnic minorities

Another issue is the special legal regime of Indigenous territories and of other ethnic minorities, and the established rights that Indigenous peoples and other ethnic minorities have over their traditional territories and natural resources. ABS laws must pay due attention to such specific legislation when regulating access to genetic materials held by Indigenous peoples and ethnic minorities, as well as access to associated traditional knowledge. The CBD establishes, in its Article 8j, that contracting parties must, subject to their national legislation, "respect, preserve and maintain knowledge, innovations and practices of Indigenous and local communities embodying traditional lifestyles relevant for the conservation and sustainable use of biological diversity." They must also "promote their wider application with the approval and involvement of the holders of such knowledge, innovations and practices and encourage the equitable sharing of the benefits arising from the utilization of such knowledge, innovations and practices," and several countries have approved national legislation recognizing these rights too. At the international level, other important instruments are the UN Declaration on the Rights of Indigenous Peoples (UNDRIP), approved in 2007 (a non-legally binding instrument), and the Convention No. 169 of the International Labour Organization (ILO), concerning Indigenous and Tribal Peoples in Independent Countries (a legally binding instrument), which was approved in 1989 and came into force in 1991.[6] In Latin America, most countries have ratified the ILO Convention No. 169. In relation to Indigenous peoples' rights over natural resources and traditional knowledge, the main elements of these instruments are:

1 UNDRIP recognizes that "respect for Indigenous knowledge, cultures and traditional practices contributes to sustainable and equitable development and proper management of the environment" (preamble). It also recognizes that Indigenous peoples have the right to maintain, control, protect, and develop their cultural heritage, traditional knowledge, and traditional cultural expressions, as well as the manifestations of their sciences, technologies, and cultures, including human and *genetic resources*, *seeds*, medicines, *knowledge of the properties of fauna and flora*, oral traditions, literatures, designs, sports and traditional games and visual and performing arts. *They also have the right to maintain, control, protect and develop their intellectual property over such cultural heritage, traditional knowledge*, and traditional cultural expressions (Article 31, emphasis added).[7]
2 Convention No. 169 of the ILO, concerning Indigenous and Tribal Peoples in Independent Countries, establishes that the rights of ownership and possession of Indigenous and tribal peoples over the lands which they traditionally occupy must be recognized, and that measures must be taken to safeguard their right to use lands not exclusively occupied by them, but to which they have traditionally had access for their subsistence and traditional activities. It adds that particular attention must be paid to the situation of nomadic peoples and shifting cultivators in this respect (Article 14). In addition, the Convention recognizes

that the rights of Indigenous and tribal peoples to the natural resources pertaining to their lands must be specially safeguarded, and that these rights include their right to participate in the use, management, and conservation of these resources (Article 15).

At the national level, the constitutions of most Latin American countries have provisions on Indigenous and other ethnic minorities' rights. In the 1990s, new constitutions were adopted in several Latin American countries, and most of them recognize themselves as multicultural and pluri-ethnic, and protect the cultural and territorial rights of their Indigenous peoples and afrodescendants, such as Colombia (1991), Paraguay (1992), Peru (1993), Bolivia (1994 and 2009), Argentina (1994), Ecuador (1998 and 2008), and Venezuela (1999). Guatemala promulgated its new constitution in 1985, and Brazil promulgated its new constitution in 1988, one year before the approval of the ILO Convention No. 169.

Thus, it can be said that, in countries that have ratified such international treaties and/or adopted national legislation recognizing the special rights of their ethnic minorities and local communities, access to genetic materials held by them is subject to their prior informed consent, and such genetic materials cannot be considered to be in the public domain and under the management and control of contracting parties to the ITPGRFA. Genetic resources held by these groups can be accessed or placed in the multilateral system only with the consent of local communities. However, local communities include not only Indigenous and other ethnic groups, but also small-scale and traditional farmers, whose rights must also be established at the national level. Very few countries have national regulations on farmers' rights (other than the so-called "farmers' privilege"), and there are many uncertainties regarding how to obtain prior informed consent and to establish bilateral benefit-sharing mechanisms in relation to PGRFA and to associated traditional knowledge held by local communities. However, it must be taken into consideration that the Treaty does recognize farmers' rights, which means that national ABS laws may also include the obligation of users to obtain the prior informed consent of farmers to access genetic resources and the traditional knowledge that they hold (Andersen *et al.*, 2010). Farmers' rights will be discussed in more detail in Chapter 8.

Brazilian access and benefit-sharing law and plant genetic resources for food and agriculture

Brazil was one of the first megadiverse countries to adopt national legislation on ABS, aimed at implementing the CBD.[8] Provisional Act[9] (Medida Provisória) No. 2186-16, of 2001, regulates access to genetic resources and to associated traditional knowledge, benefit-sharing, and the transfer of technology for the conservation and use of biological diversity. Brazil signed the ITPGRFA on June 10, 2002, and ratified it on May 22, 2006.[10] Manioc (*Manihot esculenta* only) is the only crop listed in Annex I of the Treaty and included in the multilateral system of the Treaty whose center of origin and diversity is Brazil. During the negotiations on the Treaty, the inclusion of peanuts (whose centers of origin and diversity are Brazil and Bolivia) in the multilateral system was discussed, but in the end Brazil and Bolivia agreed that this crop should be excluded.

The ABS legal regime set up under MP 2186-16/2001 was conceived for wild biodiversity, particularly for chemical, pharmaceutical, and/or other non-food/feed industrial uses, with very little consideration for the specific nature of PGRFA.[11] However, MP 2186-16/2001 sets out, in Article 19.2, that transfers of genetic resources of species that are subject to facilitated exchange international agreements signed by Brazil, including on food security, will be made available in accordance with conditions contained therein. As Brazil has ratified the ITPGRFA, the special rules of its multilateral system apply to Annex I crops. MP 2186-16/2001 has established a differentiated ABS regime for plant genetic resources for food and agriculture included in Annex I (and in the multilateral system), distinct from the ABS system for genetic resources in general. However, there is no specific ABS regime for other PGRFA (i.e., those *not* included in Annex I). The general ABS regime for genetic resources applies to all plant genetic resources which are *not* included in Annex I of the Treaty, whether wild or domesticated. We shall thus look first at the general provisions of MP 2186-16/2001, and then analyze its application to PGRFA more specifically.

In Brazil, the Genetic Heritage Management Council (Conselho de Gestão do Patrimônio Genético – CGEN) is responsible for implementing national policies on access to genetic resources and associated traditional knowledge and benefit-sharing.[12] CGEN was created in April, 2002, and is composed of 19 representatives from ministries[13] and other governmental institutions.[14] When CGEN was created, only representatives of government could participate in its sessions. In 2003, however, representatives of other stakeholders (scientific institutions, Indigenous and other traditional communities, and private industry) started to participate in the Council's sessions, being entitled to speak but not to vote. This informal practice (of allowing nonmembers of CGEN to speak at its sessions) was legally recognized only in 2007, when Presidential Decree No. 6159 established that the Council could invite "specialists and representatives from different sectors of society" to contribute and provide inputs on issues being discussed by the Council. However, nongovernmental participants still do not have the right to vote, which is heavily criticized by civil society organizations as being antidemocratic.

CGEN also develops technical and administrative rules aimed at enforcing MP 2186-16/2001, and access to genetic resources and associated traditional knowledge requires its prior authorization. Access to genetic resources located in Indigenous territories is subject to the prior consent of Indigenous communities involved,[15] and to prior consultation (*not* authorization) with the official federal agency for Indigenous affairs (called FUNAI – Fundação Nacional do Índio); access to genetic resources in private lands is subject to the prior consent of its owners; access to genetic resources in protected areas is subject to the prior authorization of the competent authority (responsible for the protected area management); access to genetic resources in areas that are "essential for national security" is subject to prior authorization of the National Defense Council; and access to genetic resources in Brazilian jurisdictional waters, on the continental shelf, and in the exclusive economic zone is subject to the prior authorization of the maritime authority. Therefore, in all above-mentioned cases, CGEN cannot grant authorization of access to genetic resources without the prior consent of these communities or competent authorities. Whenever access involves traditional knowledge held by Indigenous and local communities, their prior consent is also necessary.[16]

Indigenous peoples enjoy special territorial rights in Brazil. Indigenous territories are, according to the Brazilian Constitution (Article 231), inalienable public lands, owned by the federal state, but Indigenous peoples have the rights of permanent possession of their traditional territories, as well as exclusive rights to their natural resources (with exceptions made for mining activities and utilization of water resources, including for energy purposes).[17] The Brazilian Constitution (Article 68 of transitional constitutional provisions) also recognizes special territorial rights to *quilombola* communities, which are also ethnic minorities formed from the descendants of runaway slaves (called *maroon* in other Latin American countries). Recognition of territorial rights of the *quilombolas* takes place through the concession of a collective and indivisible property title, and such territories are inalienable. Whenever there are private property titles over the limits of *quilombolas*' territories, their prior expropriation is necessary. A total of 1,624 communities have already been recognized as *quilombolas*, but civil society organizations estimate that there are approximately 3,000 *quilombola* communities in Brazil. However, as of June 2010, only 180 communities had received their property titles from the federal state (Comissão Pró-Indio de São Paulo, 2011). There are other traditional communities living in protected areas, such as extractive reserves (which belong to a category of protected area that allows traditional populations to live inside their limits; Santilli, 2010). Not all traditional communities live in protected areas, and not all of them have officially recognized rights over their traditional territories. Decree 6040, of 2007, defines traditional communities as

> culturally differentiated groups which identify themselves as such, have their own forms of social organization, occupy and use territories and natural resources as a condition for cultural,

> social, religious, ancestral and economic reproduction, and use knowledge, innovations and practices generated and transmitted by tradition.

The Decree establishes a National Policy for the Sustainable Development of Traditional Communities, which must be implemented by a commission composed of representatives of public agencies and of traditional communities (which include rubber tappers, nut gatherers, fisherfolks, etc.). Access to genetic resources held by traditional and local communities is also subject to their prior informed consent, even though there is no consensus as to who actually belongs to a traditional or local community, as mentioned above.

The main legal instruments established by Provisional Act 2186-16/2001 are (1) authorization for access to genetic resources and associated traditional knowledge and for transfer of genetic resources' samples (or accessions) to other institutions; (2) a benefit-sharing contract, which is mandatory only when access is for commercial purposes/bioprospection; (3) a material transfer agreement, to be signed by the recipient of the genetic resources, indicating whether there was access to associated traditional knowledge. Benefit-sharing contracts signed by providers and users of genetic resources and associated traditional knowledge[18] may establish sharing of profits from the commercialization of derived products, payment of royalties, access to and transfer of technology, no-cost licensing of products and processes, capacity-building, etc. All expeditions and research projects that involve accessing genetic resources and traditional knowledge must be developed in partnership with a Brazilian institution. Foreign institutions are not allowed to develop such activities by themselves, without a Brazilian partner institution.

It is important to keep in mind that the CBD and Provisional Act 2186-16/2001 do not define any ownership or proprietary rights of genetic resources. Sovereign rights and property rights are distinct concepts which should not be confused. Exercising sovereign rights, countries may decide that certain natural resources (such as genetic resources) are public property (or state property), but not necessarily. [Decision 391 of the Andean Community and the Ethiopian Proclamation (482/2006), for example, recognize genetic resources as the property of the state.] Provisional Act 2186-16/2001 was issued as an expression of the Brazilian State's sovereign rights over its genetic resources, but it does not establish any ownership or proprietary rights of the Brazilian State over its genetic resources. A proposal (bill) for a constitutional amendment that would make genetic resources the property of the federal state is currently being discussed by the Brazilian Congress, but there is very little political support for such a proposal, and it is very unlikely that it will be approved (as of April 2011). Most stakeholders agree that genetic resources are public interest resources, and that they have social, cultural, and economic value for the whole society, and not just for the owners of the areas where they are located, whether public or private.

A number of questions have been raised over the past ten years as the Provisional Act 2186-16/2001 has been applied to specific cases. To clarify which activities are covered by the ABS requirements, CGEN published an official "technical orientation" to make it clear that accessing genetic resources is different from collecting biological material. According to the CGEN's Technical Orientation No. 01/2003, access is "the activity carried out with genetic resources with the objective of isolating, identifying or using information of genetic origin or molecules and substances arising from the metabolism of living beings and of extracts obtained from such organisms." The activity requires authorization from CGEN only when it fits this definition. Collection of biological material, without the purpose of accessing genetic resources, is not subject to access authorization. Collection of biological material in protected areas requires another type of official permit, issued by Instituto Chico Mendes de Conservação da Biodiversidade (ICMBIO, the federal agency responsible for the management of protected areas[19]), but not authorization from CGEN.

Under Provisional Act 2186-16/2001, benefit-sharing contracts are mandatory only when access to genetic resources and traditional knowledge is for commercial purposes or bioprospecting. When genetic resources are accessed in federal public protected areas (parks, wildlife reserves, etc.), the Union (federal state) must be a party to the benefit-sharing contract,[20] and the same applies to public lands owned by states or municipalities (Brazil is a federal state). In other situations where the Union is not a party, it is entitled only to a share of the benefits. If the benefit-sharing contract is signed with a private landowner (on whose land the resources are to be accessed), the benefits will go to him or her, privately. This is one of the main drawbacks of Provisional Act 2186-16/2001: there is no requirement that benefits received by private landowners will flow to initiatives aimed at conserving biological diversity. Benefits received by the Union (federal state) must be deposited in the National Environmental Fund, in the Naval Fund, or in the National Scientific and Technological Development Fund, and according to Provisional Act 2186-16/2001, they must be destined for biodiversity conservation (Article 33).[21] However, private landowners can use benefits for any purpose, not necessarily biodiversity conservation activities.

Provisional Act 2186-16/2001 adopts a very privatist stance in relation to genetic resources, which are public interest resources, independently of being found *in situ* on private or public lands. This is one of the aspects of Provisional Act 2186-16/2001 that must change, so that benefits are always applied to biodiversity conservation projects, and private landowners should receive only a share (when access takes place in their properties). CGEN's Resolution No. 8/2003 has partially mitigated such a privatist stance, by stating that, in cases of "relevant public interest," the private landowners' consent to access may not be required, but it is still not clear what this really means. The Nagoya Protocol (which Brazil has signed but not yet ratified, as of July 23, 2011), Article 9, establishes that parties must "encourage users and providers to direct benefits arising from the utilization of genetic resources towards the conservation of biological diversity and the sustainable use of its components," which is a wise orientation that must be enforced through national laws.

There is another benefit-sharing requirement that has already been changed since Provisional Act 2186-16/2001 was enacted. This Act required the signing of a benefit-sharing contract before any authorization of access to genetic resources or traditional knowledge for bioprospecting (with commercial purposes) could be granted by CGEN. That way, even before biosprospecting activities began, benefit-sharing had to be agreed on by providers and users of genetic resources, and this was extremely complex, as the parties involved could have no idea of what would be the outcomes of the bioprospecting activities. They could lead to commercially valuable products or not (as is most often the case). For both the providers and the users of genetic resources and knowledge, it was hard to agree on benefit-sharing while the results of bioprospecting were still unknown. In 2007, a Presidential Decree (No. 6159) set out that if the provider of genetic resources (or traditional knowledge) agrees, the benefit-sharing contract can be signed at a later date, as long as it is prior to the development of any new commercial product and to any claims of IP rights.

Furthermore, in August, 2006, CGEN approved Resolution 21,[22] which established that certain types of research and scientific activities are not subject to access authorizations. These are activities aimed at evaluating or elucidating the evolutionary history of a species or taxonomic group, relationships/interactions of living beings among themselves, or relationships/interactions between living beings and the environment (where they coexist) or the genetic diversity of populations; parentage testings, sexing techniques, and chromosomal or DNA analysis aimed at identifying species or specimens; epidemiological research or research aimed at identifying etiological agents of diseases, as well as measuring the concentration of substances whose presence in the organism (in certain amounts) indicate illnesses or physiologic states; and research aimed at creating collections of DNA, tissue, germplasm, blood, or serum. The exemption (from access authorization) granted to these types of research considered that they fall within the concept of "access to genetic resources" owing to the use of methodological molecular tools in a circumstantial manner, but not because their main objectives are directly connected with access to genetic resources. Therefore, the legal requirement of an access authorization is not justified.

Provisional Act 2186-16/2001 also establishes, in its Article 31, that any (natural or legal) person who claims a patent over processes or products originating from genetic resources is required to disclose the origin of such genetic resources and of associated traditional knowledge. They must also sign a declaration stating that they have met all ABS requirements set by Provisional Act 2186-16/2001, and reveal the number and date of the corresponding access authorization. This requirement applies to patents claims over products or processes originating from genetic resources accessed after June 30, 2002 (when the Provisional Act came into force), according to CGEN's Resolution 39, approved on February 12, 2009.[23] However, Article 31 refers only to "industrial" property rights (patents, trademarks, industrial designs, etc.), and not to plant breeders' rights. This means that, once more, such regulation did not consider the specific legal regime of plant genetic resources. Such regulation does not ensure disclosure of origin when plant breeders' rights are claimed, which is a serious drawback.

Since Provisional Act 2186-16/2001 is enforceable only in the Brazilian territory, it is important that countries that are primarily users (or importers) of genetic resources also adopt similar laws, requiring that claims for patents or other IP rights disclose the origin of genetic resources and associated traditional knowledge used in the development of processes and products (which are to be protected), and that ABS requirements established by providers' countries are fully respected. To date, Norway is one of the very few user countries that has adopted a domestic law to ensure compliance with ABS laws of providers' countries: according to Act No. 100, of June 19, 2009 (Nature Diversity Act),

> the import for utilisation in Norway of genetic material from a state that requires consent for collection or export of such material may only take place in accordance with such consent. The person that has control of the material is bound by the conditions that have been set for consent, and the state may enforce the conditions by bringing legal action on behalf of the person that set them.

In addition,

> when genetic material from another country is utilised in Norway for research or commercial purposes, it must be accompanied by information regarding the country from which the genetic material has been received (provider country). If national law in the provider country requires consent for the collection of biological material, it shall be accompanied by information to the effect that such consent has been obtained.

According to the Norwegian Act (Section 60), if the provider country is a country other than the country of origin of the genetic material, the country of origin must also be stated. The country of origin means the country in which the material was collected from *in situ* sources. If national law in the country of origin requires consent for the collection of genetic material, information as to whether such consent has been obtained must be provided. The Act states that the Norwegian king may make regulations prescribing that if utilization involves use of the traditional knowledge of local communities or Indigenous peoples, the genetic material must be accompanied by information to that effect. The Act also establishes that when genetic material covered by ITPGRFA is utilized in Norway for research or commercial purposes, it must be accompanied by information that the material has been acquired in accordance with the SMTA established under the Treaty.[24]

Ten years have passed since Provisional Act 2186-16/2001 came into force in Brazil, and there is a certain consensus among most stakeholders (governmental agencies, NGOs, scientific researchers, local and Indigenous communities, private sector, etc.) that this regulation needs to be revised and several points clarified because of its lack of clarity and transparency. However, there is a great deal of dispute and controversy over *how* it should be revised. Several legal bills (aimed at revising Provisional Act 2186-16/2001) have been drafted since 2003, but there is currently (as of July 2011) no official (governmental) bill, because the Brazilian ministries of the environment, agriculture and science and technology cannot come to an agreement over many ABS issues. Thus, the sending (by the federal government) of an official legal bill to the National Congress has been postponed over and over again since 2003.[25]

One of most serious drawbacks of Provisional Act 2186-16/2001 is the fact that it does not consider the numerous situations when genetic resources and associated traditional knowledge are shared among several traditional and local communities, and it may be extremely complex to determine who must grant prior informed consent and with whom benefits arising from their commercial utilization should be shared. It may also give rise to disputes between local communities about who owns the resources: this has already happened, for instance, in the case of research conducted by the Department of Psychobiology of the Federal University of São Paulo (UNIFESP), involving bioprospecting of medicinal plants traditionally used by the Krahô Indigenous people, who live in the Brazilian state of Tocantins. The researchers obtained prior informed consent from three Krahô communities, but when the research started, in 1999, other Krahô communities living in the same Indigenous territory but with historical conflicts with the other communities started to complain that they had not granted their consent to the research, and that they were also holders of such resources and traditional knowledge. The research could not go on, due to failed negotiations on ABS (Kishi, 2009; Kleba, 2009). There are no mechanisms, under Provisional Act 2186-16/2001, to address such situations.

Indigenous organizations, NGOs, and some governmental agencies have presented several proposals for the creation of a national benefit-sharing fund, divided into ecological and ethnographic regions, and managed with the participation of all stakeholders. However, such proposals have not yet been approved.

If bilateral contracts pose the above-mentioned difficulties for shared wild genetic resources, these are even more complex when PGRFA are involved, owing to their special nature, already discussed in this book. There is a great interdependency among countries and also among local farming communities within the same national territory: frequently, they are used to share and exchange plant genetic resources and agricultural knowledge and practices, through social networks and according to local rules, and it may be very complex to define who can grant prior informed consent to access to PGRFA and who should receive the benefits arising from their utilization.

In Brazil, Provisional Act 2186-16/2001 does not include special ABS requirements for PGRFA. Transfers of plant genetic resources included in Annex I of the ITPGRFA are subject to the terms and conditions of the multilateral system, according to Article 19, paragraph 2, of the Provisional Act, as Brazil has already ratified the Treaty. However, there is no specific ABS regime for other PGRFA (i.e., those *not* included in Annex I), and the general ABS regime for genetic resources (established by Provisional Act 2186-16/2001) poses some difficulties in its application to PGRFA.

Currently, CGEN requires access authorizations only for Brazilian native plant species. In the case of domesticated species, this has been interpreted as those that can be found in *in situ* conditions in the Brazilian territory. According to the CBD (which Brazil has already ratified and incorporated into national law), "in situ conditions" means "conditions where genetic resources exist within ecosystems and natural habitats, and, *in the case of domesticated or cultivated species, in the surroundings where they have developed their distinctive properties*" (emphasis added). The CBD also defines "country providing genetic resources" as "the country supplying genetic resources collected from *in situ* sources, including populations of both wild *and domesticated species*" (emphasis added).

There is currently a fierce debate on the need for access authorizations for domesticated plant species that are not native to Brazil but that have developed "distinctive properties" in the Brazilian territory as a result of natural selection or on-farm management by local and Indigenous communities. This is the case of several local (also called traditional or creole) maize varieties: although Meso-American countries (especially Mexico and Guatemala) are recognized as the centers of origin for maize, there are local (also called "creole") maize varieties found in Brazil that do not exist in Meso-America. They have become so adapted to Brazilian conditions, and have been managed on-farm by so many generations of local/traditional farmers, that some people understand that they have developed "distinctive properties" in the Brazilian territory and that access to such varieties must be subject to an access authorization, to be granted by CGEN, after the prior informed consent of local communities. According to this argument, local/creole varieties incorporate traditional knowledge in themselves, intrinsically. Others argue that maize is an exotic species, which was introduced in Brazil and, therefore, is not part of the Brazilian "genetic heritage," and no access authorization can be required for access to exotic genetic resources. CGEN has already decided, for instance, that access to genetic resources of sugarcane (*Saccharum* spp.) does not require access authorization, because it is an exotic species, but in the case of mamona (*Ricinus communis* L.), which is also an exotic species, CGEN has asked for more technical information before it decides whether the mamona varieties (to be accessed) have developed "distinctive properties" in Brazil. Some CGEN members consider that such "distinctive properties" must comprise a genetic change in relation to the original plant, and not only phenotypical or morphological differences.

There is no definitive clarification of this issue, and CGEN has been deciding on a case-by-case basis. Among others, CGEN has already granted authorization of access to genetic resources and traditional knowledge associated to maize landraces held by *quilombola* communities who live in Vale do Ribeira, in the state of São Paulo. It has also granted authorizations of access to genetic resources and traditional knowledge associated with plant varieties (of manioc/cassava, peanuts, maize, beans, pumpkins, bananas, cotton, and yams), as well as their wild relatives, held by Indigenous communities (Yawalapiti and Kayabi) who live in the Xingu National Park, in Brazil. More complex, however, is the regulation of access to local/traditional plant varieties conserved in *ex situ* collections which were collected *after* the coming into force of Provisional Act 2186-16/2001. If access to such *ex situ* collections also takes place *after* the coming into force of this Act, the user must obtain the prior informed consent of providers (local communities), as identified by the holder of the collection. However, in most cases, it is not possible to identify the provider of the genetic resource and/or associated traditional knowledge, as it is rare to find information about local/farmer communities or their traditional knowledge in genebank accessions. According to CGEN's Resolution 32/2008, in such cases, the Council must examine, on a case-by-case basis, if it is possible to exempt the user of a genetic resource conserved *ex situ* of the obligation to obtain prior informed consent and to share benefits, and what will be the destination of eventual benefit-sharing funds. When genetic resources were collected *before* the coming into force of Provisional Act 2186-16/2001, but access will take place *after* its coming into force, prior consent and benefit-sharing must be agreed upon with the institution that holds the *ex situ* collection (except when it was collected in Indigenous territories, protected areas, jurisdictional waters, on the continental shelf, or in the exclusive economic zone).

There is no consensus, among CGEN members, as to who fits into the concept of traditional or local communities. When genetic resources (and associated traditional knowledge) are accessed in Indigenous and *quilombolas*' territories, it is clear that the prior informed consent of such communities is required, in accordance with Provisional Act 2186-16/2001, as these communities have clearly established rights over their natural resources and cultural heritage. However, when other "local communities" are involved, such as local/traditional farmers, rubber tappers, fisherfolk, etc., there is no consensus in the interpretation of the law. Provisional Act 2186-16/2001 defines "local communities" as human groups that are "differentiated by their cultural conditions, organize themselves traditionally along successive generations, hold their own customs and preserve their social and economic institutions." The Federal University of Santa Catarina (southern Brazil), for instance, has asked for an authorization to access traditional knowledge associated with genetic resources of *goiaba serrana* (*Feijoa sellowiana*), a type of guava (fruit) that is native to the south of Brazil. Traditional knowledge associated to *goiaba serrana* is held by local/traditional farming communities living in the municipalities of São Joaquim, Urubici, and Urupema. CGEN understood that such communities did not fall within the concept of "local communities" for the purposes of the ABS regulation, and that no access authorization was required. Another controversial case involved an authorization of access to traditional knowledge associated with traditional/local processes of producing manioc/cassava flour; in this case CGEN also understood that the access authorization was not required.

CGEN has also edited a Technical Orientation (No. 07/2009) regulating access to genetic resources *for the specific purpose of plant breeding.* Considering that access to genetic resources for plant breeding can fall into any of the three access categories (access for scientific research, for commercial/bioprospection, or for technological development), this Technical Orientation establishes the following distinctions, for the purpose of granting access authorizations aimed at plant breeding:

1 scientific research: a set of activities aimed at selecting promising genotypes for the beginning of bioprospection;
2 bioprospecting: a subsequent stage in which the promising genotypes selected in the first research phase are tested for distinctness, uniformity, and stability (DUS) and for cultivation and use, or other equivalent tests;
3 technological development: the final stage of the breeding program, which involves obtaining genetic seeds or basic plants, in the case of vegetatively propagated species.

This orientation was issued because not all plant breeding programs lead necessarily to the development of new cultivars (plant varieties), and they have different phases and objectives, so they must request access authorizations according to the type of activity.

Many controversial issues will only be definitively clarified when Brazil has a national law that establishes a differentiated ABS regime for all PGRFA, and not only for Annex I crops included in the multilateral system of the Treaty. Instead of trying to define who is the "legitimate" holder of a certain plant genetic resource or agricultural knowledge, such a law could create collective mechanisms, such as national benefit-sharing funds, to address situations involving resources shared among different local communities. Benefit-sharing funds could be distributed to participatory breeding programs and community seedbanks or used to pay traditional and agroecological farmers for environmental services and supporting on-farm conservation of agrobiodiversity, with the participation of local farmers (Santilli, 2009).

Peruvian access and benefit-sharing law and plant genetic resources for food and agriculture

Peru is part of the Andean Community of Nations (along with Bolivia, Colombia, and Ecuador; Venezuela withdrew in 2006), and all member countries of the Andean Community have ratified the CBD. In 1996, the Andean Community of Nations approved Decision 391, which establishes a common regime on access to genetic resources. Decision 391 was approved in accordance with Article 15 of the CBD, and it was the first subregional ABS regime. It establishes a bilateral ABS system: providers and users of genetic resources and traditional knowledge must sign bilateral contracts, on a case-by-case basis, and this system applies to all genetic resources, found *in situ* and *ex situ*, and to all their by-products (or derivatives) as well as to their intangible components (traditional knowledge), for the purposes of research, biological prospecting, conservation, industrial application, and commercial use, among others. Decision 391 became legally binding for member countries of the Andean Community in 1996. However, Peru approved a national regulation on access to genetic resources only in January of 2009 (Resolución Ministerial No. 087-2008-Minam, ratified by Decreto Supremo 003-2009-Minam, officially published on January 18, 2009). Between 1996 (when Decision 391 was approved) and 2009 (when the national Peruvian regulation was approved), almost all transfers of genetic resources were suspended,[26] owing to lack of legal clarity. Access to domesticated species and to genetic material from national genebanks was authorized only for scientific purposes, through material transfer agreements signed by the INIA (Instituto Nacional de Innovación Agraria, National Institute for Agrarian Innovation, of the Peruvian Ministry of

Agriculture), in which recipients were committed to forgo any IP rights over the transferred genetic material, according to Lapeña *et al.* (2010). Decision 391 does not establish any specific ABS rules on access for PGRFA, and countries that are part of the Andean Community will have to adopt special rules to implement the ITPGRFA at the national level, as Decision 391 does not contemplate the multilateral system of ABS established by the Treaty for Annex I crops. Peru signed the ITPGRFA on October 8, 2002, and ratified it on June 5, 2003. Many questions were raised about the compatibility between the Treaty and Decision 391 in Peru, until the approval, in 2009, of Resolución Ministerial No. 087-2008-Minam, ratified by Decreto Supremo 003-2009-Minam (Lapeña *et al.*, 2010). This Regulation finally clarifies the ABS regime for PGRFA:

1 In the case of PGRFA of Annex I crops, they are included in the multilateral system of the Treaty, and access is regulated by this specialized international ABS regime (Article 5c), and not by the national regulation edited in 2009.

2 PGRFA of non-Annex I crops are regulated by the Ministerial Resolution 087-2008-Minam and by Supreme Decree 003-2009-Minam,[27] which establishes different regimes for access for scientific purposes and for commercial purposes, as well as for plant genetic resources found in *ex situ* collections:

   2.1 Access to non-Annex I crops for scientific purposes is regulated by the Ministerial Resolution 087–2008-Minam and by Supreme Decree 003-2009-Minam. Framework access contracts (*contratos marco de acesso*) must be celebrated between the competent authority (*autoridad de administracion y ejecución*) and universities, research centers or researchers, which have to be registered. Some of the main requirements of such contracts are: – participation of national professionals in the activities of collection and research on genetic resources and their derivatives; – obligation not to transfer genetic resources to third parties that are not included in the research project; – establishment of clauses regarding payments for the collection, as well as clauses relating to payments to the provider of the genetic resource; – establishment of clauses regarding IP rights on processes or products resulting from the use of genetic resources accessed or their derivatives; – compulsory deposit of duplicates of all material collected at institutions approved by the competent authority (Article 25).

   2.2 Access to non-Annex I crops for commercial purposes is also regulated by the Ministerial Resolution 087-2008-Minam and by Supreme Decree 003-2009-Minam. An access authorization must be requested to the competent authority before an access contract can be signed. The competent authorities are the INIA, for "genetic resources, molecules, combination or mixture of natural molecules, including raw extracts and other derivatives contained in the cultivated or domesticated inland species. Such content can be found in all or in part of the specimen" (Article 15); the Ministry of Agriculture, in coordination with the INIA, for "genetic resources of wild relatives of cultivated species"; the Ministry of Agriculture (Dirección General Forestal y de Fauna Silvestre), for "genetic resources, molecules, combination or mixture of natural molecules, including raw extracts and other derivatives contained in wild inland species."; and the Production Ministry (vice-ministry for fisheries), for "genetic resources, molecules, combination or mixture of natural molecules, including raw extracts and other derivatives contained in hydrobiological marine species and in species of inland waters." Access contracts are drawn up between one of the above-mentioned authorities and the recipient (user) of genetic resources. The access contract must take into consideration "the rights and interests of providers of genetic resources, of providers of biological resources that contain such resources and of providers of the intangible component (traditional knowledge)," and they must include provisions on prior informed consent, mutually agreed terms for access, and fair and equitable benefit-sharing agreements (Article 20). The same applies to "accessory contracts" (*contratos accesorios*), which are drawn up between the recipient (user) of the genetic resources and (a) the owner, possessor, or manager of the place/land where the biological resource (that contains the genetic resource) is located; (b) the *ex situ* conservation center where the genetic material is conserved; (c) the owner, possessor, or manager of the biological resource that contains the genetic resource; (d) the provider of the intangible component (knowledge, innovations, and practices) of the genetic resource (when the provider is an Indigenous people or community, the contract must respect all national and international instruments that protect traditional knowledge of Indigenous peoples and communities); or (e) the national institution (Article 21).

3 Access to non-Annex I crops that are conserved in *ex situ* collections held by the IARCs of CGIAR. As Lapeña *et al.* (2010) explain, the Peruvian Ministerial Resolution of 2009 (mentioned above) states[28] that genetic resources originating in Peru and found in *ex situ* genebanks of the IARCs of CGIAR, and which are *not* included in Annex I of the ITPGRFA, are subject to the national Peruvian regulation (and not to the terms and conditions established by the multilateral system of the Treaty). Nevertheless, at the second session of the Treaty's governing body, held from October 29 to November 2, 2007, in Rome, contracting parties (to the Treaty) decided that access to plant genetic resources *not* listed in Annex I, and *collected prior to the entry into force of the*

Treaty (on June 29, 2004), that are held by the IARCs of CGIAR should be provided pursuant to the terms and conditions set out by the SMTA used for Annex I crops included in the multilateral system. Non-Annex I material collected by the IARCs of CGIAR *after the coming into force of the Treaty*, must be available for access on terms consistent with those mutually agreed between the IARCs that receive the material and the country of origin of such resources (according to Article 15.3 of the Treaty). However, according to the governing body decision mentioned above, access to genetic resources of non-Annex I crops held *ex situ* by IARCs and *which were collected before June 29, 2004* (when the Treaty came into force) should be transferred according to the SMTA, and not according to the general Peruvian legislation on access to genetic resources, established in 2009. This contradiction between the ITPGRFA's governing body decision, issued in 2007, and the Peruvian Ministerial Resolution of 2009 is creating great uncertainty in relation to the transfer of non-Annex I crops held by IARCs, and CIP (Centro Internacional de la Papa/International Potato Center). An IARC based in Peru, has decided to suspend any international transfer of Andean roots and tubercles until this situation is clarified (Lapeña *et al.*, 2010). Some examples of non-Annex I crops held by IARC *ex situ* collections are maca, arracacha, yacón, and quinoa.

4 Access to plant genetic resources conserved in *ex situ* collections located in Peru:

  4.1 For scientific purposes: all transfers of genetic resources from *ex situ* conservation centers to national or foreign researchers are subject to the SMTA signed by providers and users of these resources, with a detailed description of the research project and its participants (Article 32). Such an agreement must necessarily include the following conditions (for the transfer of genetic resources of Peruvian origin): prohibition to claim ownership over genetic material per se or over its derived products; obligation to transfer genetic material to third parties without the authorization of the competent authority; recognition of the source/origin of the genetic material.

  4.2 For commercial purposes: such transfers are subject to an access contract (according to the rules described above) and to an accessory contract (*contrato accesorio*) drawn up between the Peruvian *ex situ* conservation center and the user/recipient of the genetic material. The material transfer agreement is considered to be a type of accessory contract (Lapeña *et al.*, 2010).

5 Ministerial Resolution 087-2008-Minam establishes, in its Article 27, that the Ministry of the Environment can limit (partially or totally) access in the following cases: access to the genetic resources of endemic, rare, or endangered species, subspecies, varieties, or races; access to vulnerable or fragile ecosystems that could be negatively impacted by

access; access that would have adverse effects on human health or the cultural identity of Indigenous peoples or undesirable environmental impacts on ecosystems that cannot be easily controlled; access that may cause danger of genetic erosion; access that breaches biosafety rules; and access to genetic resources or geographical areas considered to be strategic.

6 When access involves associated traditional knowledge, all provisions of Law 27811/2002 (on the protection of collective knowledge of Indigenous peoples associated to biological resources) must be respected.[29] Some of the main concepts and rules established by this Law are:

6.1 Indigenous people are "native people whose rights are prior to the formation of the Peruvian state, and maintain a unique culture, a territory and self-recognize as such." This concept includes people in voluntary isolation or uncontacted (by our society), as well as peasant and native communities. "Indigenous" includes and can be used as a synonym for "originary," "traditional," "ethnic," "ancestral," "native," or other denominations (Article 2).

6.2 Collective knowledge is accumulated and transgenerational knowledge, developed by Indigenous peoples and communities in relation to the properties, uses and characteristics of biological diversity. (The law regulates only collective Indigenous knowledge, and not individual knowledge, which is regulated by customary law.) Collective knowledge includes knowledge held by different Indigenous peoples, and the rights of Indigenous peoples over their collective knowledge are permanent and inalienable (Articles 2 and 10).

6.3 Prior informed consent means authorization granted by an Indigenous organization that is representative of the Indigenous peoples who hold collective knowledge, in accordance with the rules (customs and practices) that are recognized by them, for access and use of such collective knowledge, and as long as sufficient information is previously provided on the purposes, risks and implications of access, including eventual uses of collective knowledge, and if it is relevant, its value (Article 2).

6.4 A License Contract for the Use of Collective Knowledge (*contrato de licencia de uso de conocimientos colectivos*) is an express agreement between the organization representing Indigenous peoples that possess the collective knowledge and a third party (the user of such collective knowledge) that establishes the terms and conditions for the use of such collective knowledge (Article 2). Such a contract must be registered by INDECOPI (Instituto Nacional de Defensa de la Competencia y de la Protección de la Propiedad Intelectual), the public agency in charge of granting IP rights.

6.5 To access collective knowledge for scientific, commercial, or industrial purposes, it is necessary to obtain the prior informed consent of representative Indigenous organizations. As Ruiz Muller (2006) points out, representative Indigenous organizations (and not the communities themselves) are responsible for granting (or not) prior informed consent. The Indigenous organization must include the largest possible number of Indigenous peoples who hold the collective knowledge, and take into consideration their interests and concerns, especially in relation to spiritual values and religious beliefs (Article 6). When access is for commercial (or industrial) purposes, the contract must include benefit-sharing (Article 7).

6.6 A Fund for the Development of Indigenous Peoples (Fondo para el Desarrollo de los Pueblos Indígenas) was created, with the objective of financing projects and other activities aimed at promoting Indigenous peoples' development. Indigenous peoples can access such resources through their representative organizations, and they can present proposals to be funded. The fund is managed by five representatives of Indigenous organizations, two representatives of Instituto Nacional de Desarrollo de los Pueblos Andinos, Amazónicos y Afroperuanos (National Institute for the Development of Andean, Amazonian and Afroperuvian peoples, INDEPA). A percentage of not less than 10 percent of gross sales, before taxes, resulting from the commercialization of products developed from collective knowledge will be directed to this fund (Article 8).

6.7 Collective knowledge is considered to be in the public domain when it has been accessible to persons other than Indigenous peoples, through mass media such as publications, or when it refers to properties, uses, or characteristics of a biological resource which are massively known outside Indigenous peoples and communities. When such collective knowledge has entered into the public domain in the last 20 years, a certain percentage of gross sales, before taxes, resulting from the commercialization of products developed from collective knowledge will be directed to the Fund for the Development of Indigenous Peoples (Article 13).

6.8 The law defines three types of Registers of Collective Knowledge of Indigenous Peoples (Registros de Conocimientos Colectivos de los Pueblos Indígenas). Registro Nacional Público de Conocimientos Colectivos de los Pueblos Indígenas is for the registration of collective knowledge that is in the public domain while Registro Nacional Confidencial de Conocimientos Colectivos de los Pueblos Indígenas for the registration of confidential collective knowledge and which third parties cannot access. INDECOPI is responsible for these two registers. Registros Locales de Conocimientos Colectivos de los Pueblos Indígenas[30] are local registers of collective knowledge, established by the Indigenous peoples and communities themselves, according to their customs and practices (Articles 15–18). The main purposes of these registers are to preserve and safeguard the collective knowledge of Indigenous peoples and their rights over them; and to provide INDECOPI with information to protect the interests of Indigenous peoples in relation to their collective knowledge (Article 16). INDECOPI must send the information in the public register to the main patent offices of the world, so that they consider them as prior art in the examination of novelty and inventiveness of patent applications. The objective is to prevent patents over products or processes developed from collective knowledge (Article 23).

6.9 Protection granted by the law: The Indigenous peoples possessing collective knowledge must be protected against the disclosure, acquisition or use of that collective knowledge without their consent and in an unfair manner, to the extent that the collective knowledge is not in the public domain. They must also be protected against unauthorized disclosure in the event that a third party has illegitimately had access to collective knowledge covered by a safeguard (nondisclosure) clause (Article 42).

Other Peruvian laws that incorporate specific provisions on the protection of traditional knowledge, innovations and practices of Indigenous and local communities are Law 26839/2007, on the conservation and sustainable use of biological diversity, its regulation (Supreme Decree 068/2001) and the National Strategy on Biological Diversity, approved by Supreme Decree 102/2001) (Ruiz Muller, 2006). In 2004, Peruvian Law No. 28216,[31] on access to biological diversity and to associated collective knowledge of Indigenous peoples, established a National Commission against Biopiracy.[32] According to this Law, biopiracy is

> the nonauthorized and noncompensated access and use of biological resources or traditional knowledge of Indigenous peoples by third parties, without the correspondent authorization and in violation of the principles established by the Convention on Biological Diversity and other regulations on this issue. This misappropriation can take place by physical control, through property rights over products that incorporate such resources, illegally obtained, or in some cases, through their invocation.

According to Lapeña *et al.* (2010), six patent claims over plant genetic resources native from Peru have been paralyzed owing to the intervention of the Commission, including maca, sacha inchi, and camu camu. The Commission has chosen 35 biological resources of Peruvian origin to be prioritized in the identification of biopiracy cases (through patents), and 15 of them are PGRFA (Lapeña *et al.*, 2010).

The Region of Cuzco has also enacted Regional Ordnance (Ordenanza Regional) No. 048-2008, which regulates access to genetic resources and associated traditional knowledge of Indigenous and peasant communities in the Cuzco region. In 2005, Peru enacted Law No. 28477, which declares, as "natural heritage" of the Peruvian nation, a list of 35 native crops[33] (including potatoes, manioc/cassava, *kiwicha*, *maca*, *oca*, and quinoa), three native livestock (guinea pig/cuy, alpaca, and llama) and 11 wild fauna species. Peru has also edited National Directorial Resolution (Resolución Directoral Nacional) No. 1986/INC,[34] on December 23, 2009, which declared, as "national cultural heritage," the knowledge, practices, and technologies associated to the traditional cultivation of maize in the Sacred Valley of Incas, in the Andes of Peru.

A comparison between the Brazilian and the Peruvian access and benefit-sharing laws, in relation to plant genetic resources for food and agriculture

Brazil and Peru are probably the South American countries that have enacted the greatest number of legal acts on access to genetic resources and traditional knowledge and benefit-sharing. Both have complex and extensive legislation aimed at implementing the CBD at the national level, as described above. Both countries have enacted rules on access authorizations and bilateral contracts between providers and users/recipients of genetic resources and traditional knowledge. Both countries have established different procedures for access to genetic resources for scientific purposes and for commercial purposes, which is in line with the Nagoya Protocol, which determines (in its Article 8) that contracting parties must adopt "simplified measures on access for noncommercial research purposes." Neither of the two countries has yet (as of July 2011) ratified the Nayoga Protocol, but it is very likely that both of them will do so (Brazil has already signed it).

The two countries have, however, followed different paths in relation to access to traditional knowledge. Peru has enacted a specific law on collective knowledge, and only Indigenous organizations can be party to contracts, not the Indigenous communities themselves. Brazil does not have a specific law on traditional knowledge, but it is regulated by Provisional Act 2186-16/2001, in a special chapter (No. 3) that is still not fully implemented in Brazil. The Act states that a registry *may* be created, but so far (as of July 2011), this is not the case partly because of the opposition of many local communities, who feel that such a registry could make it easier for third parties to claim IP rights over their knowledge and practices, as it would be publicly accessible (Peru has created a confidential register). Brazil, in contrast, has adopted other public policies aimed at safeguarding cultural heritage, including traditional knowledge associated with biodiversity (see Chapter 11). In Brazil, Indigenous communities have to grant their prior informed consent before CGEN can issue any access authorization. In Peru, Indigenous organizations, and not the communities themselves, are the signatories of ABS contracts, which indicates that Indigenous organizations tend to be more representative than in Brazil. It would be extremely complex to require contracts to be signed directly by Indigenous organizations in Brazil. The legitimacy and representativeness of Indigenous organizations would probably be often questioned.

It must be remembered that there are important differences in the Indigenous populations of the two countries: In Brazil, there are 234 Indigenous peoples, who speak 180 different languages and have a population of 600,000 people (approximately), representing only 0.2 percent of the total national population. Although 98.63 percent of the extension of Indigenous territories is located in the Amazon region, only 60 percent of the Indigenous population live in the Amazon region (Instituto Socioambiental, 2011). In Brazil, around 68 percent of Indigenous peoples have a population of fewer than 1,000 individuals, and they are mostly scattered in small communities all over the country. Most Indigenous organizations are local, and very few have a national representativeness. In Peru, in contrast, Indigenous peoples account for around 30 percent of the national population (estimates vary from 25 percent to 48 percent, depending on the inclusion criteria), and Quechua, the main Indigenous language, is spoken by around 20 percent of the total population. (It has also been estimated that the native language of 19 percent of five-year old children is Quechua, Aymara, or another Indigenous language.[35]) Jaqaru and Kawki (which belong to the Aymara family of languages) as well as Ashaninka and Aguaruna, among others, are also important Indigenous languages. In Peru, some Indigenous organizations, such as AIDESEP (Asociación Interétnica de Desarollo de la Selva Peruana), are more representative at the national level.

Neither of the two countries has solved the issue of traditional knowledge shared among different Indigenous or local communities. The Peruvian regulation says that the Indigenous organization (that signs the ABS contract) must include the largest possible number of Indigenous peoples who hold the collective knowledge, and take into consideration their interests and concerns (Article 6), whereas the Brazilian regulation simply does not contemplate such a situation of common resource or knowledge. Peru has created a special fund for traditional knowledge, which does not exist in Brazil, as only the benefits received by the Union (federal state) are deposited in the National Environmental Fund, in the Naval Fund, or in the National Scientific and Technological Development Fund. In Brazil, there is no specific fund for ABS, nor any specific fund for traditional knowledge, and only the benefits received by the Union (federal state) are necessarily used in biodiversity conservation. According to the Peruvian regulation, access to collective knowledge that has entered into the public domain in the last 20 years is subject to benefit-sharing, deposited in the Fund for the Development of Indigenous Peoples. This means that collective knowledge in the public domain for more than 20 year does not trigger benefit-sharing. It seems extremely complicated to determine when certain knowledge has entered into the public domain, because of its dynamic and collective nature, and because such knowledge is generally transmitted orally.

Resources and traditional knowledge can be shared not only by different Indigenous peoples in the same country, but also across different countries. The Amazonian countries, for instance, have many biological resources in common, as well as traditional knowledge that is shared by Indigenous peoples living in the territories of different countries (this is the case, for instance, of the Ashaninka, also called Kampa, who live in Brazil and Peru, of the Ticuna, who live in Brazil, Peru, and Colombia, and of the Yanomami, who live in Brazil and Venezuela. According to Instituto Socioambiental (2011), 43 Indigenous peoples of Brazil have some part of their population living in other countries. A common ABS regime could be negotiated, for instance, between Amazonian countries, as recommended by the Nagoya Protocol, in its Article 11, on "transboundary cooperation." In addition, regional benefit-sharing mechanisms could be established to cover situations in which resources and traditional knowledge are common to several countries. The Nagoya Protocol refers to a global multilateral benefit-sharing mechanism, but it would be wise also to adopt regional multilateral benefit-sharing mechanisms, as negotiations among countries of the same region tend to be much easier than at a global level.

Even though both the Peruvian and Brazilian ABS laws establish that traditional exchange of genetic resources and traditional knowledge among Indigenous and local communities, according to their customs, beliefs, and practices, are not affected by their provisions, this is not so simple in practice. Expectations that benefit-sharing will generate great amounts of money can lead communities to restrict their traditional practices of sharing and exchanging resources and knowledge through their social networks, and the mere fact that a law says that this will not happen is not enough to prevent it. Collective benefit-sharing mechanisms must be prioritized, so that communities do not have to dispute, among themselves, who is the "owner" or provider of a certain resource or knowledge. After all, the conservation of biological diversity depends directly on such exchange and sharing mechanisms, and will be negatively impacted by restrictions on the circulation of genetic resources and traditional knowledge.

In relation to plant genetic resources, both countries have already enacted national regulations, but they are restricted to Annex I crops, covered by the multilateral system of ITPGRFA. So far, both countries have focused their legal acts on bilateral ABS regimes, with very little attention given to the specific nature of PGRFA. From now on, it will be important that they consider enacting legislation that is specific to these resources. They must regulate access to plant genetic resources found *in situ* in their territories, and their utilization for food and agriculture. Not only collective benefit-sharing mechanisms at the national level must be established, but also multilateral ABS systems at the regional level, involving several countries of the same geographical region.

Notes

1 According to Article 15.3 of the ITPGRFA, non-Annex I material received and conserved by IARCs of CGIAR after the coming into force of the Treaty will be available for access on terms consistent with those mutually agreed between the IARCs that receive the material and the country of origin of such resources (or the country that has acquired those resources in accordance with the CBD or other applicable law). That is, non-Annex I material collected after the entry into force of the Treaty falls outside the scope of the multilateral system in the IARCs, and is subject to the CBD regime.

2 Source: http://www.cbd.int/abs/measures/ (accessed April 18, 2011).

3 Only 24 of the world's countries have federal systems. However, their citizens make up 40 percent of the world's population. Other examples of federal states are United States, Mexico, Argentina, Australia, and India. However, there are different types of federalism, and the divisions of power between national and provincial/state governments vary greatly from one country to the other.

4 Acre State Law No. 1235, of 1997.

5 Amapá State Law No. 388, of 1997.

6 The following Latin American and Caribbean countries have ratified ILO International Convention No. 169, as of March 15, 2011 (http://www.ilo.org/ilolex/cgi-lex/ratifce.pl?C169): Argentina, Bolivia, Brazil, Chile, Colombia, Costa Rica, Ecuador, Dominica (in the Caribbean Sea), Guatemala, Honduras, Mexico, Nicaragua, Paraguay, Peru, and Venezuela. Most Latin American countries have ratified ILO Convention No. 169 (with the exceptions of Belize, El Salvador, Guiana, Surinam, and Uruguay). El Salvador, Cuba, Haiti, Panama, and the Dominican Republic have ratified the previous ILO International Convention on Indigenous and Tribal Populations Convention, No. 107, approved in 1957. Since the adoption of Convention No. 169, Convention No. 107 is no longer open for ratification, but it is still in force in 18 countries. However, Convention No. 107 is based on the assumption that Indigenous and tribal populations are temporary societies destined to disappear with "modernization," and it encourages integration. Convention No. 169 is founded on the belief that Indigenous and tribal populations are permanent societies, and on the recognition of, and respect for, ethnic and cultural diversity. Source: http://www.ilo.org/Indigenous/Conventions/no107/lang--en/index.htm (accessed March 15, 2011). Fiji (in the Australian continent, a South Pacific country) has ratified ILO Convention No. 169, as well as Nepal, and in Europe, Spain, Norway, the Netherlands, and Denmark have ratified it. The Central African Republic is the only African country to have ratified ILO Convention No. 169 (as of March 15, 2011).

7 To consult the UNDRIP, see http://www.un.org/esa/socdev/unpfii/en/drip.html (accessed March 15, 2011).

8 The CBD was signed by Brazil in 1992, during the United Nations Conference on Environment and Development, held in Rio de Janeiro. It was ratified by the National Congress through Legislative Decree 2, of February 3, 1994, and promulgated by Presidential Decree 2519, of March 16, 1998.

9 Provisional acts (*medidas provisórias*) are a type of legislation issued by the president of Brazil in cases of emergency (urgency and relevance), according to Article 62 of the Brazilian Constitution. They come into force immediately after their publication, and must be submitted to the National Congress afterwards. The National Congress may approve, amend, or reject them. Technically, provisional acts are not laws, as they are not created through a regular legislative process, but they are approved by a presidential unilateral act, and must be submitted for the approval of the National Congress after their coming into force. Provisional Act 2186-16/2001 (which is often called "the Brazilian ABS law") was first published in June 2000 and it was edited 16 times by the Brazilian federal government. In 2001, it became a permanent act through a constitutional amendment (No. 32), and it is currently the main instrument regulating ABS in Brazil. To consult an English version of Provisional Act 2186-16/2001, see http://www.mma.gov.br/estruturas/sbf_dpg/_arquivos/mp2186i.pdf (accessed March 20, 2011).

10 The Brazilian National Congress approved the text of the ITPGRFA through Legislative Decree No. 70, of April 18, 2006, which was ratified by the federal government on May 22, 2006. The Treaty came into force in Brazil on August 20, 2006. Presidential Decree No. 6476, of June 5, 2008, promulgated the Treaty. The objective of promulgation is to incorporate the international Treaty's text into national law. Source: http://www.dji.com.br/decretos/2008–006476/2008–006476.htm (accessed March 20, 2011).

11 Brazil, in spite of having between 50,000 and 55,000 vascular plant species, and the greatest biodiversity in the world, is highly dependent on plant genetic resources originating from other countries for food and agriculture. Many of the staple crops of Brazilians' diet come from other countries, such as rice, beans, wheat, corn, and sugarcane. Many native species have, nevertheless, local and regional importance, such as manioc/cassava, pineapples, cashew, *cupuaçu*, passion fruit, Brazil nuts, *guarana*, *jaboticaba*, peanuts, and some palm tree species.

12 On September 24, 2003, CGEN's Decision No. 40 gave IBAMA (Brazilian Institute of the Environment and Renewable Natural Resources) the power to grant authorizations of access to genetic resources for scientific purposes to national public and private institutions, when there is no potential for economic use and there is no access to associated traditional knowledge. If access is for bioprospecting or technological development purposes or if it involves accessing associated traditional knowledge (for scientific research, bioprospecting, or technological development), CGEN is responsible for the access authorization. On August 27, 2009, CGEN's Decision No. 246 gave CNPQ (National Council for Scientific and Technological) the same power, under the same conditions as IBAMA. "Access to genetic resources for scientific purposes" means that, a priori, the research has no potential economic use identified. Provisional Act 2186-16/2001 defines bioprospecting as any exploratory activity aimed at "identifying genetic resources and associated traditional knowledge that can potentially have a commercial use" (Article 7, VII). When the feasibility of using a functional attribute of a genetic resource to develop products or processes for industrial or commercial uses is confirmed, the commercial use is characterized (for the purpose of making a bioprospecting authorization necessary), according to CGEN's Technical Orientation No. 6/2008. Technological development

is defined in CGEN Technical Orientation No. 4/2004 as any systematic work resulting from existing knowledge aimed at producing specific innovations, developing or changing products or processes, for an economic application.

13 Decree 3945, of 2001, establishes the composition of the Genetic Heritage Management Council. It was altered by Decree 4946, of 2003, Decree 5439, of 2005, and Decree 6159, of 2007. To consult these legal acts, see http://www.planalto.gov.br/ccivil_03/decreto/2001/D3945.htm (accessed March 20, 2011).

14 The executive secretariat of the Genetic Heritage Management Council (CGEN) is part of the structure of the Ministry of the Environment (more specifically, of the Secretariat of Biodiversity and Forests).

15 CGEN Resolution No. 09/2003 and Resolution No. 19/2005 establish the procedures for obtaining prior informed consent to access genetic resources found *in situ* in Indigenous territories, private areas, or areas owned or held by local communities, or genetic resources found in sustainable use protected areas (such as extractive reserves, inhabited by traditional and local communities), for scientific research (with no commercial purpose). Resolutions No. 12/2004 and No. 22/2006 establish such procedures when access to genetic resources is for bioprospecting and commercial purposes. Resolution No. 32/2008 regulates access to genetic resources collected *in situ* and conserved in *ex situ* collections.

16 CGEN Resolutions No. 5/2003 and No. 19/2005 establish the procedures for obtaining the prior informed consent of Indigenous and local communities, when access to traditional knowledge is for scientific (noncommercial) purposes. Resolution No. 6/2003 applies when access to traditional knowledge is for bioprospecting for commercial purposes. To consult these acts, see http://www.mma.gov.br/estruturas/sbf_dpg/_arquivos/res6.pdf (accessed March 15, 2011).

17 Federal Law 6,001, approved in 1973, establishes the statute of Indigenous peoples in Brazil, but this Law is currently being revised by the National Congress, as some of its provisions became incompatible with the Brazilian Constitution approved in 1988.

18 CGEN Resolution No. 11/2004 establishes guidelines for benefit-sharing contracts involving access to genetic resources and/or associated traditional knowledge provided by Indigenous or local communities.

19 This permit is granted through SISBIO (National System of Authorizations and Information on Biodiversity) (www.icmbio.gov.br/sisbio; accessed March 11, 2011). Administrative Order No. 154/2007 regulates the collection of biological material for scientific purposes in federal protected areas.

20 Resolution No. 27/2007 establishes guidelines for benefit-sharing contracts where the Union (federal state) is a party.

21 Presidential Decree No. 6915/2009 regulates the use of benefit-sharing funds received by the Union (federal state).

22 Resolution No. 21/2006 was partially altered by Resolution 28 of 2007 and by Resolution No. 30/2008. Decree 5459/2005 establishes administrative sanctions applicable to unauthorized activities, such as fines, confiscation of products derived from unauthorized access, suspension of sales of such products, etc.

23 Resolution No. 34/2009 can be consulted at http://www.mma.gov.br/estruturas/sbf_dpg/_arquivos/res34_cons.pdf (accessed March 11, 2011).

24 For more information on user countries' ABS measures, consult Tvedt and Young (2007).

25 However, congressman Ricardo Tripoli has presented (to the National Congress) legal bills Nos. 7709/2010 and 7710/2010, which alter the composition of CGEN, including representatives of all stakeholders (private sector,

NGOs, scientists, etc.), create a national benefit-sharing fund, whose funds are destined for biodiversity conservation and sustainable use, and make biopiracy a crime (currently, biopiracy is subject only to civil and administrative sanctions, because it is not defined as a crime by law). However, as of July 5, 2011, these legal bills have not yet been approved by Congress.

26 According to Lapeña *et al.* (2010), the only ABS contract signed between 1996 and 2009, in Peru, involved the Korea Research Institute of Bioscience and Biotechnology (KRIBB) and was aimed at researching traditional medicinal plants in the Amazon. See also Pastor (2008) and Pastor and Sigüenas (2008).

27 This legal act can be consulted at http://www.wipo.int/wipolex/en/text.jsp?file_id=202176#LinkTarget_199 (accessed March 11, 2011).

28 According to the 5th Disposición Transitória (Provisional Disposition) of the Peruvian Ministerial Resolution No. 087-2008-Minam, "Genetic resources originating in Peru, which are found in *ex situ* collections and are not included in Annex I of the FAO International Treaty, held in custody by the International Agriculture Centers of the CGIAR are subject to the provisions of this regulation."

29 This law can be consulted at http://www.bvindecopi.gob.pe/legis/l27811.pdf (accessed March 11, 2011). For more information on legal protection of traditional knowledge in Peru, see Ruiz Miller (2006).

30 For more information on how to register a collective knowledge in Peru, see http://aplicaciones.indecopi.gob.pe/portalctpi/SolicitarRegistro.jsp?lng=1.

31 To access the complete text of this law, see http://www.congreso.gob.pe/ntley/Imagenes/Leyes/28216.pdf (accessed March 11, 2011).

32 For more information, see www.biopirateria.gob.p (accessed March 11, 2011).

33 For a complete list of all crops, see http://www.wipo.int/wipolex/en/text.jsp?file_id=184340 (accessed March 11, 2011).

34 To access the complete text of this resolution, see http://intranet.inc.gob.pe/intranet/dpcn/anexos/74_1.pdf (accessed March 11, 2011).

35 Sources: IWGIA – Grupo Internacional de Trabajo sobre Asuntos Indígenas (http://www.iwgia.org/sw31055.asp and http://www.unicef.org/lac/PERU_revisado.pdf) and Atlas Sociolingüístico de Pueblos Indígenas de América Latina - Fichas nacionales (http://www.unicef.org/lac/PERU_revisado.pdf) (both accessed March 11, 2011).

Bibliography

Andersen, R., Tvedt, M., Fauchald, O.K., Winge, T., Rosendal, K. and Schei, P.J. (2010) *International Agreements and Processes Affecting an International Regime on Access and Benefit Sharing under the Convention on Biological Diversity: Implications for its Scope and Possibilities of a Sectoral Approach*, Fridtjof Nansen Institute, Oslo.

Chiarolla, C. and Jungcurt, S. (2011) 'Outstanding issues on access and benefit sharing under the multilateral system of the International Treaty on Plant Genetic Resources for Food and Agriculture', a background study paper by the Berne Declaration and the Development Fund (www.evb.ch/en/p25019093.html and www.utviklingsfondet.no/filestore/ITPGRABSStudy.pdf; accessed March 11, 2011).

Comissão Pró-Índio de São Paulo. Comunidades Quilombolas no Brasil (http://www.cpisp.org.br/comunidades/; accessed March 11, 2011).

Garfoth, K., López Noriega, I., Cabrera Medaglia, J., Nnadozie, K. and Nemogá, G.R. (2005) *Overview of the National and Regional Implementation of Access to Genetic Resources and Benefit-Sharing Measures*, Centre for International Sustainable Development Law, McGill University Faculty of Law, Montreal (http://www.cisdl.org/pdf/ABS_ImpStudy_sm.pdf; accessed March 11, 2011).

Instituto Socioambiental (2011) Povos Indígenas no Brasil (http://pib.socioambiental.org/pt; accessed March 11, 2011).

Instituto Interamericano de Cooperación para la Agricultura (IICA) (2010) Programa Cooperativo de Investigación, Desarollo e Innovación Agrícola para los Trópicos Suramericanos (PROCITROPICOS), *Recursos Fitogenéticos en los Trópicos Suramericanos*.

Kamau, E.C. and Winter, G. (eds.) *Genetic Resources, Traditional Knowledge and the Law*, Earthscan, London.

Kishi, S. (2009) 'PIC in access to TK in Brazil', in Kamau, E.C. and Winter, G. (eds.) *Genetic Resources, Traditional Knowledge and the Law*, Earthscan, London, pp. 311–325.

Kleba, J. (2009) 'A socio-legal inquiry into the protection of disseminated traditional knowledge – learning from Brazilian cases', in Kamau, E.C. and Winter, G. (eds.) *Genetic Resources, Traditional Knowledge and the Law*, Earthscan, London, pp. 119–142.

Lapeña, I., Sigueñas M., López Noriega, I. and Ramirez, M. (2010) *Incentivos y Desincentivos para la Participación del Perú en el Sistema Multilateral del Tratado Internacional sobre Recursos Fitogenéticos para la Alimentación y la Agricultura*, Bioversity International, Rome.

Pastor, S. (2008) *Agrobiodiversidad Nativa del Perú y patentes*, GRPI, Sociedad Peruana de Derecho Ambiental, Lima.

Pastor, S. and Sigüeñas, M. (2008) *Bioprospección el en Perú*, GRPI, Sociedad Peruana de Derecho Ambiental, Lima.

Ruiz Muller, M. (2006) *La Protección Jurídica de los Conocimientos Tradicionales: Algunos Avances Políticos y Normativos en América Latina*, Sociedad Peruana de Derecho Ambiental, Lima, and IUCN, Gland, Switzerland.

Ruiz Muller, M. (2008) *Guía Explicativa de la Decisión 391 y una propuesta alternativa para regular el acceso a los recursos genéticos en la Sub-región Andina*, Sociedad Peruana de Derecho Ambiental, Lima.

Ruiz Muller, M. (2010) *Valoración y Protección de los Conocimientos Tradicionales en la Amazonía del Perú: Sistematización de una Experiência*, Sociedad Peruana de Derecho Ambiental, Lima.

Ruiz Muller, M. and Ferro, P. (eds.) (2005) *Apuntes sobre Agrobiodiversidad: Conservación, Biotecnología y Conocimientos tradicionales*, Sociedad Peruana de Derecho Ambiental, Lima, and IPGRI, Rome.

Santilli, J. (2009) *Agrobiodiversidade e Direitos dos Agricultores*, Peirópolis, São Paulo.

Santilli, J. (2010) 'Human-inhabited protected areas and the Law: integration of local communities and protected areas in Brazilian Law', *Journal of Sustainable Forestry*, vol. 29, no. 2–4, pp. 390–402.

Tvedt, M. and Young, T. (2007) *Beyond Access: Exploring Implementation of the Fair and Equitable Sharing Commitment in the CBD*, IUCN, Gland, Switzerland, and Fridtjof Nansen Institute, Oslo.

Verdum, R. (org.) (2009) Povos Indígenas: Constituições e Reformas Políticas na América Latina, Instituto de Estudos Socioeconômicos, Brasília.

Visser, B. and Borring, J. (2011) 'The European Regional Group: Europe's role and positions during the negotiations and early implementation of the International Treaty', in Frison, C., López, F. and Esquinas-Alcázar, J.T. (eds.) *Plant Genetic Resources and Food Security: Stakeholder Perspectives on the International Treaty on Plant Genetic Resources for Food and Agriculture*, Earthscan, London.

8

FARMERS' RIGHTS

Historical background

The realization of farmers' rights is essential to ensure the conservation and sustainable use of agrobiodiversity, and is a cornerstone of the ITPGRFA. Farmers' rights include collective rights to land and agrarian reform, water, natural resources, energy, appropriate technology and education, health care, political participation and freedom of association, among others.[1] These rights are intrinsically linked, but here we will focus on farmers' rights as set out in Article 9 and in other provisions of the Treaty, as we believe that this instrument provides an important opportunity to construct and implement farmers' rights at the national level, as it is the first legally binding international agreement that explicitly recognizes farmers' rights. Clearly farmers' rights are not limited to those recognized by the ITPGRFA – it must be stressed – but this is a starting point.

Currently, there is no international instrument that deals specifically with farmers' rights, and the responsibility for realizing farmers' rights, as set out in the ITPGRFA, rests with national governments, according to their needs and priorities. There are no legally binding international standards on farmers' rights. Via Campesina (an international movement of peasants) has already called for an International Convention on the Rights of Peasants, arguing that there are already conventions to protect other vulnerable groups, such as Indigenous peoples, women, children, and migrant workers.[2] However, no such international convention yet exists, and contracting parties to the ITPGRFA must adopt national laws and measures to protect and promote farmers' rights. Next, we will show how the concept of farmers' rights developed internationally and reached the formulation in the Treaty. Then, we will analyze how farmers' rights can be implemented at the national level, and more specifically in developing countries.

The expression "farmers' rights" was coined in the 1980s by Pat Mooney and Cary Fowler, from the NGO Rural Advancement Foundation International (RAFI),[3] to highlight the valuable contribution of farmers to the global genetic pool and food diversity. They defended the recognition of farmer's rights at the FAO Commission on Plant Genetic Resources as a measure for north–south equality and as a counterweight for plant breeders' rights, which already existed and were legally enforced. According to Mooney (2011), international NGOs insisted that "farmers' varieties were the product of farmer genius and should not be treated in any way as being less than varieties produced by the public or private sector."[4] Since then, the expression "farmers' rights" has become widely used, and is included in several international instruments, but with few concrete results.

Farmers' rights were formally recognized for the first time in 1989, when the FAO conference adopted Resolution 5/89, which endorsed farmers' rights as "rights originating from past, present and future contributions of farmers to conservation, development and availability of plant genetic resources, particularly those from centers of origin/diversity." These rights were

> vested in the international Community,[5] as trustee for present and future generations of farmers, for the purpose of ensuring full benefits to farmers, and supporting the continuation of their contributions, as well as the attainment of the overall purposes of the International Undertaking on Plant Genetic Resources for Food and Agriculture, in order to: a) ensure that the need for conservation is globally recognized and that sufficient funds for these purposes will be available; b) assist farmers and farming communities, in all regions of the world, but especially in the centers of origin and diversity of plant genetic resources, in the protection and conservation of their plant genetic resources, and of the natural biosphere; c) allow farmers, their communities, and countries in all regions, to participate fully in the benefits derived, at present and in the future, from the improved use of plant genetic resources, through plant breeding and other scientific methods.

Resolution 5/89 considers that (a) in the history of mankind, unnumbered generations of farmers have conserved, improved, and made available plant genetic resources; (b) the majority of these plant genetic resources come from developing countries, the contribution of whose farmers has not been sufficiently recognized or rewarded; (c) farmers, especially those in developing countries, should benefit fully from the improved and increased use of the natural resources they have preserved; and (d) there is a need to continue the conservation (*in situ* and *ex situ*), development, and use of the plant genetic resources in all countries, and to strengthen the capabilities of developing countries in these areas. Resolution 5/89 was the first international instrument to recognize farmers' rights, but it was not legally binding. It was adopted as an annex of the International Undertaking on Plant Genetic Resources for Food and Agriculture, along with Resolution

4/89, which also recognized plant breeders' rights as provided for under the UPOV (International Union for the Protection of New Varieties of Plant). As Andersen (2005) points out, recognition of farmers' rights (under Resolution 5/89) was

> clearly motivated by the need to create acceptance for the formulations on plant breeders' rights, particularly among developing countries. Nevertheless, opponents of plant breeders' rights gained recognition of farmers' rights in exchange for something that already existed, i.e. plant breeders' rights.

Two years later, a new resolution (Resolution 3/91) decided that farmers' rights would be implemented through an international fund on plant genetic resources, to support plant genetic conservation, particularly, but not exclusively, in developing countries. This fund, however, never materialized, which created frustration among developing countries.

In 1992, Agenda 21, a comprehensive plan of action to be taken globally, nationally, and locally to promote sustainable development, approved during the United Nations Conference on Environment and Development (UNCED), also stressed the need to strengthen the Global System on the Conservation and Sustainable Use of Plant Genetic Resources for Food and Agriculture by, *inter alia*, "taking further steps to realize farmers' rights."[6] At the Nairobi (Kenya) Conference that approved the agreed text of the CBD, on May 1992, Resolution 3 also recognized the need to seek solutions to outstanding matters concerning plant genetic resources, in particular (a) access to *ex situ* collections not addressed by the CBD and (b) farmers' rights. The CBD does not explicitly recognize farmers' rights, but sets forth, in its Article 8(j), that knowledge, innovations, and practices of local communities and Indigenous populations should be respected and that application of this knowledge should be encouraged by means of approval and participation of its holders and benefit-sharing with local and Indigenous communities. In 1996, the Global Plan of Action for Conservation and Sustainable Use of Plant Genetic Resources for Food and Agriculture (GPA) also included, among its long-term objectives, the "realization of farmers' rights, at the national, regional and international levels." In 1999, a study by the Economic and Social Council on the right to adequate food submitted to the UN Human Rights Commission held that farmers' rights should be given attention by the human rights community and promoted, "since our future supply of food, and its sustainability, depend on such rights being established on a firm footing."

Although the concept of farmers' rights has been incorporated into many international instruments, there was never consensus as to its meaning, the extent of its content, or the best way to implement it. Some arguments for the recognition of farmers' rights are listed below:

1 It would be a measure of equity between the holders of plant germplasm (farmers, especially those living in centers of diversity of agricultural crops, in tropical and subtropical countries) and the holders of agricultural biotechnology (based mainly in northern countries). There would be a moral obligation to ensure that farmers are compensated for their contribution to conservation of agrobiodiversity. Whereas IP rights – in the form of patents or plant breeders' rights – compensate breeders and encourage them to develop new commercial varieties, there is no compensation and/or support for farmers to continue to conserve and sustainably use plant genetic resources. Furthermore, IP rights tend to compensate innovations regardless of the fact that, in many cases, these innovations are only the last step in inventions and are based on knowledge accumulated throughout millennia by generations of men and women in different parts of the world.
2 It would be a way to promote the conservation of plant genetic resources and ensure current and future food security. Recognition of farmers' rights would benefit not only farmers themselves, but humankind. Some farmers' organizations argue, however, that this is a utilitarian approach to farmers' rights, as they should contribute not only to the conservation of plant genetic resources, but also to the political empowerment and to the improvement of living conditions of farmers. It is reductionism to consider traditional and local agricultural systems as mere sources of plant genetic resources, to be used by conventional breeders (Bertacchini, 2008). They represent the basis of survival for nearly 1.5 billion people worldwide, and enabling farmers to maintain, develop, and utilize crop diversity to meet their daily needs is critical to combat poverty and to eradicate hunger.
3 Farmers' rights would ensure sufficient legal space for farmers to continue saving, using, exchanging, and selling farm-saved seed and other propagating material. They would prevent seed laws and plant breeders' rights from restricting such traditional farmers' practices. However, farmers' rights are much wider in scope than the so-called "farmers' privilege," which is just an exemption to plant breeders' rights over protected varieties.
4 Recognition of farmers' rights would be, in reality, just a "formalization" or "codification" of practices, uses, and customs traditionally adopted by local farming communities. It would be just a formal recognition of agricultural practices adopted by farmers for thousands of years.

Farmers' rights were one of the most controversial issues during negotiations of the ITPGRFA. According to Egziabher *et al.* (2011), the major push for the inclusion of farmers' rights in the Treaty came from the African group, which threatened to pull out of the negotiations unless there was a clear position (from developed countries) to accept them. Negotiations on farmers' rights started with the US delegation arguing that such rights should be left out of the Treaty, as "international law protects only individual and not group rights, and trying to include group rights would destroy individual rights." However, many developing countries argued that if there were not going to be farmers' rights, there would be no access (to plant genetic resources), and that the absence of goodwill in dealing with farmers' rights would remove goodwill from access too (Egziabher *et al.*, 2011). They also called for fair and equitable benefit-sharing and for support to farmers of developing countries who maintain and enrich plant genetic diversity and contribute to food security.

In 1996, developing countries presented a (common) proposal for the recognition of farmers' rights, and the EU and the United States also presented their own proposals. These three proposals served as the basis for the negotiations that took place between 1996 and 1999, when the articles on farmers' rights were finally defined and incorporated into the (draft) text of the Treaty (which was finally approved in 2001). According to Bjornstad (2004), the proposal presented by the United States did not mention farmers' rights, but affirmed that states and regional economic integration organizations (REIOs) should take measures to promote the efforts of their farmers to conserve and use sustainably PGRFA. It also suggests measures related to plant conservation: strengthening national germplasm systems; programs that preserve and improve native germplasm; promotion of and research into crops that are not widely used; and activities that help to control the erosion of arable land. No references are made to whether conservation by farmers had benefited agricultural production. The US proposal stressed that the support to farmers' activities to conserve and use sustainable PGRFA should take place "without restricting or disturbing trade" (Bjornstad, 2004).

The proposals presented by the EU and by the developing countries had some points in common. Both proposals recognized that the enormous contribution made by farmers of all regions of the world, and particularly those in the centers of origin and crop diversity, for the conservation and development of plant genetic resources, constitute the basis for food and agriculture production throughout the world. Both included measures aimed at ensuring that farmers could continue to conserve, manage, and improve plant genetic resources. The European proposal established that contracting parties, "for the purpose of strengthening the role of farmers in conservation and sustainable use of PGRFA and ensuring fair and equitable sharing of benefits," must, as far as possible and as appropriate, *inter alia*, "subject to its national legislation, respect, preserve and maintain the knowledge, innovations and practices of farmers relevant to the conservation and sustainable use of plant genetic resources for food and agriculture." The proposal of the developing countries stated that the responsibility for realizing farmers' rights at the national level rests with both the national governments *and* the international community (Bjornstad, 2004)

Some of the main ideas of the developing countries' proposal are described by Bjornstad (2004):

- protection and promotion of the collective rights of farmers with respect to their innovations, knowledge, and practices;
- assistance to farmers, in different regions of the world, especially in centers of origin and diversity of plant genetic resources, in the conservation, improvement, and sustainable use of such resources;
- development of a *sui generis* system, at the international level and in each country, which recognizes, protects, and compensates farmers and traditional communities for their knowledge, innovations, and practices, and ensures fair and equitable sharing of benefits deriving from the use of plant genetic resources;
- recognition and protection of the rights of farmers and their communities to save, use, exchange, share, and market their seeds and any other plant reproductive material, including the right to re-use farm-saved seed;
- a guarantee that the prior informed consent of farmers and local communities is obtained before the collection of plant resources is undertaken;
- a guarantee that farmers and local communities fully participate in the definition and implementation of the measures and legislation on farmers' rights at national and international levels;
- revision, assessment, and, if appropriate, modification of IP rights systems, land tenure, and seed laws in order to ensure their harmony with farmers' rights;
- establishment and implementation of an international fund (aimed at promoting farmers' rights).

Many points in the proposal presented by developing countries are directly associated with other international instruments. The expressions "prior and informed consent," "fair and equitable sharing of benefits deriving from the utilization of plant genetic resources," as well as the application of "knowledge, innovations and practices of local communities and Indigenous populations," with the "approval and participation of its holders and benefit-sharing with local and Indigenous communities," are used in the CBD. The expression *sui generis* is employed in the WTO's TRIPS Agreement, more specifically in its Article 27.3b, which sets forth that countries must grant protection for plant varieties, through patents or an effective *sui generis* system or a combination thereof. The rights of farmers to save, exchange, share, and sell their seeds face, in many countries, restrictions imposed by seed laws and by plant breeders' rights (especially in countries that signed the 1991 UPOV Act), as Bjornstad (2004) points out. When the ITPGRFA was being negotiated, most countries were still in the process of implementing the CBD and TRIPS Agreement at the national level (some of them still are), and the concepts established in such instruments prevailed. The developing countries' proposal was based on the same bilateral instruments used by the CBD, and on the limited legal space created by the TRIPS Agreement for a *sui generis* system of protection of IP rights over plant varieties. When the proposal was presented, very few alternatives to a bilateral system to protect farmers' rights were conceived. Even though the proposal mentions the "collective" rights of farmers to their resources and knowledge, no collective benefit-sharing mechanisms are proposed. All benefit-sharing mechanisms require the identification of "providers" and "users" of genetic resources, and the signing of bilateral agreements. Furthermore, the proposal does not make a direct link between farmers' rights and the Treaty's provisions on conservation and sustainable use of plant genetic resources, which later were recognized to be closely connected with farmers' rights.

The ITPGRFA incorporated some ideas of the proposal presented by developing countries, but left many of them out of the final text. Next, we will present the main provisions on farmers' rights, as set out in the preamble and in Article 9 of the ITPGRFA:

In the preamble:

> Affirming that the past, present and future contributions of farmers in all regions of the world, particularly those in centers of origin and diversity, in conserving, improving and making available these resources, is the basis of farmers' rights;
>
> Affirming also that the rights recognized in this Treaty to save, use, exchange and sell farm-saved seed and other propagating material, and to participate in decision-making regarding, and in the fair and equitable sharing of the benefits arising from, the use of plant genetic resources for food and agriculture, are fundamental to the realization of farmers' rights, as well as the promotion of farmers' rights at national and international levels.

In Article 9:

> 9.1 The contracting parties recognize the enormous contribution that the local and Indigenous communities and farmers of all regions of the world, particularly those in the centers of origin and crop diversity, have made and will continue to make for the conservation and development of plant genetic resources which constitute the basis of food and agriculture production throughout the world.
>
> 9.2 The contracting parties agree that the responsibility for realizing farmers' rights, as they relate to plant genetic resources for food

and agriculture, rests with national governments. In accordance with their needs and priorities, each contracting party should, as appropriate, and subject to its national legislation, take measures to protect and promote farmers' rights, including:

(a) protection of traditional knowledge relevant to plant genetic resources for food and agriculture;

(b) the right to equitably participate in sharing benefits arising from the utilization of plant genetic resources for food and agriculture; and

(c) the right to participate in making decisions, at the national level, on matters related to the conservation and sustainable use of plant genetic resources for food and agriculture.

9.3 Nothing in this Article shall be interpreted to limit any rights that farmers have to save, use, exchange and sell farm-saved seed/propagating material, subject to national law and as appropriate.

There is a contradiction between the Treaty's preamble, which recognizes the need for the promotion of farmers' rights at both the national and international levels, and Article 9.2 of the Treaty, which establishes that the responsibility for realizing farmers' rights rests with national governments. Although the Treaty acknowledges that countries must adopt measures to protect farmers' rights, each country may decide which measures to adopt, and policies and actions listed in the Treaty are merely illustrative, allowing countries to adopt others. The Treaty did not establish international parameters to be necessarily adopted by contracting parties, which reflects the lack of consensus regarding how to implement farmers' rights. The Treaty could have kept some flexibility, to allow countries to adapt farmers' rights to local contexts, but it should have set minimum international standards. The Treaty merely listed illustrative measures that *may* be adopted by countries, which will make it difficult for the governing body to evaluate whether a country is implementing farmers' rights or not.

During the fourth session of the governing body of the Treaty, held in Bali (Indonesia) from March 14 to 18, 2011, many developing countries argued that they needed (international) financial assistance and technical advice for the implementation of farmers' rights. In addition, the Asian region, the Near East, and Norway supported the creation of an *ad hoc* Technical Committee on Farmers' Rights and Sustainable Use of Plant Genetic Resources for Food and Agriculture. The South West Pacific, the European region, and Canada stressed that the responsibility of realizing farmers' rights rests with national governments. The options of establishing a committee on farmers' rights or expanding the mandate of the *ad hoc* Technical Committee on Sustainable Use to include farmers' rights were discussed. A resolution establishing an *ad hoc* Technical Committee on Sustainable Use was approved by the governing body,[7] and this committee will also work on important issues related to the implementation of farmers' rights. Contracting parties were also encouraged to continue submitting views, experiences, and best practices on the implementation of farmers' rights to the secretariat, and they were invited to consider reviewing and, if necessary, adjusting their national measures affecting the realization of farmers' rights.[8]

Achieving a balance between international and national action in relation to farmers' rights is still a challenge, but progress was made, as the fourth session adopted a resolution on farmers' rights.[9] In its preamble, the resolution establishes the context within which it was adopted, in that it:

- recalls the recognition in the ITPGRFA of the enormous contribution that local and Indigenous communities and farmers of all regions of the world have made, and will continue to make, to the conservation and development of PGRFA throughout the world;
- recalls the importance of fully implementing Article 9 of the Treaty;
- recalls that, according to Article 9 of the Treaty, the responsibility for realizing farmers' rights, as they relate to plant genetic resources for food and agriculture, rests with national governments and is subject to national law;
- emphasizes the link between farmers' rights under Article 9 and the provisions on conservation and sustainable use under Articles 5 and 6 of the Treaty;
- acknowledges that there is uncertainty in many countries as to how farmers' rights can be implemented and that the challenges related to the realization of farmers' rights are likely to vary from country to country;
- recognizes that exchange of experiences and mutual assistance between contracting parties can significantly contribute to making progress in the implementation of the provisions on farmers' rights in the Treaty;
- recognizes the contribution the governing body may give in support of the implementation of farmers' rights;
- recalls that Resolution 6/2009 called for regional consultations, to be convened by the secretariat, subject to the agreed priorities of the program of work and to the availability of financial resources (Resolution 6/2009 was adopted by the governing body during its third session, held in Tunis, in 2009);
- regrets, however, that the secretariat was not able to convene the regional workshops called for in Resolution 6/2009, owing to lack of financial resources and capacity; and
- notes the results of the international consultations on farmers' rights, submitted to the secretariat by Ethiopia, that were carried out in response to the call for regional workshops in Resolution 6/2009.

The Resolution also:

- invites each contracting party to consider reviewing and, if necessary, adjusting its national measures affecting the realization of farmers' rights[10] as set out in Article 9 of the Treaty, to protect and promote farmers' rights;
- encourages contracting parties and other relevant organizations to continue submitting views, experiences, and best practices on the implementation of farmers' rights as set out in Article 9 of the Treaty, involving, as appropriate, farmers' organizations and other stakeholders;
- invites contracting parties to consider convening national and local consultations on farmers' rights with the participation of farmers and other relevant stakeholders;
- requests the secretariat to convene regional workshops on farmers' rights, subject to the agreed priorities of the work program and budget, and to the availability of financial resources, aiming at discussing national experiences on the implementation of farmers' rights as set out in Article 9 of the Treaty, involving, as appropriate, farmers' organizations and other stakeholders;
- requests the secretariat to collect the views, experiences and best practices submitted by contracting parties and relevant organizations, and the reports of the regional workshops as a basis for an agenda item for consideration by the *ad hoc* Technical Committee on Sustainable Use of Plant Genetic Resources for Food and Agriculture and to disseminate relevant information through the website of the Treaty, where appropriate;
- appreciates the involvement of farmers' organizations in the work of the governing body, as appropriate, according to the rules of procedure of the governing body;
- encourages each contracting party to closely relate the realization of farmers' rights, as appropriate and subject to national legislation, with the implementation of Articles 5 and 6 of the Treaty, in particular the measures in Articles 5.1c,d and 6.2c,d,e,f,g;
- invites contracting parties and relevant organizations to facilitate and support the participation of farmers' organizations and relevant stakeholder groups in the regional consultations on farmers' rights;
- encourages contracting parties to engage the participation of farmers' organizations and relevant stakeholders in matters related to the conservation and sustainable use of plant genetic resources through awareness raising and capacity-building.

The Global Consultation Conference on Farmers' Rights, held in Addis Ababa in November 2010, produced an input paper to the discussion on the implementation of Article 9 on farmers' rights.[11] This input paper was submitted by Ethiopia to the secretariat of the Treaty, and was presented and discussed during the fourth session of the governing body. Among its recommendations is the establishment of an *ad hoc* working group to develop voluntary guidelines on the national implementation of Article 9 and related provisions, in a transparent, participatory, and inclusive manner, with the effective involvement of farmers' organizations. The recommendation recognizes progress achieved by governments in the reform of the UN/FAO Committee on World Food Security, with significant improvements in the effective participation by farmers' organizations, and requests the governing body to consider adopting the procedures agreed by the Committee on World Food Security as a template for new procedures that will ensure the full participation of all stakeholder groups. Civil society organizations present at the fourth session also expressed satisfaction with the new procedures for civil society participation in the Committee on World Food Security, and requested that a study regarding adoption of a similar approach for small-scale farmers be prepared for consideration by the fifth session of the governing body.

Farmers' rights to save, use, exchange, and sell farm-saved seeds and other propagating materials

The preamble of the ITPGRFA refers explicitly to the rights that farmers have to "save, use, exchange and sell farm-saved seed and other propagating material." Article 9.3, however, affirms that "nothing in this article shall be interpreted to limit any rights that farmers have to save, use, exchange and sell farm-saved seed/propagating material, subject to national law and as appropriate." Although the preamble affirmatively recognizes these rights, Article 9.3 is neutral and sets forth only that the decision rests with each country, and according to its national law. Article 9.3 reflects the lack of consensus among countries that defended a positive recognition of farmers' rights to save, use, exchange, and sell farm-saved seeds and countries that were against such a positive recognition, which could lead to restrictions on plant breeders' rights that are incompatible with the 1991 UPOV Act.

Article 9.3 does not, however, make any restrictions on the options that can be adopted by countries regarding the implementation of farmers' rights at the national level, even if it includes limitations on IP rights over plant varieties. This is probably one of the most divisive issues regarding the implementation of farmers' rights. Neither the proposal submitted by the EU nor the one presented by the United States made any reference to farmers' rights to save, use, exchange, and sell farm-saved seeds. For agrobiodiversity conservation and sustainable use strategies, it is crucial to ensure sufficient legal space within seed laws, IP legislation, and ABS for farmers to continue saving, using, exchanging, and selling farm-saved seeds and propagating material. It is also important to consider that farm-saved seed includes both local traditional varieties and improved varieties that have been further adapted and developed by farmers.

As Pelegrina and Salazar (2011) describe it (emphasis added):

> Traditional agriculture depends on the constant exchange and movement of plant genetic resources to manage different biotic and abiotic stresses and to provide for the different needs of farming communities. These natural and farmers' selection pressures developed the plant genetic diversity that the world inherited today. Diverse, free and democratic management of plant genetic resources will allow greater options for climate adaptation. *The right of farmers to save, use, exchange and sell seeds is one of the most basic foundations of the farmers' system of plant genetic resources management. This is how plant genetic resources diversity is maintained and created.*

During the consultations on farmers' rights, a prime concern expressed by participants (especially those from developing countries) is the need for support from the governing body of the Treaty to develop or adjust national laws for the realization of farmers' rights, especially seed and IP laws, which tend to undermine such rights.

There is a strong pressure for developing countries to adopt the 1991 UPOV Act, which restricts such farmers' rights (see Chapter 5 for further details). Nevertheless, it is not only plant breeders' laws that impose restrictions on farmers' rights to save, exchange, use, and sell farm-saved seeds. Such restrictions apply only to IP-protected plant varieties. Seed laws, however, regulate the production, marketing, and utilization of seeds and propagating material of all plant varieties, and not only of protected varieties (there are some exceptions, which are discussed in Chapter 4). Seed laws and plant breeders' rights must be reviewed and adjusted, so that they take into account the need to:

- broaden the genetic base of crops and increase the range of genetic diversity available to farmers;
- promote the expanded use of local and locally adapted crops and varieties;
- maximize intra- and interspecific variation for the benefit of farmers, especially those who generate and use their own varieties and apply ecological principles in maintaining soil fertility and in combating diseases, weeds and pests; and
- support the wider use of diversity of varieties and species in on-farm management, conservation and sustainable use of crops.

After all, these are obligations that all contracting parties to the ITPGRFA have assumed, in accordance with Article 6, on the sustainable use of plant genetic resources, which is strongly linked to farmers' rights. Article 6 is very clear that contracting parties must "develop and maintain appropriate policy and legal measures that promote sustainable use of agrobiodiversity," and this includes, among other measures, "reviewing and adjusting breeding strategies and regulations concerning variety release and seed distribution" (Article 6.2g). Resolution 6/2009, adopted by the governing body during its third session, and the resolution on farmers' rights, adopted during the fourth session, both invite contracting parties to "consider reviewing and, if necessary, adjusting [their] national measures affecting the realization of farmers' rights as set out in Article 9 of the International Treaty, to protect and promote farmers' rights." In addition, the second resolution encourages each contracting party to "closely relate the realization of farmers' rights, as appropriate and subject to national legislation, with the implementation of Articles 5 and 6 of the International Treaty, in particular the measures in Articles 5.1(c,d) and 6.2(c,d,e,f,g)." As Article 9 is not part of the multilateral system of ABS established by the International Treaty, it refers to *all* plant genetic resources for food and agriculture, and not only to those included in Annex I.

It is important to ensure a broad genetic basis of crops, and both seed laws and plant breeders' rights should contain exceptions allowing small-scale/local farmers to save, exchange, use, and sell farm-saved seeds (of protected or public domain varieties) to other small-scale (local, traditional, and agroecological) farmers, as long as it takes place in local markets and among local farmers. The definition of what constitutes a "local" market is complex, and it must take into consideration not only administrative/political and agronomical aspects, but also sociocultural ones. Some proposals aimed at balancing IP rights and farmers' rights to save, use, exchange, and sell farm-saved seeds (of protected varieties) have already been suggested by different stakeholders, for example limiting the farmers' rights to save, use, exchange, and sell seeds (of protected varieties) to crops produced for consumption at the national level (i.e., this right would not apply to export crops) or limiting the aforementioned farmers' rights to crops used for human food or animal feed (i.e., this right would not apply, for instance, to ornamental plants). Both proposals are legally and politically feasible, and should be considered by contracting parties when they implement farmers' rights at the national level.

Use of commercial plant varieties as a source of diversity in farmers' breeding: extending the breeders' privilege to farmers

As plant breeders, farmers must also benefit from the "breeders' exemption,"[12] which allows the use of plant varieties (including those developed by the formal sector and protected by plant breeders' rights) as sources of diversity to develop new varieties. The breeders' exemption must be extended to farmers who develop and improve their plant varieties, using their own breeding methods and techniques.

According to the UPOV Convention, the authorization of the breeder is not required either for the utilization of the plant variety as an initial source of variation for the purpose of creating other varieties or for the marketing of such varieties. Such exception to plant breeders' rights must apply also to farmers/breeders. Some national laws already recognize the rights of farmers as breeders (such as the laws of India and Ethiopia). The European Seed Association (ESA) has also issued a position paper on farmers' rights in which it states that it "fully supports an open access to all genetic resources, including landraces, genebank accessions, wild relatives and protected varieties for breeding purposes by all breeders: farmers or companies alike." The ESA is against any regulation that forbids or discourages farmers from breeding or participating in plant breeding or using other ways of improving the value of their crop.[13]

Pelegrina and Salazar (2011) point out that, although modern cultivars are often used by small-scale farmers, this does not mean that small farmers have not also created diversity, using introduced cultivars as raw materials for their selection. Thus, new types of varieties or populations have emerged, selected from modern cultivars, landraces, and local varieties. Pelegrina and Salazar (2011) give some examples:

> Farmers in North Cotabato, Philippines, developed 120 farmer rice varieties in 6 years, in contrast to the national release of only 55 inbred lines, in 10 years, from public research institutions. In the Mekong Delta of Vietnam, there are more than 100 farmer varieties covering more than 100,000 hectares of rice area. In the North and Central parts of Vietnam, farmers have developed more than 150 new farmer varieties. Due to traits that fit the market and intensive systems that most farmers now practice, their new rice varieties are also nonphoto-sensitive, of short to medium duration, and are no longer tall. Furthermore, these new varieties carry adapted traits that fit the farming conditions of different macro and micro eco-systems. Saving, using, exchanging and selling seeds among themselves helped create these new cultivars. All traditional or introduced varieties constitute raw materials to be developed and adapted. If the rice varieties were protected with IP rights that would discourage farmers from exchanging and selling among themselves, these varieties would not have emerged.

That is, even where modern varieties replace landraces, farmers do not lose their ability to innovate.

In a very interesting article, Rene Salazar, Niels Louwaars, and Bert Visser (Salazar *et al.*, 2007) also show that plant varieties conserved, developed, and improved by farmers are not restricted to varieties commonly known as "local and traditional" (landraces), and that farmers continue to develop new plant varieties. These new farmers' varieties are developed and improved by farmers based on different sources of variation, which include not only landraces but also varieties developed by the formal (private and public) breeding sector. Rene Salazar *et al.* (2007) point out that, even in market-oriented and intensive production systems, farmers continue to create their own varieties. In many situations, modern varieties simply replaced local/traditional varieties as a source of diversity, but farmers' breeding was not abolished. Farmers often recognize attractive features of modern varieties, including high yields and novel resistances, but also identify various characters that are not appreciated, especially regarding taste, processing qualities, and resilience under less than optimal growing conditions, so they promote interbreeding (crossing) of varieties to produce new varieties or lines with desirable properties. Salazar *et al.* (2007) mention some examples of countries in Southeast Asia, but point out that there are also accounts of these practices in other countries. Let us look at one of the examples mentioned by Salazar *et al.* (2007).

The IR 36 rice variety developed by the International Rice Research Institute (IRRI) is one of the most widely disseminated rice varieties and is extensively used by Asian farmers. On the island of Bohol, in the Philippines, local communities prefer rice grains with red color, because this characteristic is associated with better quality and greater satisfaction after the meal. Some time after introduction of the IR 36 rice variety, new phenotypes of this variety began to appear in Bohol, with red-colored grains (in the original variety, developed by IRRI, grains were white). The Philippines Seed Board conducted molecular tests on the red grains and found that they descended from the original IR 36 rice variety. Farmers in Bohol had incorporated the preferred red pericarp trait by introgression of genes from traditional red rice varieties exhibiting this trait into the newly released Philippines Seed Board varieties. This is an interesting example not only of the use of a conventional variety as a source of variation in breeding carried out by farmers, but also of how a variety developed by the formal sector acquired a "local" trait as a result of its adaptation to conditions favored and determined by local communities. In another location, where environmental and cultural factors favor a different trait (such as a certain taste or resistance to a certain pest), the same variety could have acquired other "local" traits, setting it apart from the original variety, as a result of improvement made by farmers.

Conducting research on genetic diversity of maize varieties in the Indigenous community of Cuzalapa, in western Mexico, Dominique Louette (1999) showed that farmers do not use only strictly local varieties, and that they regularly introduce exotic varieties and exchange seeds with other farmers. According to Louette, the assumption that traditional systems are "closed" and "isolated" with respect to the flow of genetic material is clearly contradicted by the results of her research, which involved 39 farmers (one-fifth of all farmers in Cuzalapa) over a three-year period (1989, 1990, and 1991) and was performed in a center of origin and diversity for maize, close to the Sierra de Manantlán Biosphere Reserve (created especially to protect wild relatives of maize). She considers that the dynamic nature of agricultural systems precludes "freezing" local varieties into a static system, as local varieties exist as part of a dynamic system that extends beyond a single place. According to Louette, "traditional cultivars are not genetically stable populations that can be well defined for conservation purposes; rather, local varieties constitute systems that are genetically open."

It is the possibility of accessing seeds (in accordance with local customs and rules) and of using them as a source of genetic variation that gives rise to high genetic diversity. Restrictions on the exchange of seeds imposed by the 1991 UPOV Act (which prevents exchange and sales of seeds, even in local markets) and by seed laws (conceived only for seeds released by the formal sector) make experiences such as those described above almost impossible. Thus, there must be no legal restrictions on farmers' use of protected varieties as a source of genetic diversity and on the exchange of seeds among themselves. Otherwise, they will not be able to innovate and develop new varieties using both landraces and modern varieties, to the detriment of conservation and sustainable use of agrobiodiversity.

Farmers' rights to save, exchange, use, sell, develop, and improve seeds of local and/or commercial varieties must be ensured as fundamental conditions for on-farm conservation and management of agricultural diversity. If these rights are not ensured, agrobiodiversity conservation actions and policies will have limited impact, as they will always face legal restrictions imposed on local and traditional practices that are crucial for the sustainability of agricultural systems. How will a public policy aimed at promoting agrobiodiversity be successful if seed laws forbid seed exchanges among farmers, through their social networks? How can sustainable agricultural systems be encouraged if sales of seeds that are adapted to certain socioenvironmental conditions in local markets are forbidden by law? How can public policies support traditional practices that are considered illegal by seed and IP legislations?

Protection of traditional knowledge and collective benefit-sharing mechanisms

The ITPGRFA, in its Article 9, also protects farmers' rights to traditional knowledge relevant to PGRFA, which includes innovations, practices, and knowledge related to seeds and agricultural systems. Traditional knowledge associated with agrobiodiversity includes cultivation practices, biological pest and disease control, selection, development and improvement of locally adapted varieties, maintenance of soil fertility, etc. Local varieties, developed by farmers and traditional communities, incorporate such associated knowledge.

The distinction between tangible (or material) components (plant genetic resources) and intangible ones (associated knowledge) of agrobiodiversity tends to be artificial, as it is hard to dissociate local plant varieties from associated knowledge, which is incorporated in the biological resource itself.[14] As Emperaire and Santilli (2006) put it:

> Traditional knowledge associated with a plant which is domesticated and selected by local communities is expressed in the existence of the biological object itself, which is the plant. Without the agronomical knowledge of local communities, their techniques and experimentations in selection and conservation, these objects would not exist, whether they are plants used for food, medicine, ornamental or others. Agricultural diversity is in itself an expression and materialization of traditional knowledge.

Both protection of traditional knowledge and the right to equitably participate in benefits arising from the utilization of PGRFA are recognized as farmers' rights under Article 9 of the ITPGRFA. These rights have inspired different proposals for their protection, which are sometimes defended separately and other times jointly. Some proposals aim to ensure benefit-sharing for farmers (for their knowledge and genetic resources) through the recognition of IP rights for plant varieties developed by farmers, in the same way that these rights are recognized for plant varieties developed by the formal sector (private or public). According to such proposals, farmers should receive royalty payments for their plant varieties in the same way that commercial breeders receive royalties for their protected varieties. In addition, IP rights would avoid misappropriation of farmers' varieties and agricultural knowledge, and oblige users of such resources and knowledge to share benefits with farmers.

There are many difficulties in implementing such a *sui generis* system of IP. IP rights mean exclusion and monopolies over seeds, both of which tend to discourage the free circulation of agricultural resources and knowledge, undermining the bases of local and traditional agricultural systems. Not only would IP rights prevent farmers from using genetic resources of commercial/modern varieties, but monopolistic rights would also exclude farmers from benefiting from new varieties developed by other farmers. Furthermore, this regime would deny the collective and cumulative character of innovations achieved by farmers, and it would be complex to define the holders of these rights, considering that exchanges performed by local communities take place through complex social networks and according to local rules and institutions. Carlos Corrêa (2000) notes that it would be illogical to protect farmers' rights through the IP system, as it was precisely this system that created the problems that the concept of farmers' rights seeks to solve.

There are other proposals based on bilateral ABS systems, in accordance with the principles of the CBD. According to them, access to plant genetic resources conserved *in situ*/on-farm by farmers, as well as to traditional knowledge, should be subject to prior and informed consent and sharing of benefits arising from their utilization. Bilateral contracts should be drawn up between providers and users of seeds and traditional knowledge.

Andersen (2006) identifies two approaches to the understanding of farmers' rights: the *ownership approach* and the *stewardship approach*. She argues that there is a latent conflict between these two, and that the stewardship approach must prevail if farmers' rights are to be realized within the framework of the ITPGRFA. According to Andersen (2006), the *ownership approach* refers to the right of farmers to be rewarded for genetic material obtained from their fields and which is used in commercial varieties and/or protected with IP rights. The idea is that such a reward system is necessary to enable equitable sharing of benefits arising from the use of agrobiodiversity and to establish an incentive structure for continued maintenance of this diversity. Access and benefit-sharing legislation and farmers' IP rights are suggested as central instruments. The *stewardship approach* refers to the rights that farmers must be granted in order to enable them to continue as stewards of agrobiodiversity. The idea is that the legal space required for farmers to continue this role must be upheld and that farmers involved in the maintenance of agrobiodiversity – on behalf of our generation, for the benefit of all humankind – should be rewarded and supported for their contributions (Andersen, 2006). She proposes the following working definition:

> farmers' rights consist of the customary rights that farmers have had as stewards of agrobiodiversity since the dawn of agriculture to save, grow, share, develop and maintain plant varieties, of their legitimate right to be rewarded and supported for their contribution to the global pool of genetic resources as well as to the development of commercial varieties of plants, and to participate in decision making on issues that may affect these rights.

Collective benefit-sharing mechanisms must be developed, and farmers' rights must be recognized as essentially collective, and not individual, rights. The most important benefit for farmers is probably the creation of legal space for farmers to save, use, exchange, produce, and sell their seeds, free of legal obstacles and restrictions which do not consider the specificities of local/traditional systems (and are imposed by seed laws, IP rights, and, in some cases, ABS laws). Other collective benefit-sharing mechanisms must be promoted, for example public policies aimed at valuing and strengthening local and traditional agricultural systems and at adding environmental and cultural value to local/regional products. This could be achieved by, for example:

- promoting geographical indications and certifications;
- allowing farmers to participate in all decision-making processes (that affect agrobiodiversity) at the international, national, regional, and local levels;
- strengthening local/traditional farmers' capacity to participate in local and national agricultural markets;
- promoting participatory plant breeding, seed fairs and local seedbanks, managed by farmers' communities, and seed exchange networks;
- recognizing agrobiodiverse traditional/local agricultural systems as biocultural heritage (Argumedo *et al.*, 2011) and as cultural landscapes, with the adoption of policies and measures to safeguard such systems, at the national and international levels; and
- establishing national benefit-sharing funds to support on-farm conservation and sustainable use of agrobiodiversity (which would complement the international benefit-sharing fund created by the treaty);
- introducing a collective payment for environmental services;
- creating a special category of protected area aimed at conserving agrobiodiversity.

Actions and policies aimed at supporting on-farm conservation and management of agrobiodiversity are probably among the most effective and fruitful ways of sharing benefits arising from the use of plant genetic diversity. Such policies must include farmers' knowledge and practices, and ensure the continuity of biological, social, and cultural processes which generate agrobiodiversity. Links between agrobiodiversity conservation and sustainable local development must be promoted and strengthened. On-farm conservation actions and policies must focus on the agricultural system as a whole, with all its components, and not only on specific species or varieties, and consider its dynamic nature. The entire sociocultural system supporting agrobiodiversity must be considered, including local perceptions and values associated with genetic resources. Articles 5 and 6 of the ITPGRFA establish contracting parties' obligations in relation to conservation and sustainable use of agrobiodiversity, and they are directly related to the implementation of farmers' rights. On-farm conservation accomplishes several functions, in addition to conservation in itself, such as the political and social empowerment of local communities and the improvement of their living conditions.

Plant varieties developed by farmers must not be protected by an IP system, not even a *sui generis* one. Sharing of benefits with farmers should not be linked to the sales of products developed through genetic material accessed in *ex situ* collections or collected in *in situ*/on-farm conditions, as the role of farmers in conservation of agrobiodiversity would be greatly underestimated: after all, farmers have conserved and managed agricultural resources for thousands of years, and considering that their contribution is limited to the genetic material used in the development of commercial varieties is a gross underestimation of their contribution to the global gene pool. Besides, as Ramanna (2006) points out, rewarding and recognizing farmers for their contribution goes beyond tracking the extent to which their innovations/material have been a part of patents or other IP rights. Another danger of defining farmers' rights as IP rights, according to Borowaik (2004), is that it may end up helping to legitimize asymmetries by creating the impression that there is parity among competing rights: breeders and farmers have parallel rights platforms to get their fair shares. However, as Borowaik notes, reality is much more asymmetric, and such systems could promote a further shift away from farmer-centered agriculture (Borowaik, 2004). Benefit-sharing must benefit all farmers, not only those who hold varieties that are used by plant breeders or which are protected by IP rights. (National benefit-sharing funds are discussed in Chapter 7.)

Participatory plant breeding

According to Article 6.2c of the ITPGRFA, contracting parties must promote plant breeding efforts which, with the participation of farmers, particularly in developing countries, strengthen the capacity to develop varieties particularly adapted to social, economic, and ecological conditions, including in marginal areas. Participatory plant breeding can also be considered a benefit-sharing mechanism.

Participatory plant breeding had already been included in the GPA, adopted during the fourth International Technical Conference on Plant Genetic Resources, held in Leipzig, Germany, in 1996, and that is currently being updated. According to paragraph 184d, of the GPA, governments, and their national agricultural research systems, supported by the IARCs and other research and extension organizations must "explore and, in appropriate circumstances, make use of decentralized and 'participatory' plant breeding strategies to develop plant varieties specifically adapted to local environments." At the international level, several participatory plant breeding programs are being (or have been) developed, and the Working Group on Participatory Plant Breeding, established by CGIAR in 1996, has documented more than 80 participatory plant breeding programs worldwide.

But what is participatory plant breeding? How does it differ from conventional plant breeding? According to the definition of CIAT (International

Center for Tropical Agriculture), one of the CGIAR research centers, participatory plant breeding is

> the systematic and regular involvement of farmers as decision-makers in all stages of a plant breeding program. Farmer involvement in plant breeding can take many forms, including: definition of breeding goals and priorities; selection or provision of germplasm sources; hosting of trials; selection of lines for further crossing; evaluation of results; planning for the following year's activities; suggestion of methodological changes; and multiplication and commercialization of the seed of selected lines.[15]

According to CIAT, participatory varietal selection is the most familiar form of farmer participation in plant breeding. Participatory varietal selection traces its origin back to the farming systems research of the 1970s, with farmers becoming involved in the breeding process itself in the 1990s. In participatory varietal selection, farmers are involved in evaluating a range of stable lines and selecting those most appropriate for their own uses for subsequent independent testing. Participatory plant breeding involves a significantly higher and more complex degree of farmer involvement in decision-making at earlier and more fundamental stages of the varietal development process. With this higher level of participation comes much greater potential for farmer empowerment and for bringing about improvements in the livelihoods of rural people. Participatory plant breeding and participatory varietal selection raise farmers' awareness of regulatory frameworks and pave the way for involvement in efforts to influence these, particularly when existing frameworks limit farmers' opportunities to access benefits from the genetic materials that they helped to develop. Participation of farmers in plant breeding programs offers important opportunities to safeguard and strengthen farmers' rights (Vernooy, 2003; Ceccarelli and Grando, 2007; Halewood *et al.*, 2007).

According to Brazilian geneticist Altair Toledo Machado (see Machado and Machado, 2007; Machado *et al.*, 2008), participatory plant breeding was developed as an alternative to conventional plant breeding, and is used mainly in developing countries and in marginal areas that are under environmental, social, and economic stresses. Farmers of these areas did not benefit from conventional plant breeding programs and their improved varieties, which often require heavy doses of fertilizer and other chemicals, that most poor farmers can't afford. Besides, conventional breeding tends to focus on broad adaptability, and not on adaptability to specific environmental, social and cultural conditions, and it also tends to prioritize crops with a high commercial value.

As Gerry Toomey (1999) explains: professional breeders, often working in relative isolation from farmers, have sometimes been unaware of the multitude of preferences – beyond yield and resistance to diseases and pests – of their target farmers. Ease of harvest and storage, taste and cooking qualities, how fast a crop matures, and the suitability of crop residues as livestock feed are just a few of the dozens of plant traits of interest to small-scale farmers. Despite this wealth of knowledge, in many cases farmers' participation in conventional breeding programs has been limited to evaluating and commenting on a few advanced experimental varieties just prior to their official release. A goal of the participatory plant breeding is to build on farmers' knowledge, which involves clearly identifying farmers' needs and preferences and the reasoning behind them.

According to Machado (see Machado and Machado, 2007; Machado *et al.*, 2008), participatory plant breeding includes the knowledge, skills, experiences, and practices of farmers as essential components, and takes place in a decentralized manner, with farmers participating in all stages of the breeding process. In conventional breeding, it is the (professional) breeder who defines the objectives and conducts all selection and evaluation processes – only evaluation of the genetic material is sometimes carried out with the participation of farmers – and the organization is fully centralized. According to Machado, participatory plant breeding has broader objectives than conventional breeding. It is aimed not only at high yield and productivity, but also at conserving biodiversity, developing locally adapted varieties, evaluating them in a participatory way, and diversifying systems and seed production.

Machado also describes the participatory plant breeding program developed in the farming community called Sol da Manhã, in the municipality of Seropédica, in the state of Rio de Janeiro (Brazil). The program lasted from 1986 to 2000 and its main objective was to characterize and select nitrogen-efficient maize varieties, and to increase yields, so that farmers could survive and produce in areas where soils are predominantly sandy, with low levels of organic matter and fertility. The program included the rescue, characterization, selection, and conservation of maize varieties, with the participation of farmers in all stages of the process. A new maize variety was developed, and named Sol da Manhã. It is highly efficient in its use of nitrogen, and yield increased from 2,000 to 4,000 kilograms per hectare (Machado and Machado, 2007). This experience motivated the creation, in 1990, of a Maize Network, with the participation of public institutions (EMBRAPA), the PTA Network (a network of NGOs active in the areas of agroecology and small-scale family farming, which acted in 12 Brazilian states), and farmers. Its objective was to implement participatory strategies to use and conserve maize genetic diversity in farmers' communities. This network lasted until 1996, but several organizations, such as the Advisory Services for Alternative Agriculture Projects (AS-PTA), the Center for Alternative Technology of Zona da Mata Center (CTA), from Minas Gerais, and the Center for Alternative and Popular Technology

(CETAP), to name a few, continued their activities, always adopting a participatory and agroecological approach. In western Santa Catarina State, in southern Brazil, the town of Anchieta became known as the "Capital of Creole Maize." Starting in 1996, the Union of Family Farmers of Anchieta (Sintraf/Anchieta), with support from the municipal government and some civil society organizations, started promoting actions for rescue, use, and conservation of local varieties of many species, especially maize, which also inspired other initiatives. The first State Festival (Fair) of Creole Maize took place in 2000, promoted by the aforementioned union, in partnership with the Movement of Small Farmers (MPA) and the municipal government, and 5,000 people participated in this festival. The first National Festival (Fair) of Creole Maize was held in 2002, when 943 varieties of plant species, of which 228 were creole maize, were displayed. Approximately 20,000 people participated in the festival, which was also held in Anchieta. The same success has been achieved in subsequent festivals, which usually take place every two years (Vogt *et al.*, 2007).

In Central America, one of the pioneer participatory plant breeding programs is the Programa Colaborativo de Fitomejoramiento Participativo en Mesoamérica (FPMA), developed through partnerships between governmental organizations and NGOs, farmers' organizations, and national and international agricultural research centers. The program works with small-scale farmers from Guatemala, El Salvador, Honduras, Nicaragua, Costa Rica, Cuba, and Mexico, and is aimed at conserving, characterizing, and improving varieties of maize, beans, sorghum, and other plant species. Its first and second phases ran from 2000 to 2004 and from 2005 to 2009, respectively. In its third phase (2010–2014), the main focus is on the management, conservation, and development of agrobiodiversity, using a participatory plant breeding approach. Sustainable production of food and seeds is its main target. The program has already produced some significant results: in Costa Rica, nine varieties of beans have been released; in Cuba, a maize variety has been released; in Honduras, 12 varieties of beans and four of maize have been released; and in Nicaragua, six varieties of beans, two of maize, and four of sorghum have already been released.[16] Another important initiative of the program is the creation of the Seed Community Reserve of Quilinco (Reserva Comunitaria de Semillas de Quilinco), based in Chiantla, in Sierra de los Cuchumatanes,[17] in western Guatemala. This reserve contains seeds of strategic local plant varieties, and maintains collections of maize, beans, fava beans, wheat, oats, and medicinal plants, managed by farmers.

Participatory plant breeding is not regulated in most countries. ABS and IP legislation do not have specific provisions on participatory plant breeding. However, participatory plant breeding programs are spreading worldwide, and they are increasingly being adopted by agricultural research organizations and NGOs. Such programs combine agricultural science and local knowledge and benefit small-scale farmers who have been bypassed by the Green Revolution. As already mentioned, the ITPGRFA recognizes participatory plant breeding as an important tool to promote sustainable use of agobiodiversity. National laws must also create special regulations to encourage and support participatory plant breeding programs, that is, seed, IP, and ABS legislation must consider the specificities of participatory plant breeding, and how it differs from conventional breeding, in order to avoid legal obstacles to such programs.

Ownership of IP rights over new varieties developed by participatory plant breeding programs has been discussed in several forums. If new varieties (developed through participatory plant breeding programs) meet the legal requirements to be protected by IP rights (especially plant breeders' rights), who will be the holder of these rights? Farmers and/or professional breeders who have worked together? Two options are identified: not protecting such varieties (through IP rights) or establishing co-ownership of rights for all participants in the participatory plant breeding program, including scientists, farmers, and others involved. However, a large number of co-owners of IP rights would create serious difficulties for the exercise of such rights, as any activities related to the use of the new varieties would need the prior approval of all co-owners. In addition, the prior informed consent of all co-holders of such resources and knowledge would be necessary for access authorizations, and benefit-sharing contracts would have to involve all co-holders too (in countries that have approved ABS laws). Seed laws may also impose restrictions on the sale and production of seeds of such varieties, when they impose strict homogeneity and stability requirements, which may or not be met by varieties released by participatory plant breeding programs.

It would be better not to protect (through IP rights) new plant varieties developed by participatory plant breeding programs. However, it is also important to ensure that they remain in the public domain, and are not misappropriated by third parties. This will depend, however, on the interest and willingness of local farmers and professional breeders to keep such plant varieties in the public domain, and the compatibility between the public domain and local norms regulating the exchange and circulation of agricultural resources and knowledge. Another possibility is to make use of copyleft licenses, similar (but adapted) to those used by the open source software movement. Copyleft licenses allow uses for specific (noncommercial, for instance) purposes, restrict others, and prevent misappropriation of resources and knowledge by third parties (for further discussion, see Chapter 10).

Farmers' political participation

Another important farmers' right is the right to participate in all decision-making processes that impact the conservation and sustainable use of PGRFA. This includes participation of small-scale/family/local farmers' representatives in all political forums (national councils, commissions, committees, hearings, etc., and in the work of the governing body of the ITPGRFA) that are responsible for agricultural and agrarian policies and legislation, to ensure that their needs and rights are duly respected and promoted. This includes all instances of power (international, federal, state, provincial, and municipal) and all policy areas impacting farmers' rights (land reform, rural credits, subsidies and insurance policies, food and nutritional security, agricultural technology and research priorities, local/rural development policies, environmental policies, water management and use policies, etc.). The farmers' right to participate in decision-making must be interpreted in a broad and inclusive manner, including any political decisions with impacts on local agricultural systems and PGRFA.

Farmers' representatives must also be allowed to participate in decisions regarding the implementation of national ABS laws. They must participate not only in decisions related to access authorizations to plant genetic resources and traditional knowledge, held by farming communities, but also in the definition of national policies on ABS for plant genetic resources. It is important that national authorities involve farmers' representatives in the implementation of the CBD and of the ITPGRFA at the domestic level. In many countries, implementation of the CBD is the responsibility of the ministry of the environment, whereas responsibility for the implementation of the Treaty typically lies with the ministry of agriculture, which often creates conflicts. However, not only representatives of governmental organizations must participate in such implementation processes, but also other stakeholders, such as farmers and their representative organizations.

In addition, representatives of small-scale/family/local farmers must also participate in decisions regarding objectives and priorities of agricultural research. They must not only participate in participatory plant breeding programs, but also take part in decisions on the priorities of conventional plant breeding, so that these take into consideration the needs of small-scale/family/local farmers, and give more attention to locally adapted crops, varieties, and underutilized species.

Farmers must also participate in the elaboration and application of seed laws and regulations, which establish requirements and criteria for the utilization, production, and marketing of seeds. They must also participate in decision-making processes related to variety release and registration, including criteria for the tests and assessments regarding value for cultivation and use and genetic homogeneity and stability, which are usually established in a top-down manner by ministries of agriculture and other official/technical agencies.

Such decision-making processes must occur in a collective and democratic way, and collective spaces must be created, such as commissions or committees on variety release and registration. In many countries, such decisions are taken unilaterally by official authorities, without any social participation, which is very antidemocratic. It is more inclusive and transparent to create seed commissions/committees with the participation of all stakeholders, including farmers' representatives. This is the only way to ensure that such decision-making processes take into consideration the interests and needs of all agricultural stakeholders. It is also important to guarantee a balanced composition of such commissions, and to ensure that they include representatives of all categories of farmers, as well as representatives of governmental and nongovernmental sectors. A democratic and inclusive participation of representatives of small-scale/family/local farmers could result, for instance, in the elaboration of specific regulations for the utilization, production, and marketing of local/traditional seeds. Or perhaps in the exemption of some plant species and varieties from certain legal requirements, such as genetic homogeneity and stability. Some plant varieties could be excluded from mandatory registration and specific requirements could be established for the registration of other varieties. Or a special registry for local/traditional varieties could be created. The same flexibility/adaptation should apply to the determination of the cultivation and use value of local/traditional plant varieties. After all, if the main objective of seed laws is to ensure that farmers have access to good-quality seeds, it is essential that they also participate in decision-making processes regarding standards and criteria for production, marketing, and use of these propagating materials. Besides, conservation and sustainable use of agrobiodiversity would greatly benefit from the creation of a wider legal space for plant heterogeneity.

Some identity and quality standards that usually apply to the whole territory of a country could also become more flexible, and regionalized, especially in large countries, such as Brazil, where there are great social, cultural, and environmental differences among geographical regions. Regional regulations could enable production and marketing of varieties adapted to conditions of specific geographical regions, even if they do not meet the standards for countrywide/national distribution.

Some initiatives for the implementation of farmers' rights at the national level are discussed next.

India's Protection of Plant Varieties and Farmers' Rights Act and the new Indian Seeds Bill

India's Protection of Plant Varieties and Farmers' Rights Act of 2001 is probably the most far-reaching legislation with regard to establishing rights for farmers to save, use, exchange, and sell farm-saved seed[18] (see also 'The Brazilian seed law and traditional, local, and creole plant varieties' in Chapter 4). However, such rights are threatened by a new Indian Seeds Bill, of 2004, that is currently being discussed at the Indian Parliament (as of July 15, 2011). The Seeds Bill aims to regulate the sales of seeds in India, and to replace the Seed Act of 1966 (currently in force). However, it contradicts some provisions of the Protection of Plant Varieties and Farmers' Rights Act, of 2001, and has been heavily criticized by farmers' organizations because it "compromises" the right of farmers to grow, sow, save, use, exchange, share, or sell their farm seeds.[19]

India's Protection of Plant Varieties and Farmers' Rights Act combines aspects of the UPOV Convention, regarding plant breeders' rights, with the principles established in the CBD,[20] on ABS. India is a member of the WTO, and has signed the TRIPS Agreement, but has adopted a national law which corresponds to neither of the UPOV Acts. India created a *sui generis* system for the protection of plant varieties, allowed by the TRIPS Agreement.[21]

Agriculture is an essential economic activity in India: nearly 70 percent of its population depends on agriculture for their livelihood, and agriculture is the principal contributor to India's economic output, accounting for nearly 25 percent of India's gross domestic product (GDP). Although India has a large public sector system involved in various aspects of agriculture production, including breeding and research, the majority of Indian farmers depend on the informal agricultural system of trading, exchanging, and re-using seeds. Approximately 80 percent of farmers rely on the informal seed system or farm-saved seed (Rammana and Smale, 2004; Soni, 2007).

The first Indian bill (the Protection of Plant Varieties and Farmers' Rights Act, PPVFR Act) was presented in 1993–1994 and aimed solely at the protection of plant breeders' rights over commercial varieties, which led to countless protests from NGOs that feared the effects of privatization of seeds on small-scale/traditional farmers. In 1994, when the TRIPS Agreement was approved, protests became more intense throughout the country.[22] The bill was revised five times prior to its approval in 2001, and Indian civil society organizations demanded that the country adopt a *sui generis* system for protection of plant varieties that recognized and protected farmers' rights as well as professional breeders' rights. They also demanded the creation of a registry of farmers' varieties and the participation of farmers' representatives in the implementation of the law. However, the recognition of IP rights over farmers' plant varieties, even if by means of a *sui generis* system, ended up, in some way, legitimizing the position of the private seed sector in favor of these rights (over commercial varieties). Farmers, nonetheless, obtained the legal recognition of the following rights in the approved version of the Act:

- The Act defines "breeder" as a "person or group of persons or a farmer or group of farmers or any institution which has bred, evolved, or developed any plant variety." According to the Act, a farmer who has bred or developed a new variety is entitled to registration and other protection in the same manner as a breeder of a variety. That is, farmers are recognized not only as users, but also as breeders and innovators. The

Act also establishes that a farmer who is engaged in the conservation of genetic resources of landraces and wild relatives of plants and in their improvement is entitled to recognition and reward from the National Gene Fund (provided that material selected and preserved by the farmer has been used as donors of genes in registrable varieties).

- It recognizes farmers' rights to save, use, sow, resow, exchange, share, or sell their farm produce, including seeds of protected varieties, in the same manner as they were entitled to do before the coming into force of the Act. However, farmers cannot sell "branded" seed of protected varieties. "Branded seed" means any seed put in a package or any other container and labeled in a manner indicating that such seed is of a protected variety. That is, farmers can sell both farm-produced seeds and seeds of protected varieties, as long as they are not sold in packages with labels indicating that they are of protected varieties. Furthermore, the legal possibility of using a protected variety as an initial source of diversity for creating other varieties (the breeders' exemption) is extended to "any person" (a professional or farmer/breeder).
- The above-mentioned rights apply to all farmers, and not only to specific categories of farmers. A farmer is defined as "any person who cultivates crops by cultivating the land himself, or cultivates crops by directly supervising the cultivation of land through any other person, or conserves and preserves, severally or jointly, with any other person, any wild species or traditional varieties or adds value to such wild species or traditional varieties through selection and identification of their useful properties."
- No suit, prosecution, or other legal proceeding may be brought against farmers for any violation of plant breeders' rights, if it was done or intended to be done in good faith. This provision aims to protect farmers who were not aware of plant breeders' rights when they violated them.
- When any seed of a registered variety is sold to a farmer (or group of farmers, or farmers' organizations), the breeder of the variety must disclose the expected performance under given conditions. If such propagating material fails to perform as described, farmers may claim compensation. The competent authority, after giving notice to the breeder of the variety and providing him an opportunity to file opposition, may determine that the breeder of the variety pays compensation to farmers.
- Both professional (public and private) breeders and farmers/breeders may apply for registration of varieties before the National Register of Plant Varieties. A certificate of registration for a variety confers an exclusive right to the breeder or his or her successor, his agent or licensee, to produce, sell, market, distribute, import, or export the variety.
- Varieties can be registered only if they conform to the criteria of novelty, distinctiveness, uniformity, and stability. For "extant" varieties, novelty

is not required. According to the Act, an "extant variety" means a variety available in India which is notified under Section 5 of the Seeds Act, 1966,[23] *or* a farmers' variety *or* a variety about which there is common knowledge, *or* any other variety that is in public domain.

- A "farmers' variety" means a variety that has been traditionally cultivated and evolved by the farmers in their fields, *or* is a wild relative or landrace *or* a variety about which the farmers possess common knowledge. Farmers are exempt from the payment of registration fees.
- Certificates of registration of varieties are published by the competent authority, in order to invite claims of benefit-sharing in the registered variety. Any person or group of persons (as long as they are citizens of India), firms, governmental organizations or NGOs (formed or established in India) may submit their claims of benefit-sharing in relation to the variety. On receiving such a claim, the competent authority will send a copy to the breeders of the registered varieties, who may submit their opposition to the claim. The authority will then indicate the amount of benefit-sharing (if any) to which the claimant will be entitled, taking into consideration the following: the extent and nature of the use of genetic material in the development of the variety relating to which the benefit-sharing has been claimed; and the commercial utility and demand in the market of the variety relating to which the benefit-sharing has been claimed. Such benefit-sharing funds will be deposited by the breeder of the variety in the National Gene Fund.
- Any plant breeder or other person applying for registration of a variety must disclose information regarding the use of genetic material conserved by any tribal or rural families in the breeding or development of such variety. Failure to disclose such information may result in the registrar, if satisfied that the applicant has willfully and knowingly concealed such information, rejecting the application for registration. Any person or group of persons (whether or not actively engaged in farming) or any governmental organization or NGO may, on behalf of any village or local community in India, file a claim affirming that the people of the village or community significantly contributed to the evolution of the variety, and are entitled to benefit-sharing (through the National Gene Fund).
- The authority responsible for the implementation of the act is called the Protection of Plant Varieties and Farmers' Rights Authority, and consists of a chairperson and 15 members. In addition to representatives from governmental organizations, the following members participate: one representative from a national or state-level farmers' organization, one representative from a tribal organization, one representative from the seed industry, one representative from an agricultural university, and one representative from a national or state-level women's organization associated with agricultural activities. All these representatives are nominated by the central government.

The Indian law tends to adopt an ownership approach, with all its limitations and barriers (discussed above). One of its main achievements is the explicit recognition of farmers' rights to save, use, sow, resow, exchange, share, or sell their farm produce, including seeds of protected varieties. The recognition of farmers as breeders and innovators is also important, as is the participation of farmers' representatives in the national authority set up to implement the Act.

However, a new seed bill being discussed in the Indian Parliament may restrict some farmers' rights. The new seed bill states that (emphasis added):

> Nothing in this act shall restrict the right of the farmer to save, use, exchange, share or sell his farm seeds and planting material, except that he shall not sell such seed or planting material under a brand name *or which does not conform to the minimum limit of germination, physical purity, genetic purity.*

That is, if the new seed bill comes into force, farmers will be able to sell their seeds only if they are registered and meet the same minimum standards prescribed for commercial seeds. The PPVFR Act, of 2001, only prevents farmers from selling seeds of protected varieties in branded packages, and there are no other requirements for farmers to sell seeds. Registration under the PPVFR Act was voluntary, but the new seed bill makes registration compulsory for all seeds. According to the new seed bill, every seed producer and dealer, and horticulture nursery, has to be registered with the state government. Farmers also become subject to the regulations provided for commercial producers, processors, and stockers of seeds, who need to meet specifications regarding infrastructure, equipment, and qualified manpower. Any person who contravenes any provisions of the seed bill or imports, sells or stocks seeds deemed to be misbranded or not registered, can be punished by large fines, including farmers. Under the PPVFR Act, the registration of a variety requires disclosure of the pedigree of the variety and the geographical origin of the parental material used. However, the new seed bill does not establish any obligation to disclose the pedigree of the variety under registration.[24]

Farmers' rights in the African Model Law and in the Ethiopian Proclamations

The African Model Law for the Protection of the Rights of Local Communities, Farmers and Breeders, and for the Regulation of Access to Biological Resources was approved by the African Unity Organization in July, 1998, in Ougadougou (Burkina Faso) and re-endorsed in July, 2001, in Lusaka (Zambia). The African Unity Organization was replaced by the African Union in July 2002.[25] The African Model Law is supposed to be followed by African countries to implement the CBD and the TRIPS Agreement, particularly the provisions on protecting plant varieties. The African Model Law also incorporates some components of the International Undertaking on Plant Genetic Resources (which was still in force when the African Model Law was approved; the undertaking was later replaced by the ITPGRFA). As the ITPGRFA had not been adopted when the African Model Law was being elaborated, it is based mainly on the principles of the CBD. It is a model law for protecting farmers' and plant breeders' rights and community rights,[26] as well as for regulating access to biological resources and to associated traditional knowledge.

According to the African Model Law, farmers' rights are

> recognized as stemming from the enormous contributions that local farming communities, especially their women members, of all regions of the world, particularly those in the centres of origin or diversity of crops and other agrobiodiversity, have made in the conservation, development and sustainable use of plant and animal genetic resources that constitute the basis of breeding for food and agriculture production.

Such rights are recognized and protected to enable farmers "to continue making these achievements" (Articles 24 and 25). According to the African Model Law, farmers' varieties and breeds are recognized and must be protected under the rules of practice as found in, and recognized by, the customary practices and laws of the local farming communities concerned, "whether such laws are written or not." This is a very important recognition of the legal diversity existing in African societies, an expression of their cultural diversity. The recognition of the existence of multiple local legal systems within the same territory is known as "legal pluralism" and is opposed to legal monism, which only recognizes one state/official legal system. Many African societies are pluralistic, and the recognition of local legal institutions by the African Model Law is a significant step toward the development of more legitimate legal systems, rooted in the culture and livelihoods of local communities.

The African Model Law[27] also establishes that a plant variety with specific attributes identified by a community must be granted IP through a variety certificate that does not have to meet the criteria of distinction, uniformity, and stability. This variety certificate entitles the community to exclusive rights to multiply, cultivate, use, or sell the variety, or to license its use, without prejudice to the farmers' rights set out in the law (Article 25). Farmers' rights must, with due regard for gender equity, include the rights to (Article 26):

- protection of their traditional knowledge relevant to plant and animal genetic resources;

- obtain an equitable share of benefits arising from the use of plant and animal genetic resources;
- participate in making decisions, including at the national level, on matters related to the conservation and sustainable use of plant and animal genetic resources;
- save, use, exchange, and sell farm-saved seed/propagating material of farmers' varieties, as well as the new plant varieties protected under breeders' rights, although farmers may not sell farm-saved seed/propagating material of a breeder's protected variety in the seed industry on a commercial scale;
- collectively save, use, multiply, and process farm-saved seed of protected varieties;
- use a new breeders' variety protected under the law to develop farmers' varieties, including material obtained from genebanks or plant genetic resource centres (breeders' rights on a new variety are subject to restrictions with the objective of protecting food security, health, biological diversity and any other requirements of the farming community for propagation material of a particular variety).

In addition, according to Article 27 of the African Model Law, any product derived from the sustainable use of a biological resource must be granted a certificate or label of recognition, and a certificate of fair trade must be granted to a product derived from a biological resource or knowledge or technology, in which case a significant part of the benefits derived from the product are returned to the local community. These measures are aimed at adding environmental and social value to the products of African cultural and biological diversity.

Plant breeders' rights are also recognized by the African Model Law. According to Article 28, these rights "stem from the efforts and investments made by persons and institutions for the development of new varieties of plants" and are "the basis for providing recognition and economic reward." Plant breeders' rights include the exclusive right to sell, including the right to license other persons to sell plants or propagating material of that variety, as well as the exclusive right to produce, including the right to license other persons to produce, propagating material of that variety for sale (Article 30). However, plant breeders' rights are subject to the conditions provided in the provisions on farmers' rights contained in the African Model Law (described above). The exemptions to the rights of plant breeders are established, so that any person or farmers' community may (Article 31):

- propagate, grow, and use plants of that variety for purposes other than commerce;
- sell plants or propagating material of that variety as food or for another use that does not involve the growing of the plants or the propagation of that variety;

- sell within a farm or any other place at which plants of that variety are grown any plants or propagating material of that variety;
- use plants or propagating material of the variety as an initial source of variation for the purpose of developing another new plant variety, except where the person makes repeated use of plants or propagating material of the first mentioned variety for the commercial production of another variety;
- sprout the protected variety as food for home consumption or for the market;
- use the protected variety in further breeding, research, or teaching;
- obtain, with the conditions of utilization, such a protected variety from genebanks or plant genetic resources centers.

In addition, farmers are free to save, exchange, and use part of the seed from the first crop of plants that they have grown for sowing in their own farms to produce a second and subsequent crop, according to conditions specified in the provisions on farmers' rights contained in the African Model Law (described above).

Namibia[28] and Uganda have drafted legal bills inspired by the African Model Law, which are still being discussed in the national parliaments. Zimbabwe, Malawi, and Zambia have also discussed legal bills and public policies based on the African Model Law. In 2006, Ethiopia adopted Proclamation 482/2006, which regulates access to genetic resources and community knowledge, and community rights and which draws on the African Model Law. This Proclamation provides communities with the right to receive 50 percent of the share that the Ethiopian state obtains in monetary form from the use of genetic resources. According to this Proclamation, communities have the right to make decisions about access to their knowledge, whereas the state has the authority to make decisions about access to genetic resources – on behalf of the communities. Communities do, however, have the right to disagree in cases where access to genetic resources affects their culture and their livelihood, according to Regassa Feyissa (2006).

Ethiopia also adopted a Proclamation on Plant Breeders' Rights (Proclamation 481/2006), which protects both improved and farmers' varieties and is inspired by the African Model Law. These are described as varieties having specific attributes, which have been "discovered, bred, developed or nurtured by Ethiopian farming communities, or a wild relative of a variety about which the Ethiopian farming communities have common knowledge." According to Article 28, farmers have the following rights (in relation to the use of plant varieties): (a) to save, use, exchange, and sell farm-saved seed or propagating material of farmers' varieties; (b) to use protected varieties including material obtained from genebanks or plant genetic resource centers to develop farmers' varieties; (c) to save, use, multiply, exchange, and sell farm-saved seed or propagating material of protected varieties. However, farmers may not sell farm-saved seed or propagating material of a protected variety in the seed industry as a certified seed. According to the Ethiopian Seed Proclamation 2006/2000, farmers can produce and sell certified seed to other farmers, but cannot engage in large-scale seed sales without being certified by the National Seed Agency.

Notes

1 According to Pelegrina and Salazar (2011), in 2003, farmers and farmer groups in the Philippines defined farmers' rights to comprise 38 elements covering sociopolitical, economic, and cultural rights.

2 Via Campesina, *Declaration of Rights of Peasants – Women and Men* (http://viacampesina.net/downloads/PDF/EN-3.pdf; accessed March 15, 2011).

3 Pat Mooney, Cary Fowler, and Hope Shand began working on the "seeds" issue in 1977. In 1984, the three co-founded RAFI (Rural Advancement Foundation International), whose name was changed to the ETC Group – Action Group on Erosion, Technology and Concentration – in 2001. The ETC Group is a small international NGO addressing the impact of new technologies on vulnerable communities. In 1983, Pat Mooney wrote *The Law of the Seed* (Mooney, 1983), and in 1994 Cary Fowler published *Unnatural Selection: Technology, Politics and Plant Evolution* (Fowler, 1994). Both publications are referential. Mooney's more recent work has focused on geoengineering, nanotechnology, synthetic biology, and global governance of these technologies as well as corporate involvement in their development.

4 For a detailed and interesting description of this international process, see Mooney (2011).

5 Vesting farmers' rights in the international community, as "trustee" for present and future generations of farmers, made it unclear and ambiguous who were the holders of farmers' rights. The ITPGRFA, in its Article 9, clearly granted such rights to farmers themselves.

6 Chapter 14 of Agenda 21 deals with the promotion of sustainable rural and agricultural development.

7 Source: *Earth Negotiations Bulletin*, vol. 9, no. 550, March 21, 2011 (http://www.iisd.ca/vol09/enb09550e.html).

8 Source: Andersen (2011).

9 Source: *Earth Negotiations Bulletin*, vol. 9, no. 550, March 21, 2011 (http://www.iisd.ca/vol09/enb09550e.html).

10 These consultations were organized by the Fridtjof Nansen Institute, in Norway, and the process included an email-based survey conducted between July and September 2010 and a conference held together with the Institute of Biodiversity Conservation, Ethiopia, in Addis Ababa, on November 23–25, 2010. The consultation process was supported by the Swedish International Biodiversity Programme (SwedBio), the Norwegian Agency for Development Cooperation (NORAD), the Norwegian Ministry of Agriculture and Food, the Development Fund, Norway, and the Spanish Agency for International Development Cooperation (AECID). An informal international consultation on farmers' rights was also held in Lusaka, Zambia, from September 18 to 20, 2007, jointly organized and co-hosted by the Ministry of Agriculture and Food and the Fridtjof Nansen Institute, both of Norway, and the Zambian Agriculture Research Institute.

11 Andersen, R. and Winge, T. (2011) *The 2010 Global Consultations on Farmers' Rights*, The Farmers' Rights Project, Fridtjof Nansen Institute, Lysaker, Norway (http://www.fni.no/doc&pdf/FNI-R0211.pdf; accessed March 15, 2011).
12 It is also called the breeder's privilege.
13 European Seed Association (ESA), *Position on Farmers' Rights*, Brussels, November 28, 2008 (http://www.euroseeds.org/position-papers/biodiversity/ESA_08.0773.pdf; accessed March 25, 2011).
14 In Brazil, for example, authorization is required for access to traditional, local, or creole varieties, owing to the associated traditional knowledge incorporated into their genetic material.
15 Source: *Participatory Plant Breeding* (http://www.ciat.cgiar.org/ourprograms/Climate_Capacity/prga/Documents/PPB%20brief(1).pdf; accessed March 15, 2011).
16 Source: http://www.programa-fpma.org.ni/index.php?option=com_content&view=article&id=54&Itemid=61 (accessed March 25, 2011).
17 Sierra de los Cuchumatanes is the highest nonvolcanic mountain range in Central America.
18 Source: 'Success stories from the realization of the right to save, use, exchange and sell farm-saved seed. India's Protection of Plant Variety and Farmers' Rights Act', The Farmers' Rights Project, Fridtjof Nansen Institute, Lysaker, Norway (http://www.farmersrights.org/bestpractices/success_seed.html; accessed April 10, 2011).
19 Sources: Jigeesh, A.M. 'Anti-Farmer Seeds Bill has the left up in arms against government', *India Today*, April 20, 2010 (http://indiatoday.intoday.in/site/story/'Anti-Farmer'+Seeds+Bill+has+the+Left+up+in+arms+against+govt/1/93642.html; accessed April 10, 2011); Chaudhary, P.K. 'Seeds Bill to hit farmers, agriculture: Nitish', *Times of India*, March 15, 2011 (http://articles.timesofindia.indiatimes.com/2011-03-15/patna/28691032_1_seeds-bill-hybrid-seeds-seed-production; accessed April 10, 2011); and 'India's new Seed Bill', GRAIN, with contributions from Dr. Devinder Sharma (http://www.grain.org/seedling/?id=338; accessed April 10, 2011).
20 The National Biodiversity Act, 2002, based on the CBD, regulates access to and use of genetic resources in India. This act also focuses on benefit-sharing, protection of traditional knowledge, and prior informed consent.
21 According to Article 27.3b of the TRIPS Agreement, member countries may exclude from patentability plants and animals other than microorganisms, and essentially biological processes for the production of plants and animals, other than nonbiological and microbiological processes. However, members must provide for the protection of plant varieties either by patents or by an efficient *sui generis* system, or by any combination thereof. India applied to become a member of UPOV in 2002, but as of July 15, 2011, this request has not been approved, which shows that UPOV is not likely to accept a *sui generis* system different from the one established in its Convention.
22 Sources: www.genecampaign.org; www.navdanya.org; www.sristi.org/cms/en and http://www.mssrf.org and www.sawtee.org (accessed March 30, 2009).
23 According to the Seeds Act of 1966, only varieties notified by the government need to be registered. However, if the seeds bill of 2004 is approved, all seeds for sale will have to be registered.
24 Sources: Madhavan, M.R. and Sanyal, K. (2006) 'Seeds Bill 2004' (http://www.indiatogether.org/2006/jun/law-seeds.htm); and Bala, Ravi, S. (2009) 'The conflict between Seed Bill and PPVFR Act of India' (http://www.sawtee; http://agricoop.nic.in/seeds/seeds_bill.htm; accessed March 15, 2011).

25 A model law is created as a suggested pattern for law-makers in national governments to consider adopting as part of their domestic legislation.

26 Community rights are defined as "those rights held by local communities over their biological resources or parts or derivatives thereof, and over their practices, innovations, knowledge and technologies." Community knowledge or Indigenous knowledge is defined as "the accumulated knowledge that is vital for conservation and sustainable use of biological resources and/or which is of socioeconomic value, and which has been developed over the years in Indigenous/local communities." According to the African Model Law, states must recognize the rights of communities over their biological resources; the right to collectively benefit from the use of their biological resources; their innovations, practices, knowledge, and technologies acquired through generations; the right to collectively benefit from the utilization of their innovations, practices, knowledge, and technologies; their rights to use their innovations, practices, knowledge, and technologies in the conservation and sustainable use of biological diversity; and the exercise of collective rights as legitimate custodians and users of their biological resources. Any access to a biological resource, innovation, practice, knowledge, or technology is subject to the prior informed consent of the community or communities concerned, ensuring that women fully and equally participate in decision-making.

27 The complete text of the African Model Law for the Protection of the Rights of Local Communities, Farmers And Breeders, and for the Regulation of Access to Biological Resources can be consulted at: http://www.opbw.org/nat_imp/model_laws/oau-model-law.pdf (accessed March 18, 2011).

28 For further information about the legal bill proposed by Namíbia, see Dhar (2002).

Bibliography

Adhikari, K. (2008). *Intellectual Property Rights in Agriculture: Legal Mechanisms to Protect Farmers' Rights in Nepal*, Forum for Protection of Public Interest and South Asia Watch on Trade, Economics & Environment, Kathmandu, Nepal.

Andersen, R. (2005) *The History of Farmers' Rights: A Guide to Central Documents and Literature*, The Farmers' Rights Project, Fridtjof Nansen Institute, Lysaker, Norway (http://www.fni.no/doc&pdf/FNI-R0805.pdf; accessed March 10, 2011).

Andersen, R. (2006) *Realising Farmers' Rights under the International Treaty on Plant Genetic Resources for Food and Agriculture*, The Farmers' Rights Project, Fridtjof Nansen Institute, Lysaker, Norway (http://www.fni.no/doc&pdf/FNI-R1106.pdf; accessed March 10, 2011).

Andersen, R. (2007) 'Towards a common understanding of the contents of farmers' rights', in Informal International Consultation on Farmers' Rights, September 18–20, 2007, Lusaka, Zambia, co-hosted by the Ministry of Agriculture and Food and the Fridtjof Nansen Institute, Norway, and the Zambia Agriculture Research Institute of the Ministry of Agriculture, Food and Fisheries, pp. 49–56 (http://www.fni.no/doc&pdf/farmers_rights_lusaka_consultation_final_report.pdf; accessed March 10, 2011).

Andersen, R. (2011) *Farmers' Rights in the ITPGRFA: The Fourth Session of the Governing Body*, The Farmers' Rights Project, Fridtjof Nansen Institute, Lysaker, Norway (http://www.farmersrights.org/about/fr_in_itpgrfa_5.html; accessed March 10, 2011).

Andersen, R. and Winge, T. (2008) *Success Stories from the Realization of Farmers' Rights Related to Plant Genetic Resources for Food and Agriculture*, The Farmers' Rights Project, Fridtjof Nansen Institute, Lysaker, Norway (http://www.fni.no/doc&pdf/FNI-R0408.pdf; accessed March 15, 2011).

Andersen, R. and Winge, T. (2009) *The Plant Treaty and Farmers' Rights: Implementation Issues for South Asia*, South Asia Watch on Trade, Economics & Environment, Kathmandu, Nepal (http://www.farmersrights.org/pdf/SAWTEE_ITPGRFA%20and%20Farmers'%20Rights_2009.pdf; accessed March 10, 2011).

Argumedo, A., Swiderska, K., Pimbert, M., Song, Y. and Pant, R. (2011) 'Implementing farmers' rights under the FAO International Treaty on PGRFA: the need for a broad approach based on biocultural heritage'. IIED, Asociación ANDES (Peru), Centre for Chinese Agricultural Policy and Ecoserve (India). Paper prepared for the Fourth Session of the Governing Body of the International Treaty of PGRFA, Bali, March 14–18, 2011 (http://pubs.iied.org/G03077.html; accessed March 22, 2011).

Bertacchini, E. (2008) 'Coase, Pigou and the potato: whither farmers' rights?', *Ecological Economics*, no. 68, pp. 183–193.

Bjornstad, S.I.B. (2004) 'Breakthrough for "the South"? An Analysis of the Recognition of Farmers' Rights in the International Treaty on Plant Genetic Resources for Food and Agriculture', The Farmers' Rights Project, Fridtjof Nansen Institute, Lysaker, Norway (http://www.fni.no/doc&pdf/FNI-R1304.pdf; accessed March 15, 2011).

Boef, W.S. de and Ogliari, J.B. (2007) 'Seleção de variedades e melhoramento genético participativo', in Boef, W.S. de, Thijssen, M. H., Ogliari, J.B. and Sthapit, B. (orgs.) (2007) *Biodiversidade e Agricultores: Fortalecendo o Manejo Comunitário*, L & PM, Porto Alegre, Brazil, pp. 77–93.

Borowaik, C. (2004) 'Farmers' rights: intellectual property regimes and the struggle over seeds', *Politics & Society*, vol. 32, no. 4, pp. 511–543.

Brahmi, P., Saxena, S. and Dhillon, B.S. (2004) 'The Protection of Plant Varieties and Farmers' Rights Act of India', *Current Science*, vol. 86, no. 3, pp. 392–398.

Brush, S. (org.) (1999) *Genes in the Field: On-Farm Conservation of Crop Diversity*, International Plant Genetic Resources Institute, International Development Research Centre, and Lewis Publishers, Washington, DC.

Ceccarelli, S and Grando, S. (2007) 'Decentralized-participatory plant breeding: an example of demand driven research', *Euphytica*, vol. 155, pp. 349–360.

Ceccarelli, S., Guimarães, E.P. and Weltizien (2009) *Plant Breeding and Farmer Participation*, FAO, Rome.

Christinck, A., Weltzien, E. and Hoffman, V. (eds.) (2005) *Setting Breeding Objectives and Developing Seed Systems with Farmers: A Handbook to Practical Use in Participatory Plant Breeding Projects*, Margraf Publishers, Weikersheim, Germany.

Corrêa, C. (2000) *Options for the Implementation of Farmers' Rights at the National Level*, South Centre, Geneva.

Desclaux, D. (2006) 'Sélection participative: specificités et enjeux', in Gasselin, P. and Clèment O. (coords.). Quelles variétés et semences pour des agricultures paysannes durables?, *Dossiers de L'Environnement de L'INRA*, no. 30, pp. 119–127.

Dhar, B. (2002) *Sui Generis Systems for Plant Variety Protection: Options under TRIPS*, Quaker United Nations Office, Geneva.

Egziabher, T.B.G., Matos, E. and Mwila, G. (2011) 'The African regional group: creating fair play between north and south', in Frison, C., López, F. and Esquinas-Alcázar, J.T. (eds.) *Plant Genetic Resources and Food Security: Stakeholder Perspectives on the International Treaty on Plant Genetic Resources for Food and Agriculture*, Earthscan, London.

Ekpere, J.A. (2000) *The OAU's Model Law: The Protection of the Rights of Local Communities, Farmers and Breeders, and the Regulation of Access to Biological Resources*, an explanatory booklet, Organization of African Unity, Scientific, Technical and Research Commission, Lagos.

Emperaire, L. and Santilli, J. (2006) 'A agrobiodiversidade e os direitos dos agricultores indígenas e tradicionais', in Ricardo, B. and Ricardo, F. (eds.) *Povos Indígenas no Brasil: 2001–2005*, Instituto Socioambiental, São Paulo, pp. 100–103.

Feyissa, R. (2006) *Farmers' Rights in Ethiopia*, The Farmers' Rights Project, Fridtjof Nansen Institute, Lysaker, Norway (http://www.fni.no/doc&pdf/FNI-R0706.pdf; accessed March 10, 2011).

Fowler, C. (1994) *Unnatural Selection: Technology, Politics, and Plant Evolution*, Gordon and Breach Science Publishers, Yverdon, Switzerland.

Fowler, C. and Mooney, P. (1990) *Shattering: Food, Politics, and the Loss of Genetic Diversity*, University of Arizona Press, Tucson.

Gopalakrishna, N.S. (2001) 'An "effective" *sui generis* law to protect plant varieties and farmers' rights in India: a critique', *Journal of World Intellectual Property*, vol. 4, no. 1, pp. 157–172.

Halewood, M. (ed.) (2012) *Farmers' Crop Varieties and Farmers' Rights: Challenges in Taxonomy and Law*, Earthscan, London.

Halewood, M., Deupman, P., Sthapit, B., Vernooy, R. and Ceccarelli, S. (2007) *Participatory Plant Breeding to Promote Farmers' Rights*, Bioversity International, Rome.

Kameri-Mbote, P. (2003) *Community, Farmers' and Breeders' Rights in Africa: Towards a Legal Framework for Sui Generis Legislation*, International Environmental Law Research Centre, Geneva (www.ielrc.org/content/a0302.pdf; accessed March 10, 2011).

Letty, B., Noordin, Q., Magagi, S. and Waters-Bayer, A. (2011) 'Farmers take the lead in research and development', in *2011 State of the World: Innovations that Nourish the Planet*, World Watch Institute, Washington, DC, pp. 51–58.

Louette, D. (1999) 'Traditional management of seed and genetic diversity: what is a landrace?', in Brush, S. (org.) *Genes in the Field: On-farm Conservation of Crop Diversity*, International Plant Genetic Resources Institute, Rome, and International Development Research Centre, Ottawa, pp. 109–142.

Machado, A.T. (2007) 'Manejo dos recursos vegetais em comunidades agrícolas: enfoque sobre segurança alimentar e agrobiodiversidade', in Nass, L. (ed.). *Recursos Genéticos Vegetais*, EMBRAPA Recursos Genéticos e Biotecnologia, Brasília, pp. 719–744.

Machado, A.T. and Machado, C. (2007) 'Melhoramento participativo de cultivos no Brasil', in Boef, W.S. de, Thijssen, M.H., Ogliari, J.B. and Sthapit, B. (orgs.) *Biodiversidade e Agricultores: Fortalecendo o Manejo Comunitário*, L & PM, Porto Alegre, Brazil, pp. 93–102.

Machado, A.T., Santilli, J. and Magalhães, R. (2008) *A Agrobiodiversidade com Enfoque Agroecológico: Implicações Conceituais e Jurídicas*, EMBRAPA Informação Tecnológica, Brasília.

Mooney, P. (1979) *Seeds of the Earth: A Private or Public Resource?* Inter Pares for the Canadian Council for International Cooperation and the International Coalition for Development Action, Ottawa, Canada.

Mooney, P. (1983) 'The law of the seed: another development and plant genetic resources', *Development Dialogue*, Dag Hammarskjöld Foundation, Uppsala, v. 1–2.

Mooney, P. (1996) 'The parts of life: agricultural biodiversity, Indigenous knowledge, and the role of the third system', *Development Dialogue*, Dag Hammarskjöld Foundation, Uppsala.

Mooney, P. (2011) 'The hundred year (or so) seed war: seeds, sovereignty and civil society: a historical perspective on the evolution of "the law of the seed', in Frison, C., López, F. and Esquinas-Alcázar, J.T. (eds.) *Plant Genetic Resources and Food Security: Stakeholder Perspectives on the International Treaty on Plant Genetic Resources for Food and Agriculture*, Earthscan, London.

Morris, M. and Bellon, M. (2004) 'Participatory plant breeding research: opportunities and challenges for the international crop improvement system', *Euphytica*, vol. 136, no. 1, pp. 21–35.

Nagarajan, S., Yaday, S.P. and Singh, A.K. (2008) 'Farmers' variety in the context of the Protection of Plant Varieties and Farmers' Rights Act, 2001', *Current Science*, vol. 94, no. 6, pp. 709–713.

Nnadozie, K., Lettington, R., Bruch, C., Bass, S. and King, S. (2003). *African Perspectives on Genetic Resources: A Handbook on Laws, Policies and Institutions Governing Access and Benefit-sharing*, Environmental Law Institute, Washington, DC.

Pelegrina, W.R. and Salazar, R. (2011) 'Farmers' communities: a reflection on the Treaty from small farmers' perspectives', in Frison, C., López, F. and Esquinas-Alcázar, J.T. (eds.) *Plant Genetic Resources and Food Security: Stakeholder Perspectives on the International Treaty on Plant Genetic Resources for Food and Agriculture*, Earthscan, London.

Ramanna, A. (2006). *Farmers' Rights in India: A Case Study*, The Farmers' Rights Project, Fridtjof Nansen Institute, Lysaker, Norway (http://www.fni.no/doc&pdf/FNI-R0606.pdf; accessed March 10, 2011).

Ramanna, A. and Smale, M. (2004) 'Rights and access to plant genetic resources under India's New Law', *Development Policy Review*, vol. 22, no. 4, pp. 423–442 (http://papers.ssrn.com/sol3/papers.cfm?abstract_id=558802; accessed March 10, 2011).

Sahai, S. (2001) 'Plant Variety Protection and Farmers' Rights Law', *Economic and Political Weekly*, vol. 36, no. 35, pp. 3338–3342.

Sahai, S. (2003) 'India's Plant Variety Protection and Farmers' Rights Act, 2001', *Current Science*, vol. 84, no. 3, pp. 407–412.

Salazar, R., Louwaars, N.P. and Visser, B. (2007) 'Protecting farmers' new varieties: new approaches to rights on collective innovations in plant genetic resources', *World Development*, vol. 35, no. 9, pp. 151–158.

Scurrah, M., Andersen, R. and Winge, T. (2008) *Farmers' Rights in Peru: Farmers' Perspectives*, The Farmers' Rights Project, Fridtjof Nansen Institute, Lysaker, Norway (http://www.fni.no/doc&pdf/FNI-R1608.pdf; accessed March 10, 2011).

Seshia, S. (2002) 'Plant variety protection and farmers' rights: law-making and cultivation of varietal control', *Economic and Political Weekly*, vol. 37, no. 27, pp. 2741–2747.

Smolders, H. (ed.) (2006) *Enhancing Farmers' Role in Crop Development: Framework Information for Participatory Plant Breeding in Farmer Field Schools*, Centre for Genetic Resources, Wageningen, the Netherlands.

Soni, M.K. (2007) 'Enforcing farmers'rights on uncharted territory: the role of IPRS for development in India', Washington College of Law of the American University (http://www.wcl.american.edu/pijip_static/documents/MunmeethKSoni.pdf; accessed March 10, 2011).

Thijssen, M.H., Bishaw, Z., Beshir, A. and Boef, W.S. de (eds.) (2008) *Farmers, Seeds and Varieties: Supporting Informal Seed Supply in Ethiopia*, Wageningen International, Wageningen, the Netherlands.

Toomey, G. (1999) *Farmers as Researchers: The Rise of Participatory Plant Breeding*, International Development Research Center, Ottawa (http://web.idrc.ca/ev_en.php?ID=5559&ID2=DO_TOPIC, accessed March 11, 2011).

United Nations Economic and Social Council Commission on Human Rights (1999) *The Right to Adequate Food and to be Free from Hunger*, updated study on the right to food, submitted by Mr. Asbjørn Eide in accordance with Sub-Commission Decision 1998/106. Sub-Commission on Prevention of Discrimination and Protection of Minorities, Geneva ((http://www.unhchr.ch/Huridocda/Huridoca.nsf/0/ff220c7e04411faa802567c90039c745?Opendocument; accessed March 15, 2011).

Vernooy, R. (2003) *Seeds that Give: Participatory Plant Breeding*, International Development Research Center, Ottawa.

Vogt, G.A., Canci, I. and Canci, A. (2007) 'Uso e manejo de variedades locais de milho em Anchieta (SC)', *Agriculturas: Experiências em Agroecologia*, vol. 4, no. 3, pp. 36–39.

Zerbe, N. (2005) 'Biodiversity, ownership, and Indigenous knowledge: exploring legal frameworks for community, farmers, and intellectual property rights in Africa', *Ecological economics*, vol. 53, no. 4, pp. 493–506.

Zerbe, N. (2007) 'Contesting privatization: NGOs and farmers' rights in the African Model Law', *Global Environmental Politics*, vol. 7, no. 1, pp. 97–119.

9

ANIMAL GENETIC RESOURCES FOR FOOD AND AGRICULTURE

Access and benefit-sharing and livestock keepers' rights

Agrobiodiversity is usually associated with crops and plant varieties, and seldom to animal breeds and strains, and there is very little awareness, among policy-makers, of the importance of conserving farm animal genetic diversity. However, animal genetic resources are just as important as plants for global food security and nutrition, sustainable development, and the livelihoods of millions of people. Approximately 70 percent of the world's rural poor depend on livestock for their livelihoods, and livestock contribute about 30 percent of agricultural gross domestic production in developing countries, this figure projected to increase to 39 percent by 2030 (FAO, 2007a). Animals serve countless uses and functions for humans: food (meat, eggs, milk, cheese etc.), production of fiber, skins, hides, and pelts, agricultural inputs, transport, fuel, scientific experiments, and food for other animals (fish bait, for instance), etc. Livestock represent important sources of protein for many vulnerable groups (children, convalescents, and pregnant and nursing women).

Livestock and their products also fulfill a wide variety of social and cultural functions[1] – they are important elements of many religious rituals, weddings, funerals, and other social gatherings, and contribute to sporting and leisure activities. Animals are also used for the production of traditional meals and for medicinal practices. In developing countries, many farmers rely on animals to provide inputs to crop production (draught power and manure). Where modern financial institutions are inaccessible, keeping animals that can be sold in times of need provides many households with the equivalent of savings and insurance services. In many livestock-keeping societies, exchange of animals also helps to reinforce social relationships and networks that can be drawn upon in times of need. Livestock also provide key agro-ecosystem functions, such as nutrient cycling, seed dispersal and habitat maintenance (FAO, 2007a).

In recent decades, there has been a rapid decline in livestock diversity, especially of small and highly adapted local breeds.[2] Only 14 of the more than 30 domesticated mammalian and bird species provide 90 percent of human food supply from animals, and the five main livestock species, cattle, sheep, goats, pigs, and chickens, provide the majority of food production. According to *The State of the World's Animal Genetic Resources*, the first global assessment of livestock biodiversity, published by FAO in 2007 (FAO, 2007a),[3] a number of threats to animal genetic diversity can be identified, but the most significant is the marginalization of traditional production systems and associated local breeds, driven mainly by the rapid spread of intensive livestock production, often large-scale and utilizing a narrow range of breeds. Global production of meat, milk, and eggs is increasingly based on a limited number of high-output breeds – those that are most profitably utilized in industrial production systems. Public policies tend to promote the development of large-scale animal production systems, and high-output animals, intensively bred to supply uniform products under controlled management conditions. They tend not to value local livestock breeds, which generally perform better than exotic breeds under low-input conditions, climatic stresses, and especially during times of drought. In addition, the local breeds are usually more tolerant to tick-borne diseases and tend to experience lower mortality. Thus, they provide many advantages to small-scale farmers, herders, and pastoralists, who are the guardians of much of the world's livestock biodiversity but receive very little support to play such a role, and are often marginalized by policies. A few community-based conservation and breeding programs have begun to address these issues, but they are still not sufficient and they need to be further developed. Threats to animal genetic resources also include inappropriate development policies and management strategies, disease outbreaks (such as avian influenza), and control programs, as well as various types of disasters and emergencies (droughts, floods, military conflicts, etc.) (FAO, 2007a).

FAO's Global Databank for Animal Genetic Resources for Food and Agriculture[4] contains information on a total of 7,616 livestock breeds, of which 6,536 are local breeds and 1,080 are transboundary breeds. Among the transboundary breeds, 523 are regional transboundary breeds[5] and 557 are international transboundary breeds. A total of 1,491 (20 percent) reported breeds are classified as at risk. Cattle is the species with the highest number of breeds reported as extinct (209). Large numbers of extinct pig, sheep, and horse breeds are also reported. Of even greater concern is that during the last six years 62 breeds have become extinct – amounting to the loss of almost one breed per month. These figures present only a partial picture of genetic erosion, as population data are unavailable for 36 percent of all breeds. In addition, breed inventories, and particularly surveys of population size and structure at breed level, are inadequate in many parts of the world. In Latin America and the Caribbean, for example, 68 percent and 81 percent of mammalian and avian breeds, respectively, are classified as being of unknown risk status. The figures for Africa are 59 percent for mammals and 60 percent for birds (FAO, 2007a).

Some of the key findings of the *State of the World's Animal Genetic Resources* are summarized below:

1 Today's livestock biodiversity is the result of thousands of years of human intervention. Animal husbandry, human-controlled breeding, combined with the effects of natural selection, gave rise to great genetic diversity among the world's livestock populations. Human migration, trade, military conquest, and colonization spread livestock from their original homelands, exposing them to new agroecological zones, new cultures, and new technologies, contributing to the diversification of livestock.
2 The countries and regions of the world are interdependent in the utilization of animal genetic resources. This is clear from evidence of historic gene flows and current patterns of livestock distribution. FAO (2007a) provides some examples: today, the world's most widespread cattle breed, the Holstein–Friesian, is found in at least 128 countries. Among other livestock species, large white pigs are reported in 117 countries, Saanen goats in 81 countries, and Suffolk sheep in 40 countries. International transfer of genetic material now occurs on a very large scale, both within the developed world (north–north exchanges) and from developed to developing countries (north–south). There is also some exchange between developing (south–south) countries, notably the transfer of South Asian cattle to Latin America. However, for the last century, relatively little movement of livestock germplasm has occurred from the south to the north (Hoffman, 2010). Most gene flows are focused on a limited number of breeds and have the potential to narrow the genetic resource base of the world's animal production.
3 Decision-making in the field of animal genetic resources management is often characterized by a lack of attention to multiple functions and roles played by different animal breeds in different production conditions. The value of local multipurpose animal breeds tends to be underestimated, and only some elements of livestock's overall contribution to human well-being are taken into consideration. Valuable characteristics of local livestock breeds, such as resistance or tolerance to disease, tend to be undervalued. The Global Databank for Animal Genetic Resources for Food and Agriculture contains many reports of breeds that are resistant to particular diseases, but many have not been subject to scientific investigation to explore their potential. When breeds become extinct before their disease resistance qualities have been identified, genetic resources that could greatly contribute to improving animal health and productivity are no longer accessible.

4 The growth of large-scale industrialized production in many parts of the developing world is the most economically significant trend in the global livestock sector. The industrialization process involves intensification, increase in scale, and geographical and social concentration of production. The focus is on maximizing the output of a specific product. A narrow range of breeds is used, and within-breed genetic diversity may also be reduced. Geographical concentration and the separation of livestock and crop production present a number of environmental problems, particularly related to the management of livestock wastes. Small-scale landless livestock production can be found both in and around cities and in rural areas. This type of production is less globally significant than industrial systems in terms of meeting the growth in demand for animal products. However, its important contribution to household-level food security and livelihoods needs to be taken into consideration.

5 Despite the growth of large-scale industrialized livestock production, the world's livestock production systems remain very diverse. This is particularly true for the smallholder and pastoral systems of the developing world. Locally adapted livestock remain important to the livelihoods of a large proportion of the world's poor. It is vital that policies affecting the livestock sector consider the needs of these livestock keepers and of the animal genetic resources on which they depend.

6 Small-scale farmers and pastoralists are the custodians of much of the world's animal genetic diversity. Ensuring that they are not denied the opportunity to continue performing this role will often require giving attention to policies and legal frameworks, such as those that affect access to land and water resources.

7 Genetically diverse livestock populations are an important resource to be drawn upon as production systems change and develop. Newly emerging market trends and policy objectives are continually placing new demands on the livestock sector. The prospect of future challenges such as adapting to global climate change underlines the importance of retaining a diverse portfolio of livestock breeds.

8 Institutional and technical capacity needs to be reinforced in developing countries. *In vivo* and *in vitro* conservation programs are lacking in many countries where there are significant threats to valuable animal genetic resources. A number of approaches to conservation are available, including a range of *in vivo* methods (zoos, farms, protected areas, and payments or other support measures for livestock keepers who maintain animals within their production environments), as well as *in vitro* conservation of genetic material in liquid nitrogen (semen, embryos).[6]

9 In the future, animal genetic resources from any part of the world may prove vital to breeders and livestock keepers elsewhere. There is a need for the international community to accept responsibility for the management of these shared resources. Support for developing countries and

countries with economies in transition to characterize, conserve, and utilize their livestock breeds is necessary. Wide access to animal genetic resources, for farmers, herders, breeders, and researchers is essential to sustainable use and development. Equitable frameworks for access, and for sharing the benefits derived from animal genetic resources, need to be put in place at both national and international levels.

One of the conclusions of the *State of the World's Animal Genetic Resources* is that inadequate regulatory frameworks hamper the effective management of animal genetic resources, and that legislation specifically aimed at promoting and regulating animal breeds conservation must be developed. Legal aspects of the exchange, use, and conservation of animal genetic resources have been discussed by several specialists (Ingrassia *et al.*, 2005; Hiemstra *et al.*, 2006; Anderson and Centonze, 2007; Tvedt *et al.*, 2007; Ivanković, 2008; Biber-Klemm and Temmerman, 2010). They point out the need to consider the specific nature of animal genetic resources in international and national legal frameworks because of the significant differences between animal and plant genetic resources, which include (among others):

- The generation interval in plants is low, usually less than one year, and in animals is high, ranging from one to eight years, depending on the species.
- The number of offspring is high in plants and small in most animal species (with the exception of pigs and chickens).[7]
- The economic value of an individual or germplasm is low in plants, and moderate to high in animals. In addition, the main resource for genetic change in animal genetic resources is genetic variation within the animal populations.

These authors also point out relevant differences between animal and plant genetic resources, from a legal point of view:

1 According to the WTO's TRIPS Agreement, Article 27.3(b), members may exclude from patentability "plants and animals other than micro-organisms." However, they must provide for the protection of plant varieties either by patents or by an efficient *sui generis* system, or by any combination thereof. Therefore, members are obliged to establish IP rights (patents or a *sui generis* system) on plant varieties, but NOT on animal breeds. The TRIPS Agreement leaves the choice over the patentability of animal breeds to national legislation, but the vast majority of countries do not allow patenting of animal breeds, nor of essentially biological processes for the production of animals.[8] There is a specific

(or *sui generis*) legal regime to protect IP rights over plant varieties: the UPOV Convention and national plant variety protection acts grant exclusive ownership rights over plant varieties that fulfill the criteria of novelty, distinctiveness, uniformity, and stability. They also ensure the plant breeders' exemption (also called privilege), which allows plant breeders to freely use protected varieties as a source of variation for the purpose of breeding other plant varieties. The farmer's right to use farm-saved seed of protected varieties is also protected (the level of protection depends on the UPOV Acts of 1978 or 1991). There is no specific regime to protect animal breeder's innovations, nor exemptions for livestock breeders or keepers.

2 The ITPGRFA establishes a multilateral system of ABS for crops included in its Annex I, as well as farmers' rights. There is no international agreement dealing specifically with ABS in relation to animal genetic resources, and livestock keepers' rights are not formally recognized in any legally binding international instrument. There is no multilateral system regulating international exchange of animal genetic resources, such as the one that exists for plants. Under the ITPGRFA multilateral system, the SMTA is used for transfers and exchanges of plant genetic resources, but there is no international SMTA designed for the transfer of animal genetic resources. Exchange and trade of animal genetic resources occur mainly through private contracts (bilateral agreements between sellers and buyers), and animals are usually considered to be in private ownership, and seldom in the public domain. In communal systems, sharing breeding animals is regulated by local and communal rules. There is no international network *of ex situ* collections of animal genetic resources, such as the CGIAR (currently comprising 15 international agricultural research centers, 11 of which have germplasm banks), and only a few public *ex situ* collections of animal germplasm and breeding programs. The international legal framework covering animal health-related restrictions on trade is the WTO's Agreement on Sanitary and Phytosanitary Measure, under which standards are set by the World Organization for Animal Health.

3 Access to plant genetic resources has been regulated by international agreements since 1983, when the International Undertaking on Plant Genetic Resources was approved (and later replaced by the ITPGRFA). The first internationally agreed instrument for the conservation and use of animal genetic resources is the Global Plan of Action for Animal Genetic Resources (GPA), adopted during the first International Technical Conference on Animal Genetic Resources for Food and Agriculture, held in Interlaken (Switzerland), from September 3 to 7, 2007. On the same occasion, the Interlaken Declaration on Animal Genetic Resources was also approved. Such instruments are very recent, and only a few

countries have specific regulations on animal genetic resources and policies aimed at their conservation and sustainable use.

4 The CDB dedicates very little attention to domesticated biodiversity, and even less attention to farm animal genetic resources, and most national ABS laws focus mainly on (wild) plant genetic resources. Specific laws on animal genetic resources are very rare. However, the CBD establishes sovereign rights over "genetic resources," which are defined as "genetic material of actual or potential value." "Genetic material" is defined as "any material of plant, animal, microbial or other origin containing functional units of heredity." This means that animal genetic resources are included in the ABS regime established by the CBD, and that the CBD's regulations on conservation and sustainable use of genetic resources also apply to animal genetic resources. However, it is important to keep in mind that the CBD does not define any ownership or proprietary rights of genetic resources (whether plant or animal). Sovereign rights and property rights are distinct concepts that should not be confused. In the exercise of their sovereign rights, countries may decide that certain natural resources (such as genetic resources) are public property (or state property), but this is not necessarily the case. Genetic resources are public interest resources, and they have social, cultural, and economic value for the whole of society, not just for the owners of the animals or plants that carry them. Therefore, there is no conflict between the CBD and national laws that establish that farm animals or their genetic resources are private property. Some authors understand that there is a conflict between the CBD's ABS regime and national laws establishing private ownership of animals or their genetic resources, but this is not necessarily the case. Ivanković (2008), for instance, believes that private ownership of animal genetic resources can be a challenge for fair and equitable benefit-sharing (in accordance with the CBD), as current practice is that only the owner of the animal may decide to sell the animal and with it the animal genetic resources involved, and under which conditions. This would mean that access to animal genetic resources is governed only by, and might be hampered by, the private ownership of animals and, as a consequence, free/facilitated access to animal genetic resources would require that the animals concerned be publicly owned.

5 However, the use and exchange of genetic resources (of animals or plants), whether privately or publicly owned, must be subject to the public interest. The sharp duality between private and public resources has been increasingly overcome by environmental law, as many environmental resources (such as water, land, forest resources, etc.) have their utilization restricted by environmental regulations, issued in the public interest, independently of being privately or publicly owned.

State sovereignty over genetic resources means that public interest rules may be established on the access, use, and transfer of such resources.

6 It must also be understood that genetic resources cannot be confused with the animals (or plants) that contain them. National laws can establish different ownership regimes for animals (or plants), considered in themselves (as biological materials), and their genetic resources. Private ownership of an animal does not necessarily mean private ownership of its genetic resources. It is generally assumed that the owner of an animal has the right to decide whether to use its genetic material, but it does not have to necessarily be that way. National laws may establish that certain public interest uses of animal genetic resources are not solely at the discretion of the animal's owner, for instance in the case of rare breeds, or of breeds threatened with extinction, where there is a public interest in conserving genetic material *in vitro*. In other cases, public breeding programs may need to access certain animal genetic resources (even if they are in the private domain), and such access must be ensured by law, and cannot be prevented by private owners of animals. Unlike plant genetic resources, only a very small number of local livestock breeds have been improved by breeding programs.

7 It is also important that a global common pool of animal genetic resources be established, through networks of public *ex situ* collections (currently, there are very few, but they could be created) and public support for on-farm conservation by livestock keepers. The bilateral regime of ABS established by the CBD for wild species does not take into consideration the specific nature of domesticated species, whether plant or animal, but other ABS regimes may be established. Collective benefit-sharing mechanisms, benefiting livestock keepers and breeders, as well as pastoral communities, could be established, as well as special funds for the conservation and sustainable use of livestock diversity. Collective rights must be ensured to local communities, and livestock diversity is part of their cultural heritage, and must be legally treated as such. Participatory animal breeding, as well as community-based management of animal genetic resources, especially in developing countries, must be promoted and supported, as well as public breeding programs aimed at attending the needs of small-scale livestock keepers and pastoralists (private breeding programs tend to focus mainly on high-output breeds, and the needs of small-scale livestock production systems are left unattended by public policies). These are just some examples of collective benefit-sharing mechanisms that could be promoted by international agreements. The recently approved Nagoya Protocol, which is the instrument for the implementation of the benefit-sharing provisions of the CBD, explicitly allows the possibility that other specialized international ABS instruments be developed. So far, the ITPGRFA is

> the only specialized ABS international regime, but an international ABS regime for animal genetic resources could also be established, tailored to the specific needs of the livestock sector. Some authors (Hiemstra *et al.*, 2006) have suggested the development of voluntary instruments to strengthen national policies and implementation of action at national levels, which would facilitate reaching international consensus on the conservation and sustainable use of animal genetic resources, as well as the fair and equitable sharing of benefits derived from their use, for present and future generations. Such an instrument (voluntary or binding) would be necessary also to ensure effective implementation of many objectives of the Global Plan of Action for Animal Genetic Resources, such as conservation, sustainable use, and development of animal genetic resources, as well as the development of adequate policies, institutions and capacities. This will become even more important as exchanges of animal genetic resources tend to increase as a result of climate change, globalization, and/or biotechnology (Hiemstra *et al.*, 2006). However, the debate on the integration of animal genetic resources into the ABS system is still in its initial stages (Biber-Klemm and Temmerman, 2010).

In recent years, the concept of livestock keepers' rights has been advocated by NGOs, livestock keepers, pastoralist[9] associations, and scientists who support community-based conservation of local breeds. They formed LIFE (Local Livestock for Empowerment of Rural People) Network. Köhler-Rollefson *et al.* (2010) explain that the term livestock keepers' rights was first coined and promoted by civil society organizations during the World Food Summit held in 2002. The expression was an allusion to farmers' rights, which had just been legally enshrined in the ITPGRFA. Livestock keepers' rights are based on the rationale that many breeds in developing countries disintegrate owing to the loss of the traditional rights of livestock keepers to sustain their livestock on common property resources, as well as policies that are adverse to small-scale livestock keepers. They constitute a set of principles aimed at supporting and encouraging livestock keepers to continue making a living from their breeds and thereby achieve the combined effect of conserving diversity and improving rural livelihood opportunities. They are also seen as a potential tool for protecting the rights of livestock keepers at a time when scientists and industries are making increasing use of IP rights to protect their advances in breeding and associated technologies (Tvedt *et al.*, 2007). According to Köhler-Rollefson *et al.* (2010), a series of consultations and workshops with representatives of livestock keeping communities were organized by the LIFE Network from 2003 to 2007, and resulted in a "Declaration of Livestock Keepers' Rights," which contains three principles and five rights:

Principles

1 Livestock keepers are creators of breeds and custodians of animal genetic resources for food and agriculture.
2 Livestock keepers and the sustainable use of traditional breeds are dependent on the conservation of their respective ecosystems.
3 Traditional breeds represent collective property, products of Indigenous knowledge and cultural expression of livestock keepers.

Rights

Livestock keepers have the rights to:

1 Make breeding decisions and breed the breeds they maintain.
2 Participate in policy formulation and implementation processes on animal genetic resources for food and agriculture.
3 Appropriate training and capacity-building and equal access to relevant services enabling and supporting them to raise livestock and to better process and market their products.
4 Participate in the identification of research needs and research design with respect to their genetic resources, as is mandated by the principle of prior informed consent.
5 Effectively access information on issues related to their local breeds and livestock diversity.

The principles and rights contained in the Declaration of Livestock Keepers' Rights are still not entrenched in legally binding international and national instruments. Livestock keepers' rights are still not as explicitly recognized as farmers' rights (in the ITPGFA), but some international instruments support, directly or indirectly, livestock keepers' rights. The Interlaken Declaration on Animal Genetic Resources, approved during the first International Technical Conference on Animal Genetic Resources for Food and Agriculture, held in Interlaken (Switzerland), in 2007, recognizes the following:

- The genetic resources of animal species most critical to food security, sustainable livelihoods, and human well-being are the result of both natural selection and directed selection by small-holders, farmers, pastoralists, and breeders, throughout the world, over generations.

- Maintaining the diversity of animal genetic resources for food and agriculture is essential to enable farmers, pastoralists, and animal breeders to meet current and future production challenges resulting from changes in the environment, including climate change; to enhance resistance to disease and parasites; and to respond to changes in consumer demand for animal products.
- Local and Indigenous communities and farmers, pastoralists, and animal keepers of all regions of the world have made, and will continue to make, an enormous contribution to the sustainable use, development, and conservation of animal genetic resources for food and agriculture.
- All persons engaged in animal husbandry and who have molded animal genetic resources to meet societal needs have made an enormous historic and relevant contribution. It is their ownership and management of animal genetic resources that has enabled them to make these important contributions and it is this ownership and management that should be ensured for future societal benefits. Furthermore, such persons "should participate in the fair and equitable sharing benefits arising from the utilization of animal genetic resources for food and agriculture."
- It is desirable, subject to national legislation, to respect, preserve, and maintain traditional knowledge relevant to animal breeding and production as a contribution to sustainable livelihoods, and there is a need for the participation of all stakeholders in making decisions, at the national level, on matters related to the sustainable use, development, and conservation of animal genetic resources.

The Interlaken Declaration affirms countries' commitment to the implementation of the GPA, and to ensuring that the world's livestock biodiversity is utilized to promote global food security and remains available for future generations. The GPA also recognizes that:

- All animal genetic resources for food and agriculture are the result of human intervention: they have been consciously selected and improved by pastoralists and farmers since the origins of agriculture, and have co-evolved with economies, cultures, knowledge systems, and societies. Unlike most wild biodiversity, domestic animal resources require continuous active human management, sensitive to their unique nature.
- Pastoralists, farmers, and breeders, individually and collectively, and Indigenous and local communities, play a crucial role in *in situ* conservation and development of animal genetic resources. It is important to better understand and support their role in a context of rapid economic and social change, so that they can play an effective function in *in situ* management, and share fairly and equitably in the benefits arising from the utilization of these resources. A number of actors and stakeholders can assist livestock keepers and their communities in playing this role:

researchers, extension agencies, the private sector, NGOs, and local cooperatives.

- A main objective is "to meet the needs of pastoralists and farmers, individually and collectively, within the framework of national law, to have nondiscriminatory access to genetic material, information, technologies, financial resources, research results, marketing systems and natural resources, so that they may continue to manage and improve animal genetic resources, and benefit from economic development."
- Strategic priorities include "support Indigenous and local production systems and associated knowledge systems, of importance to the maintenance and sustainable use of animal genetic resources." Specific actions (aimed at targeting such strategic priority) are:
  - assess the value and importance of Indigenous and local production systems, and identify trends and drivers of change that may affect the genetic base, and the resilience and sustainability of the production systems;
  - support Indigenous and local livestock systems of importance to animal genetic resources, including through the removal of factors contributing to genetic erosion, (support may include the provision of veterinary and extension services, delivery of micro-credit for women in rural areas, appropriate access to natural resources and to the market, resolving land tenure issues, the recognition of cultural practices and values, and adding value to their specialist products);
  - promote and enable relevant exchange, interaction, and dialogue among Indigenous and rural communities and scientists and government officials and other stakeholders, in order to integrate traditional knowledge with scientific approaches;
  - promote the development of niche markets for products derived from Indigenous and local species and breeds and strengthen processes to add value to their primary products.

Notes

1 Animal uses vary in different cultures. In countries such as China, Vietnam, and Korea, for example, dog meat is used for human food (it is even believed to enhance sexual performance). In China, there are five species of turtles that are raised in farms and sold in regional markets for food and production of medications. In Singapore, fried black scorpion is a delicacy (the high cooking temperatures neutralize its poison). Kangaroo meat is served even in pizza parlors in Australia, and fried ants (especially içá and saúva) are part of the menu of some Amazonian Indigenous communities. In Argentina, barbecued bulls' testicles are widely eaten; in Bolivia llama meat is a typical dish; and roasted guinea pigs are often found in restaurants in Bolivia and Colombia.

2 The definition of a breed is complex and varies a lot. According to Marguerat-König (2010), breed is a descriptive term: a group of animals of similar genetic background showing similar phenotypes. A breed is identifiable as a subset of a

species, the members of which are identifiable by shared phenotypic (and hence genetically encoded) characteristics. Neighboring populations, massive mutual gene flow, and close relationship make it difficult, from a genetic point of view, to distinguish between breeds. FAO gives the following definition of a breed: a group of animals for which geographical and/or cultural separation from phenotypically similar groups has led to acceptance of its separate identity. In developed countries breeds are characterized by "clear" definitions, physical characteristics, and strict definition of purity of pedigree regulated by a breeding society, backed by law. In developing countries, breeds are defined by local traditions, by identifying physical characteristics, by a geographical location, or by ethnic groups.

3 *The State of the World's Animal Genetic Resources for Food and Agriculture* draws on 169 country reports, contributions from a number of international organizations, 12 specially commissioned thematic studies, and expert knowledge to provide the first global assessment of these resources.

4 FAO's Global Databank for Animal Genetic Resources for Food and Agriculture is the most comprehensive global information source for livestock genetic diversity.

5 Under the new system of breed classification developed for *The State of the World's Animal Genetic Resources for Food and Agriculture*, the primary distinction is between breeds that occur in only one country, which are referred to as "local" breeds, and those that occur in more than one country, which are referred to as "transboundary" breeds. Within the transboundary breed category, a further distinction is drawn between "regional" transboundary breeds – those that occur in more than one country within a single region – and "international" transboundary breeds – those that occur in more than one region. There are some regional differences in terms of the relative importance of the different breed categories. In most regions – Africa, Asia, Europe and the Caucasus, Latin America and the Caribbean, and the Near and Middle East – local breeds make up more than two-thirds of all breeds. Conversely, international transboundary avian and mammalian breeds dominate in the Southwest Pacific and North America. Regional transboundary mammalian breeds are relatively numerous in Europe, the Caucasus and Africa, and to a lesser extent in Asia, whereas it is only in Europe and the Caucasus that there are many regional transboundary avian breeds. The Caucasus is a geopolitical region at the border of Europe and Asia, comprising today the states of Georgia, Armenia and Azerbaijan.

6 *Ex situ–in vitro* conservation involves maintenance of endangered animal genetic resources outside their traditional environment (cryoconservation of gametes, embryos, or somatic cells that have the potential to reconstitute live animals). *Ex situ–in vivo* conservation involves the maintenance of living animals outside the area where they evolved or are normally found, for example research stations or zoos. *In situ* or on-farm conservation requires continued use of a breed by livestock keepers in the agroecosystem in which the breed evolved or is now normally found. This includes both farms and pastoral production systems. Source: Hiemstra *et al.* (2006).

7 Schiere (2007) points out, however, that most people compare cows with maize. In fact, the generation intervals in maize are much shorter than in cows. However, the same does not happen if one compares rabbits and fruit trees.

8 Patents have entered the field of animal genetic resources only recently, with the advent of genetic engineering. Transgenic animals exist for medicinal purposes, but are still scarce for food and agriculture. At present, the main application of biotechnology in animal genetic resources is in selection processes comprising

certain biotechnological steps such as marker selection. The markets for poultry and pig are often cited as examples of strong concentration without the use of IP rights. This is the opposite situation to what happens with plants and seeds, where IP rights are essential tools for monopolistic positions of multinationals such as Monsanto. Secrecy and hybridization have played a large role in animal genetic resources for food and agriculture (Biber-Klemm and Temmerman, 2010). To be patented, domestic legislation must allow it and animals must meet the criteria of patentability: novelty, inventiveness, and industrial application. Patents on animal breeds would change the common rule that ownership of animals include the right to the subsequent generations. Patent holders would have to license the reproduction of patent-protected animals, and this would affect access to animal genetic resources, including for breeding purposes. Ownership over progeny would be detached from the female animal and would have to be negotiated with the patent holder, as long as patented characteristics are expressed (Biber-Klemm and Temmerman, 2010).

9 According to the League for Pastoral Peoples and Endogenous Livestock Development, pastoralists are people who depend for their living primarily on livestock. They inhabit those parts of the world where the potential for crop cultivation is limited owing to lack of rainfall, steep terrain, or extreme temperatures. In order to optimally exploit the meager and seasonally variable resources of their environment and to provide food and water for their animals, many pastoralists are nomadic or seminomadic. The type of livestock pastoralists keep varies according to area, and includes sheep, goats, cattle, and camels, but also yaks and horses in Central Asia, buffalo in South Asia, llamas and alpacas in South America, and reindeer in the Palearctic region. An important characteristic of pastoralists is their close relationship with their animals. The identity of pastoralists is based on the close association with their livestock that forms a key component of their social and ritual life. By keeping animals under conditions that are close to the wild, but giving them the benefit of protection and health care, pastoralists represent a cultural counterpoint to industrialized animal production in the West. There is no reliable information available on the number of pastoralists worldwide. According to one estimate, there are around 17.3 million pastoralists in Africa, 3.4 million in the Middle East and South Asia, and no more than 2 million in Central Asia. It is widely recognized by ecologists that pastoralism represents a sustainable method of utilizing certain types of ecosystems, such as deserts, steppes, and certain mountain areas. In fact, continued utilization of the world's arid lands very much depends on viable pastoral systems. Nevertheless, pastoralists have come under pressure worldwide for a variety of reasons that include population growth, environmental degradation, and unsound development and trade policies. In particular, encroachment of agriculture on their grazing territories and the privatization of former communally owned land is undermining their existence. Source: www.pastoralpeoples.org/pastoralists.htm (accessed April 10, 2011).

Bibliography

Anderson, S. and Centonze, R. (2007) 'Property rights and the management of animal genetic resources', *World Development*, vol. 35, no. 9, pp. 1529–1541.

Biber-Klemm, S. and Temmerman, M. (2010) *IPR regimes and Animal Genetic Resources: Situation and Possible Impacts*, presentation at the International Technical Expert Workshop Exploring the Need for Specific Measures for Access and Benefit-sharing of Animal Genetic Resources for Food and Agriculture, December 8–10, 2010, Center for Genetic Resources, Wageningen, the Netherlands.

Biber-Klemm, S. and Temmerman, M. (orgs.) (2010) *Rights to Animal Genetic Resources for Food and Agriculture: Notes from an Interdisciplinary Workshop*, World Trade Institute, Berne, Switzerland, http://www.nccr-trade.org/ip-9-past/rights-toanimal-genetic-resources.html; accessed April 10, 2011.

CGIAR System-wide Genetic Resources Programme (SGRP) and Bioversity International (2006) *Options and Strategies for the Conservation of Farm Animal Genetic Resources*, report of an International Workshop and Presented Papers (November 7–10, 2005, Montpellier, France) (CD-ROM) (http://www.sgrp.cgiar.org/?q=node/568; accessed April 10, 2011).

Correa, C. (2010) *Intellectual Property and Animal Genetic Resources: Developing Countries' Perspectives*, presentation at the International Technical Expert Workshop Exploring the Need for Specific Measures for Access and Benefit-sharing of Animal Genetic Resources for Food and Agriculture, December 8–10, 2010, Center for Genetic Resources, Wageningen, the Netherlands.

FAO (2007a) *The State of the World's Animal Genetic Resources for Food and Agriculture*, FAO, Rome (ftp://ftp.fao.org/docrep/fao/010/a1260e/a1260e00.pdf; accessed April 10, 2011).

FAO (2007b) *Interlaken Declaration on Animal Genetic Resources for Food and Agriculture*, FAO, Rome.

FAO (2007c) *Global Plan of Action for Animal Genetic Resources for Food and Agriculture*, FAO, Rome.

FAO (2007d) *People and Animals. Traditional Livestock Keepers: Guardians of Domestic Animal Diversity*, FAO, Rome. http://www.fao.org/docrep/010/a1057e/a1057e00.htm; accessed April 10, 2011).

FAO (2009a) Commission on Genetic Resources for Food And Agriculture. *The Use and Exchange of Animal Genetic Resources for Food and Agriculture* (http://www.lgi.lt/gencentras/teisine_info/FAOnaudojimas_kaita.PDF; accessed April 10, 2011).

FAO (2009b) *Livestock Keepers – Guardians of Biodiversity*, Animal Production and Health Paper no. 167, FAO, Rome.

FAO (2010) Commission on Genetic Resources For Food And Agriculture. Intergovernmental Technical Working Group on Animal Genetic Resources for Food and Agriculture, *Status of Trends of Animal Genetic Resources – 2011* (http://www.fao.org/docrep/meeting/021/am131e.pdf; accessed April 10, 2011).

GTZ – People and Biodiversity. *Conserving Local Livestock Breeds: Political Strategies and Legal Regulations*. Eschborn, Germany (http://www.conservation-development.net/Projekte/Nachhaltigkeit/CD1/LaenderDesSuedens/Themenblaetter/PDF/AgrobiodivConservingLocalLivestockBreeds.pdf; accessed April 10, 2011).

Hiemstra, S.J., Drucker, A.G., Tvedt, M.W., Louwaars, N., Oldenbroek, J.K., Awgichew, K., Abegaz Kebede, S., Bhat, P.N. and Da Silva Mariante, A. (2006) *Exchange, Use and Conservation of Animal Genetic Resources: Policy and Regulatory Options*, Center for Genetic Resources, Wageningen, the Netherlands.

Hoffmann, I. (2010) *International Flows of Animal Genetic Resources – Historical Perspective, Current Status and Future Expectations*, presentation at the International Technical Expert Workshop, Exploring the Need for Specific Measures for Access and Benefit-Sharing of Animal Genetic Resources for

Food and Agriculture, December 8–10, 2010, Center for Genetic Resources, Wageningen, the Netherlands.

Ingrassia, A., Manzella, D. and Martyniuk, E. (2005) *The Legal Framework for the Management of Animal Genetic Resources*, FAO, Rome.

Ivanković, M. (2008) *Analysis of Applicability of Access and Benefit-Sharing Principles on Animal Genetic Resources*, Center for Genetic Resources, Wageningen, the Netherlands (http://documents.plant.wur.nl/cgn/literature/reports/reportIvankovic_ABS_AnGR_final2008.pdf; accessed April 10, 2011).

Köhler-Rollefson, I.U (2004) *Farm Animal Genetic Resources: Safeguarding National Assets for Food Security and Trade.* A summary of workshops on animal genetic resources held in the Southern African Development Community (SADC), Federal Ministry for Economic Cooperation and Development of Germany, GTZ – People and Biodiversity, FAO and Technical Centre for Agricultural and Rural Cooperation (CTA), Eschborn, Germany.

Köhler-Rollefson, I.U. (2010) 'Livestock keepers'rights: a rights-based approach to invoking justice for pastoralists and biodiversity conserving livestock keepers', *Policy Matters*, vol. 17, pp. 113–115 (www.pastoralpeoples.org/docs/policymatters.pdf; accessed April 10, 2011).

Köhler-Rollefson, I.U., Mathias, E., Singh, H., Vivekanandan, P. and Wanyama, J. (2010) 'Livestock keepers' rights: the state of discussion', *Animal Genetic Resources*, vol. 47, pp. 119–123.

League for Pastoral Peoples and Endogenous Livestock Development (LPP) and LIFE Network (Local Livestock for Empowerment of Rural People) (2010) *Biocultural Community Protocols for Livestock Keepers*, Lokhit Pashu-Palak Sansthan (LPPS), Sadri, Rajasthan, India (www.pastoralpeoples.org/docs/BCP_for_livestock_keepers_web.pdf; accessed April 10, 2011).

LIFE Network (Local Livestock for Empowerment of Rural People) (2010) *Declaration on Livestock Keepers' Rights* (www.pastoralpeoples.org//Declaration_on_LKRs_with_ initial%20signatories_6.pdf; accessed April 10, 2011).

LIFE Network (Local Livestock for Empowerment of Rural People) (2010) *Supporting Livelihoods and Local Livestock Breeds: Guidelines for Putting Livestock Keepers' Rights into Practice* (www.pastoralpeoples.org/docs/LKR_Guidelines.pdf; accessed April 10, 2011).

Marguerat-König, C. (2010) 'Traditional animal breeding and property rights', in Biber-Klemm, S. and Temmerman, M. (orgs.) (2010) *Rights to Animal Genetic Resources for Food and Agriculture: Notes from an Interdisciplinary Workshop*, World Trade Institute, Berne (http://www.nccr-trade.org/ip-9-past/rights-toanimal-genetic-resources.html; accessed April 10, 2011).

Mariante, A. S. (2006) *Animais do Descobrimento: Raças Domésticas da História do Brasil; Animals of the Discovery: Domestic Breeds in the History of Brazil* (bilingual edition, in Portuguese and English), EMBRAPA Recursos Genéticos e Biotecnologia, Brasília, Brazil.

Mathias, E. (ed.) (2010) *Adding Value to Livestock Diversity: Marketing to Promote Local Breeds and Improve Livelihoods*, FAO Animal Production and Health Paper no. 168 (http://www.fao.org/docrep/012/i1283e/i1283e00.pdf; accessed April 10, 2011).

Mathias, E., Mundy, P. and Köhler-Rollefson, I.U (2010) 'Marketing products from local livestock breeds: an analysis of eight cases', *Animal Genetic Resources*, vol. 47, pp. 59–71

Pilling, D. (2010) 'Threats to animal genetic resources for food and agriculture – approaches to recording, description, classification and analysis, *Animal Genetic Resources*, vol. 47, pp. 11–22 (http://www.fao.org/docrep/013/i1823t/i1823t03.pdf; accessed April 10, 2011).

Schiere, H. (2007) 'Perda da diversidade de espécies e de raças de animais domésticos: um tema quase esquecido', in Boef, W.S. de, Thijssen, M.H., Ogliari, J.B. and Sthapit, B. (orgs.) (2007) *Biodiversidade e Agricultores: Fortalecendo o Manejo Comunitário*, L & PM, Porto Alegre, Brazil, pp. 53–59.

Steane, D. (1999) 'Biodiversity in domesticated animals', in Wood, D. and Lenné, J. M (eds.). *Agrobiodiversity: Characterization, Utilization and Management*, Cabi Publishing, Reading, UK, pp. 59–85.

Tvedt, M.W., Hiemstra, S.J., Drucker, A.G., Louwaars, N. and Oldenbroek, K. (2007) *Legal Aspects of Exchange, Use and Conservation of Farm Animal Genetic Resources*, Fridtjof Nansen Institute, Lysaker, Norway (www.fni.no/doc&pdf/fni-r0107.pdf; accessed April 10, 2011).

Tvedt, M.W. (2007) 'Patent protection in the field of animal breeding', *Acta Agriculturae Scandinavica, Section A – Animal Sciences*, vol. 57, no. 3, pp. 105–120 (http://www.fni.no/doc&pdf/MWT-AASSA-2007.PDF; accessed April 10, 2011).

10

THE OPEN SOURCE SOFTWARE MOVEMENT, THE COMMONS MOVEMENT AND SEEDS

What they have in common – biological open source and protected commons

The open source software movement started in the late 1970s and early 1980s, when computer programmers (also called "hackers"[1]) began reacting to restrictions imposed by copyrights[2] over free sharing of software source codes as a result of commercial and competitive interests. Computer programmers saw proprietary restrictions on access to software source codes (imposed by private companies) as barriers to their creativity and the possibility of cooperative work. They also believed that the scope for innovation would be far greater if software development were decentralized and source codes open and freely accessible. The free flow of information was an important value of the "hacker" culture, essential to new software development.

In order to modify a computer program, or to use parts of its source code in another computer program, access to the source code is necessary. Whenever a copyrighted computer program is bought, the source code is kept secret, and its owner has access only to an executable version, which prevents (or at least makes it very difficult) for him or her to copy or modify the computer program. Hence, software source codes are usually protected in two different ways: (1) through the software itself, which contains mechanisms that prevent access to its source code or (2) through copyright, which grants the holders exclusive rights to reproduce, copy, distribute, and modify computer programs. With open source software, the exact opposite takes place: the developer of a software program permits any user to access the source code and use, modify, and create derivative works based on it. The user, however, cannot prevent the free distribution of the software program, which would be considered a violation of the original developer's rights, as explains Ronaldo Lemos (2005, p. 72).

One of the leaders of the free software movement was Richard Stallman, from the Massachusetts Institute of Technology's Artificial Intelligence

Laboratory, a focal point for the "hacker" community through the 1960s and 1970s (Hope, 2008). In 1984, Stallman started to work on the development of a new free-access operating system,[3] and sent out a manifesto to other computer programmers, requesting their participation in his free software development project, as well as encouraging them to share the source codes of their own computer programs. He called his free software development project "GNU"[4] and in 1985 he founded the Free Software Foundation.[5] For Stallman, the word "free" referred not to price, but to free access and use of computer programs, that is, "free as in free speech, not as in free beer" (Stallman, 1999; Hope, 2008). Free does not mean noncommercial: free software programs can be distributed at no charge or with a high price, because users will always have the freedom to run the program, change the program, and redistribute the program with or without changes.

Stallman created a type of license that he called copyleft, as opposed to copyright, with the General Public License (GPL[6]) as its most common form. The GPL is a standard (template) license that any computer programmer may use to allow third parties to access his/her programs' source code, and to use it for any purpose, to adapt it to specific needs, to freely distribute copies, and to develop new products based on it, without the need to obtain the prior consent of the original developer. However, if the user decides to copy or distribute any new versions or derivatives of the program, he or she can only do so in accordance with a copyleft license, that is, third parties must be allowed to use and modify the derived products freely as well. Thus, everyone benefits from access to a wide variety of source code that is permanently enhanced by new innovations. Eric Raymond, in his book *The Cathedral and the Bazaar* (Raymond, 2000), demonstrates the difference between these two models of innovation: the free software movement, which is decentralized and cooperative (which he compares to a "bazaar"), and the conventional model, which is centralized and hierarchical (which he calls a "cathedral").

In 1991, Linus Torvalds, then a graduate student at the University of Helsinki (Finland), released Linux, an operating system kernel built using tools made available by the Free Software Foundation. He used a copyleft license and invited all computer programmers to contribute to the development and improvement of the system, which brought in thousands of programmers from around the world. According to Janet Hope (2008, p. 12), by the end of the decade GNU/Linux (that is, the Linux kernel together with other operating system elements supplied by the GNU project) became not only a market and technological phenomenon, but also a flagship for an entire technosocial revolution. The GNU/Linux operating system is distributed under a copyleft license, which is "viral," that is, anyone who downloads a copy of GNU/Linux is bound by its terms. Any modifications one makes to the program are not proprietary but rather are subject to the same terms of the copyleft license. Paradoxically, the license, a private contractual instrument, keeps GNU/Linux public and freely available (Aoki, 2009). In 1998, the Open Source Initiative was established as a certification body for open source licenses, and it became responsible for reviewing and approving licenses as in conformance with the "open source definition."[7]

Currently, over 150,000 projects are estimated to be in development worldwide, involving over 1.5 million developers, and the open source software[8] movement has spread many of its founding concepts to other areas of knowledge (Hope, 2008, p. 13). Some biologists have started defending open source biology and other scientists have proposed bringing biotechnology and open source closer together, as they believe the current IP system has created difficulties rather than encouraged scientific innovation because it is highly restrictive, expensive, and surrounded by secrets and impediments. Many research tools and mechanisms are protected by countless and often overlapping IP rights. In some cases, the development of more complex products depends on the use of several research tools, which forces scientists to spend large amounts of time, energy, and resources on negotiations over patents belonging to others. These negotiations delay research or, in some cases, make research impossible, particularly those without commercial purposes. This has led many researchers to propose that the same concepts used in open source software should be employed in biological sciences.[9]

One of the open source biology projects under way is called the Tropical Disease Initiative, developed by the Goldman School of Public Policy of the University of California at Berkeley, the Department of Biopharmaceutical Sciences and Pharmaceutical Chemistry of the University of California in San Francisco, and Duke University's Law School. Scientists intend to employ open source principles to produce medication for tropical diseases, such as malaria, cholera, dengue fever, and Chagas disease, which affect over half a million people worldwide. All scientists will have free and permanent access to research tools and databases, new discoveries and inventions will be shared, and products will not be protected by patents. Scientists thus intend to reduce costs, accelerate research and development of new medications, and devote their efforts to a type of research that is not in the commercial interests of large laboratories and the pharmaceutical industry, as it involves diseases that affect mainly the populations of poor countries who cannot afford the high costs of patents (Maurer *et al.*, 2004, pp. 180–183).[10]

Another initiative is Biological Innovation for Open Society (BIOS), coordinated by molecular geneticist Richard Jefferson, Director of the Centre for the Application of Molecular Biology to International Agriculture (CAMBIA), located in Canberra, Australia.[11] With the goal of extending open source concepts to biotechnology, Jefferson developed the so-called BIOS (Biological Open Source) license, aimed at overcoming some of the difficulties involved in accessing scientific knowledge and research tools protected by IP rights. Upon agreeing with the terms of the BIOS license, scientists have access to the whole portfolio of vectors and technologies developed by CAMBIA,[12] among which the "transbacter" stands out. "Transbacter" is a vector that facilitates the transfer of genes into plant cells, providing a technological alternative to other vectors protected by IP rights.[13] The BIOS license is among the most thoroughly implemented open source initiatives in the life sciences.[14]

These initiatives converge, in varying degrees, to what has become known as the commons movement. Commons are goods that are not subject to the specific control of any one person, and their use may be shared. Ronaldo Lemos (2005, pp. 18–19) points out that commons have always been present in human life throughout history, and life in society depends essentially on some goods being maintained this way, such as squares, streets, intellectual works in public domain, etc. However, what defines whether a good is common is not the intrinsic possibility (resulting from its own nature) of being shared, but rather the (social and legal) regime to which it is subjected. Thus, some goods are intrinsically noncompetitive (i.e., their use by one person does not prevent someone else from using them) but are not legally treated as commons. Good examples are intellectual works with no physical support, such as songs and literary works published on the internet, which can be used by many people at the same time. In other words, many people can access an intangible intellectual product without physically taking anything away from other users. However, owing to copyright laws, they become exclusive property of certain people. There is a monopoly of intellectual creations that would otherwise be free, according to Ronaldo Lemos (2005, pp. 18–19).

New collaborative forms of work require new legal frameworks, distinct from conventional IP rights. One of the answers was the creation of the creative commons model,[15] by Lawrence Lessig, from Stanford University, California (Lessig, 2001, 2004). The objective of creative commons is to develop license models that can be used by any individual or organization to allow third parties to distribute, copy, and use their creations (songs, movies, photos, texts, or any intellectual work).[16] In general, intellectual works can be used only with the permission of their authors, as authors have all rights reserved, and creative commons licenses allow authors to reserve only selected rights. Authors can choose from a variety of creative commons licenses models, and what they have in common is that the user must always acknowledge the original author. In some license models, the author authorizes free copying and distribution of the creation, but forbids its use for the creation of derivative works, that is, the original production cannot be altered without prior consent from the author (called the "no derivatives" license).[17] There are also license models in which the author allows the use of his or her work for noncommercial purposes, but does not allow its use for commercial purposes (called the "noncommercial" license). The "noncommercial license," however, does not mean that the work cannot be used commercially. It means that users receive no commercial rights via the Creative Commons license, but can ask the author (licensor) directly if he or she is interested in licensing the commercial rights through other agreements. Furthermore, the term "noncommercial use" does not apply to the author (the licensor), but only to others who wish to use the author's work (the licensees). The author/licensor may at any time decide to use his or her work commercially. In other Creative Commons license models, the author authorizes remixes and derivative works based on his or her creative work, as long as they are distributed under the same Creative Commons license under which the original work was published (called the "share alike" license). Essentially, a Creative Commons-licensed work is authorized by its author to be used consistent with its license terms. Some known users of Creative Commons licenses are Al Jazeera Creative Commons Repository, Google, Massachusetts Institute of Technology Open Course Ware, Public Library of Science, Wikipedia, etc.

Creative Commons licenses represent an ingenious twist on the proprietary approach: they do not deny IP rights; on the contrary, they are based on copyright, and they apply to all works that are protected by copyright law. Creative Commons licenses give everyone, from individual creators to large companies, a standardized way of keeping their copyright while allowing certain uses of their work – a "some rights reserved" approach to copyright. Third parties are allowed certain uses of the work but not others. Thus, Creative Commons-licensed works are "protected commons," not public domain goods, as the latter can be appropriated (directly or indirectly) by third parties through IP rights. Public domain goods have no effective protection against misappropriation.[18] Creative Commons-licensed works cannot be appropriated by third parties, as there is a legal protection against it, conferred by copyright law and by the Creative Commons license, which is based on copyright law.

What initiatives such as free software, open source biology, and Creative Commons licenses seek is a balance between the monopoly promoted by IP law and the protection of common resources. All of these alternatives respond, in some way, to the arguments defended by Garrett Hardin in his famous work *The Tragedy of the Commons*, in which he argues that goods that are not privately appropriated but kept as common resources are destroyed and exploited inappropriately, as no one has ownership and takes responsibility for their care (Hardin, 1968, pp. 1243–1248). Michael Heller refuted these arguments in his article "The tragedy of the anticommons: property in the transition from Marx to markets" (Heller, 1998) and later in the same year, in an article co-authored with Rebecca Eisenberg, asked "Can patents deter innovation? Anticommons in biomedical research" (Heller and Eisenberg, 1998). In these articles, the authors show the effects of the opposite situation, called the "anticommons": if there are too many owners and all have the right to exclude others, and the use of a good or service depends on complex negotiations with all owners, the good or service tends to be underused. Overlapping IP rights over a particular good or resource often result in blocking of its use and failure to develop innovations potentially useful to society. In agricultural biotechnology, an often-cited example is that of Goldenrice, a genetically engineered rice variety developed using approximately 70 different patented technologies (Hope, 2008).

And what do advocates of open source software and the commons movement have to do with farmers and their seeds? Both groups want free access to knowledge and innovations and defend sharing, rather than exclusion. Both have suffered the effects of excessive proliferation of IP rights over resources and knowledge which are fundamental to them, as well as the lack of mechanisms to protect common resources. Like the computer programmers who started the free software movement, many farmers are outraged with laws passed top-down that conflict with local rules of access and sharing of resources and knowledge. They want not only protection against misappropriation of their resources and knowledge by third parties, but also enough legal space to maintain biological, social, and cultural processes that are essential for their food security, such as seed saving and exchange through social networks.

Farmers, as well as researchers and breeders from public institutions, have also felt the effects of restrictions imposed by IP rights, which can create an atmosphere that is antagonistic to cooperation and to exchange of knowledge and germplasm. Plant breeding (by farmers or scientists) is essentially dependent on access to a wide variety of plant genetic resources, and innovations in this domain are always cumulative. It was precisely for this reason that plant breeders created a *sui generis* system to protect new plant varieties, supposedly distinct from patents because it allows the use of a protected plant variety as a source of variation in plant breeding. In its original version, the UPOV system had some (not all) elements of an open source system, because access to plant genetic resources was considered free for research and breeding purposes. The UPOV system is gradually becoming more like the patent system (especially after the adoption of the 1991 Act of the UPOV Convention), with successive restrictions imposed on breeders' exemptions and farmers' rights. What many plant breeders seek, in reality, is to rescue some of the principles of the original UPOV. Farmers, on the other hand, want to guarantee their rights to re-use, exchange, and sell seeds and plant varieties, which are becoming more and more legally restricted.

In order to achieve this, some scientists have proposed the creation of a "Biolinux" model of the GPL for plant germplasm. According to this proposal, genetic material transfer agreements would set out rules regarding the use of genetic materials similar to those of copyleft licenses, that is, anyone who receives genetic materials would agree that, if new plant varieties are developed through the use of received genetic materials, these new plant varieties will be freely available. Thus, the plant breeder can freely use the materials received in the research and in the development of new varieties, but new varieties must also be freely available to third parties. Farmers' rights to freely use, save, exchange, and develop new varieties would also be legally insured, as the protection granted by the copyleft license is precisely what insures free circulation and use. Several scientists have defended proposals for use of the open source model and copyleft licenses for agricultural seeds and varieties, such as Sylvain Desmoulière (2001), Krishna Ravi Srinivas (2002, 2006), Margareth Kipp (2005), Roberto Verzola (2005), Janet Hope (2008), Jack Kloppenburg (2008), and Keith Aoki (2009).[19]

Tom Michaels, a bean breeder, presented his proposal of a "General Public License for Plant Germplasm" (GPLPG) in 1999 (Hope, 2008; Kloppenburg, 2008). His intention was to keep plant germplasm broadly accessible for research and plant breeding. According to his proposal, scientists would supply plant germplasm to other parties accompanied by a material transfer agreement that would include provisions allowing further development and recombination of the transferred germplasm, but requiring that any new plant variety derived from the received germplasm should likewise be made available to others without any restriction for use in research and breeding, and according to the same terms of the GPLPG. In Michaels' proposal, the equivalent of a software "derivative work" would be any new plant germplasm descended from germplasm distributed under the GPLPG (Hope, 2008; Kloppenburg, 2008).

Jack Kloppenburg (2008, pp. 15ff.) proposes that an open source license (a "Biolinux for seeds" or a GPLPG) be used to create a "protected commons" space in which sharing of plant genetic resources among farmers and plant breeders is unimpeded but is also protected from appropriation by monopolists. The GPLPG would allow the sharing of plant germplasm among those who will reciprocally share, but would exclude those who will not. Wide dissemination of plant genetic resources would be stimulated and facilitated, rather than restricted. According to Kloppenburg (2008, pp. 17–20), the use of a GPLPG would be more efficient and cheaper than other measures intended to prevent inappropriate patenting, such as community registration and defensive patenting, and could potentially:

- prevent the patenting of plant genetic material and biopiracy;
- prevent the use of farm-derived genetic resources in proprietary breeding programs;
- develop a legal/institutional framework that recognizes farmers' collective sovereignty over seeds and allows farmers to freely exchange, save, improve, and sell seeds;

- develop an institutional framework in which farmers cooperate with plant scientists in the development of new plant varieties that contribute to a sustainable food system;
- develop a framework for marketing of seed that is not patented or use restricted.

For Ravi Srinivas (2006), the Biolinux model could be used to protect the biological commons in agricultural biodiversity, particularly germplasm and seeds. According to Srinivas, a Biolinux license could govern the transfer of germplasm, as part of a material transfer agreement (MTA), and it would indicate that the germplasm in the seeds is freely available to farmers and plant breeders for further development. Farmers would also have the right to use the received germplasm for development of new plant varieties, but they should offer the new varieties under a similar Biolinux license to others. The Biolinux license would prohibit the farmer or plant breeder from restricting the right of others in a manner that is inconsistent with such a license. Srinivas (2006) goes on to explain that in the Biolinux model germplasm is akin to code. According to him, "one is free to use and develop varieties as long as one does not make proprietary claims on the germplasm to prevent others from using it, and one is expected to offer the variety with a similar condition" (Srinivas, 2006). That way, a material transfer agreement that incorporates a Biolinux license or open source principles would prevent enclosure of the germplasm in the form of patents on the derivatives of germplasm, argues Srinivas (2006). He adds that a Biolinux license model would be based on the logic that farmers are both users and innovators, and suggests that it be used to protect new varieties developed through participatory plant breeding programs (Srinivas, 2002, pp. 325ff.).

All of the above-cited proposals have much in common, but they will have to overcome the difficulties presented by the fact that the legal status of plant genetic resources and of traditional knowledge is still ambiguous in most countries. Although the multilateral system of ABS established by the ITPGRFA tries, in some ways, to define and maintain a global common pool of plant genetic resources, its coverage is limited to the 64 crops and forages listed in its Annex 1, that are on the public domain and are stored *ex situ*, in germplasm banks (see Chapter 6). This means that all other plant genetic resources are, in principle, subjected to the bilateral ABS system established by the CBD, and to national ABS laws, which vary widely (many countries still do not have ABS laws, and others have very strict ones). Although the sovereign rights of states over their genetic resources are recognized by the CBD, this does not necessarily mean that they are publicly owned by states, and in many countries it is not clear whether plant genetic resources are public, public interest, or private resources. The status of traditional knowledge associated with plant genetic resources is even less clear, and there is also the fact that most countries have signed the WTO's TRIPS Agreement, committing themselves to plant variety protection through IP or *sui generis* regimes (see Chapter 5). Even the 64 crops and forages included in the multilateral system of the ITPGRFA are subjected to an ambiguous status, as far as the possibility of IP rights being claimed over them is concerned. The Treaty says that (Article 12.3d) "recipients (of plant genetic resources) shall not claim any IP or other rights that limit the facilitated access to the plant genetic resources for food and agriculture, or their genetic parts or components, in the form received from the multilateral system." It is not clear what "in the form received from the multilateral system" exactly means, and interpretations may vary greatly.

The unclear legal status of plant genetic resources poses a serious problem for the creation and use of a GPLPG. As previously mentioned, open source licenses are based on copyright law, and apply only to works that are protected by copyright law (such as software and intellectual works). In the case of the BIOS license, it is usually applied to biotechnological tools protected by patents, another form of IP. Thus, these licenses do not deny IP rights; on the contrary, they are based on the premise that IP rights holders relinquish some of their (private) rights over their creations. If it is unclear who the owners of plant genetic resources and associated knowledge are, it will also be uncertain (from a legal point of view) who will be able to relinquish any rights over them, through open source licenses. If plant genetic resources are considered to be public and owned by states (as a result of their "sovereign rights," recognized under the CBD, although this is not always the case[20]), then only states, as holders of rights over their genetic resources, would be able to allow the free use and exchange of genetic resources through open source licenses (at a national or international level). If farmers and local communities are considered to be holders of rights (property, custodianship, etc.) over their seeds and associated traditional knowledge (under Article 8j of the CBD or farmers' rights, which are still not recognized in most countries), they will be the licensors, and not the states. Whenever IP rights are recognized over plant varieties, only the holders of such rights will be able to license their use to third parties through an open source license, which is very unlikely to happen because private companies will want to protect their commercial interests. If IP rights over plant varieties are held by public institutions, however, they could license their use through open source licenses. Thus, the creation of open source licenses for plant genetic resources depends on a previous and clear definition (from a legal point of view) of who owns or holds any rights over plant germplasm and associated knowledge and who can relinquish rights over them through open source or Biolinux-model licenses.

On the other hand, it is important to distinguish "open access commons" from "socially regulated commons," as Preston Hardison (2006) points out. He calls attention to the fact that the commons movement has been inadequately taking into account the rights and aspirations of Indigenous peoples and local communities, by failing to consider the many different commons that exist. He points out that the concept of commons is not monolithic and cannot be applied indistinctively to all situations where resources and knowledge are shared and collectively developed. According to Hardison, some civil society organizations consider that there are only two antagonistic movements – represented by those who promote a hyperprivatization of "our culture," on the one hand, and those who defend public domain and free access to culture, on the other. Hardison emphasizes that there is, in reality, a vast plurality of local commons, defined by local rules and institutions, and that there are over 6,000 Indigenous peoples around the world, and many more local communities, each having its own unique set of beliefs (Hardison, 2006).

Although Indigenous peoples and other traditional and local communities may not share our concept of property, this does not mean that they do not have their own social rules to regulate use and exchange of knowledge. They often establish restrictions on who can use certain knowledge, when, under what circumstances, and with what objectives, as Hardison (2006) explains. What is intended is not the creation of a public domain regime in which everything is freely accessible to everyone under any circumstances, but rather that local rules and institutions for regulation of collective and individual rights be respected, protecting what Hardison calls "socially regulated commons."

It is for these reasons that a possible use/adaptation of open source systems for seeds and local plant varieties should always take into consideration the local rules and institutions that regulate access to and circulation of genetic materials and associated agricultural knowledge, be the holders Indigenous peoples, local communities, or small-scale farmers. Nonetheless, creative commons licenses have been sufficiently flexible to allow authors to adapt them to their needs.

There is no sense in establishing any licenses, be they copyleft or any other form, to regulate relations among members of the same Indigenous or local community, or among traditional, small-scale, or agroecological farmers, as these relations are already ruled by local institutions and networks. Such models can, however, be employed whenever relations with external third parties are involved and there is an intent of authorizing specific uses, restricting others, and preventing misappropriation of resources and knowledge by third parties. Experiences involving participatory plant breeding, carried out through partnerships between scientific researchers and farmers, for instance, could make use of copyleft licenses whenever new plant varieties are developed, provided that the status of plant genetic resources and associated knowledge is already clearly defined by the legislation of the country in which they are located.

Notes

1 In this context, "hacker" does not mean someone who cracks a security system, but a computer programmer or developer.
2 Source codes of computer programs can be protected under copyrights or as trade secrets.
3 An operating system is the core program without which a computer cannot run.
4 See www.gnu.org/philosophy/philosophy.html (accessed December 10, 2010).
5 See www. fsf.org/about/ (accessed December 10, 2010).
6 Initially, it was called GNU Public License.
7 See www.opensource.org/docs/osd; accessed 10 December, 2010.
8 As Janet Hope points out (Hope 2009, p. 177), though based on "free software," this phenomenon is now more commonly referred to as "open source." Those who coined the term "open source" wanted to see nonproprietary software more widely adopted, including in commercial settings, and considered the language of freedom to be confusing (some people think that "free" means noncommercial).
9 See www.sciencecommons.org and www.freedomofscience.org (accessed December 10, 2010).
10 Other examples of open source science are (1) Biobricks Foundation, a nonprofit organization dedicated to promoting and coordinating the production of standardized modules or "biobricks" that can be assembled into functional synthetic biological systems; (2) the International HapMap Project, a private–public collaboration to create a haplotype map of the human genome, the HapMap, which intends to describe the common patterns of human DNA sequence variation. The HapMap is expected to be a key resource for researchers to find genes affecting health, disease, and responses to drugs and environmental factors. See www. hapmap.ncbi.nlm.nih.gov/abouthapmap.html (accessed December 10, 2010).
11 See www.cambia.org and www.bios.net; www.thesynapticleap.org (accessed December 10, 2010) and Berthels (2009).
12 Both CAMBIA and BIOS are supported by the Rockfeller Foundation.
13 The Public Intellectual Property Resource for Agriculture (PIPRA) is a clearinghouse institution that is designed to integrate, through collaborative management, the portfolio of IP technological innovations in agriculture (this technology portfolio is usually fragmented across a large number of public institutions). PIPRA's goal is to mobilize technologies from a wide range of public/nonprofit technology providers to address specific projects for the improvement of subsistence and specialty crops that are not being addressed by commercial seed and agricultural biotechnology companies. See www.pipra.org and Bennett and Boettiger (2009).
14 Janet Hope points out, however, that BIOS license cannot be considered a "truly" open source license, because it gives the licensor (of the biological tool or technology) too much control over downstream development of the licensed IP and technology, and there is insufficient freedom to "fork the code" – that is, to reject the licensor's leadership of the ongoing process of cumulative innovation (Hope, 2009, p. 191). According to Hope (2008, p. 318), BIOS licenses also contain a grant-back structure that differs from most copyleft licenses in ways that could limit their adoption. Licenses are granted from CAMBIA to a licensee, and all improvements are granted back to CAMBIA, which may then license them to other researchers. This network, with CAMBIA at the center, is enforced by restrictions on licensees sublicensing the technology.

15 See www.creativecommons.org (accessed December 10, 2010).
16 Creative Commons does not recommend the use of Creative Commons licenses for software. It recommends the use of software licenses made available by the Free Software Foundation or listed at the Open Source Initiative, as these licenses were designed specifically for use with software.
17 A derivative work is based on another work but is not an exact, verbatim, copy. Some examples are the translation from one language to another or a film version of a book. Creative Commons licenses allow the user to exercise the rights permitted under the license in any format or media, even in the "no derivatives" license.
18 In October 2010, Creative Commons announced the release of its Public Domain Mark, a tool that enables works free of known copyright restrictions to be labeled in a way that allows them to be easily discovered over the internet. The Public Domain Mark is to be used for marking works already free of copyright, and it complements Creative Commons' CC0 public domain dedication, which enables authors to relinquish their rights prior to the expiration of copyright (http://creativecommons.org/about/pdm; accessed December 10, 2010).
19 In Brazil, the International Free Software Forum created, in 2006, a "Free Seedbank," with the objective of providing Indigenous and other traditional communities in Rio Grande do Sul with seeds free from genetic modifications and from IP rights. Source: Evangelista, R. *Banco de Sementes Livres (*http://wiki.softwarelivre.org/Sementes/Noticia20060111215216; accessed December 10, 2010).
20 In Brazil, for instance, a proposal for constitutional amendment is being discussed at the National Congress, and if it is approved, it will turn genetic resources public federal property, a legal status similar to that of mineral resources. This shows that there is no certainty that "sovereign rights" over genetic resources necessarily means "public ownership."

Bibliography

American Society of Plant Biologists (2007) 'The freedom to innovate: a privilege or a right?' *Plant Cell*, vol. 19, pp. 1433–1434 (www.plantcell.org/cgi/content/full/19/5/1433; accessed February 19, 2009).

Aoki, K. (2009) 'Free seeds, not free beer: participatory plant breeding, open source seeds and acknowledging user innovation in agriculture' (http://ssrn.com/abstract=1390273; accessed December 10, 2010).

Bellivier, F. and Noiville, C. (2009) *La Bioéquité: Batailles Autour du Partage du Vivant*, Autrement, Paris.

Bennett, A. and Boettiger, S. (2009) 'The Public Intellectual Property Resource for Agriculture (PIPRA): a standard licence public sector clearinghouse for agricultural IP', in Van Overwalle, G. (ed.) *Gene Patents and Collaborative Licensing Models*, Cambridge University Press, New York, pp. 135–150.

Berthels, N. (2009) 'Cambia's Biological Open Source Initiative (BIOS)', in Van Overwalle, G. (ed.) *Gene Patents and Collaborative Licensing Models*, Cambridge University Press, pp. 194–203.

Brewster, A., Chapman, A. and Hansen, S. (2005) 'Facilitating humanitarian access to pharmaceutical and agricultural innovation', *Innovation Strategy Today*, vol. 1, no. 3, pp. 203–216 (www.biodevelopments.org/innovation/index.htm; accessed February 19, 2009).

Carneiro da Cunha, M. (2007) *'Culture' and Culture: Traditional Knowledge and Intellectual Rights*, Prickly Paradigm Press, Chicago.

Desmoulière, S. (2001) 'Approche ethnobotanique de la diversité varietale du manioc en Amazonie centrale: gestion et perspectives de conservation', PhD thesis, Muséum National d'Histoire Naturelle, Laboratoire d'Ethnobiologie-Biogéographie, Paris (http://yepca.org/thz/pdf/0-Sommaire.pdf; accessed February 19, 2009).

Feldman, R. (2004) 'The open source biotechnology movement: is it patent misuse?', *Minnesota Journal of Law, Science & Technology*, vol. 6, pp. 117–16 (http://mjlst.umn.edu/uploads/CS/bW/CSbWvh00RsLS5YM4raBkYg/feldman_a2.pdf; accessed February 19, 2009).

Halewood, M. (2010) 'Governing the management and use of pooled microbial genetic resources: lessons from the global crop commons', *International Journal of the Commons*, vol. 4, no. 1, pp. 404–436 (http://www.thecommonsjournal.org/index.php/ijc/article/viewArticle/152; accessed January 10, 2011).

Hardin, G. (1968) 'The tragedy of the commons', *Science*, vol. 162, pp. 1243–1248 (http://www.garretthardinsociety.org/articles/art_tragedy_of_the_commons.html; accessed February 19, 2009).

Hardison, P. (2006) 'Indigenous peoples and the commons' (http://wo.ala.org/tce/wp-content/uploads/2008/10/ip-and-the-commons.pdf; accessed February 19, 2009).

Hecker, F. (2000) *Setting Up Shop: The Business of Open-Source Software*, Mountain View, CA (http://hecker.org/writings/setting-up-shop; accessed February 19, 2009).

Heller, M. (1998) 'The tragedy of the anticommons: property in the transition from Marx to markets', *Harvard Law Review*, vol. 111, no. 3, pp. 621–688 (http://www.unc.edu/courses/2007fall/geog/804/001/Heller%201998%20Tragedy%20of%20the%20Anticommons.pdf; accessed February 19, 2009).

Heller, M. and Eisenberg, R. (1998) 'Can patents deter inovation? Anticommons in biomedical research', *Science*, vol, 280, pp. 698–701 (http://www.sciencemag.org/content/280/5364/698.full; accessed February 19, 2009).

Hope, J. (2008) *Biobazaar: The Open Source Revolution and Biotechnology*, Harvard University Press, Cambridge, MA.

Hope, J. (2009) 'Open source genetics: conceptual framework', in Van Overwalle, G. (ed.) *Gene Patents and Collaborative Licensing Models*, Cambridge University Press, pp. 171–193.

Instituto Socioambiental (2010) Final report of the Project "Traditional Knowledge. Innovate to Advance: Proposing New Forms of Safeguarding Collective Rights of Indigenous Peoples." This project focuses on the following cases: registration of traditional Wanano dances and Baniwa songs, from Indigenous people from the Negro River region in the Brazilian Amazon, and the registration of songs by the Yudjá and Ikpeng Indigenous peoples, in the Xingu region of Brazil (http://ct.socioambiental.org/relatoriofinal; accessed December 10, 2010).

Kipp, M. (2005) 'Software and seeds: open source methods', *First Monday*, vol. 10, no. 9 (http://131.193.153.231/www/issues/issue10_9/kipp/index.html; accessed February 19, 2009).

Kloppenburg, J. (2008) 'Seeds, sovereignty and the Via Campesina: plants, property, and the promise of open source biology' (http://www.drs.wisc.edu/documents/

articles/kloppenburg/2008%20Seeds%20and%20Sovereignty.pdf; accessed February 19, 2009).

Lemos, R. (2005). *Direito, Tecnologia e Cultura*, Fundação Getúlio Vargas, Rio de Janeiro.

Lessig, L. (2001) *The Future of Ideas: The Fate of the Commons in a Connected World*, Random House, New York (www.the-future-of-ideas.com/download; accessed February 19, 2009).

Lessig, L. (2004) *Free Culture: The Nature and Future of Creativity*, Penguin Books, New York.

Levy, S. (2001) *Hackers: Heroes of the Computer Revolution*, Penguin Books, New York; orig. pub. 1984 (http://www.dvara.net/HK/LevyStevenHackers1&2.pdf; accessed February 19, 2009).

Maurer, S.M., Rai, A. and Sali, A. (2004) 'Finding cures for tropical diseases: is open source an answer?', *PLoS Medicine*, vol. 1, no. 3, pp. 180–183 (www.tropicaldisease.org/documents/MauRaiSal_PLOS2004.pdf; accessed February 19, 2009).

Pollack, A. (2005) 'Open-sources practices for biotechnology', *The New York Times*, New York, 10/2/2005, http://www.nytimes.com/2005/02/10/technology/10gene.html; accessed February 19, 2009).

Raymond, E. (2000) *The Cathedral and the Bazaar* (www.catb.org/~esr/writings/cathedral-bazaar; accessed February 17, 2009).

Srinivas, K.R. (2002) 'The case for Biolinuxes and other pro-commons innovations', in Sarai Reader: *The Cities of Every Day Life*, Center for the Study of Developing Societies, New Delhi (http://www.sarai.net/publications/readers/02-the-cities-of-everyday-life/09biolinux.pdf; accessed February 19, 2009).

Srinivas, K.R. (2006) 'Intellectual property rights and bio commons: open source and beyond', *International Social Science Journal*, vol. 58, no. 188, pp. 319–334 (http://papers.ssrn.com/sol3/papers.cfm?abstract_id=1288552; accessed February 19, 2009).

Stallman, R. (1999) 'The GNU operating system and the free software movement', in Dibona, C., Ockman, S. and Stone, M. (eds.). *Open Sources: Voices from the Open Source Revolution*, O'Reilly, Cambridge, MA (www.oreilly.com/catalog/opensources/book/stallman.html; accessed February 19, 2009).

Toomey, G. (2007) *Sharing the Fruits of Science*, University Affairs, Ottawa, Ontario (www.universityaffairs.ca/sharing-the-fruits-of-science.aspx; accessed February 19, 2009).

Verzola, R. (2005) 'Software and seeds: lessons in community sharing', *Seedling*, pp. 13–17 (http://www.grain.org/seedling/?id=410; accessed February 19, 2009).

Weber, S. (2004) *The Success of Open Source*, Harvard University Press, Cambridge.

11

AGROBIODIVERSITY AND CULTURAL HERITAGE LAW

Cultivated plants as cultural artifacts: "agriculture"

To Carl Sauer (1986), cultivated plants are "cultural artifacts," and to Laure Emperaire (2008), they are "biological objects in nature, but cultural in essence." Culture is even present in the term "agriculture," and the word "culture," historically, has also meant cultivation of the land. Culture and agriculture are, therefore, intimately related, and we can use legal instruments aimed at safeguarding cultural heritage to recognize and promote agrobiodiversity-rich agricultural systems and all their elements, both tangible (such as agroecosystems and cultivated plants) and intangible (agricultural techniques, practices, and knowledge).

Safeguarding traditional foodways and dietary diversity is also an important way to promote agrobiodiversity and food security. The conservation of crop genetic diversity cannot be dissociated from crop use, and promoting traditional diets, which are generally more diverse, safeguards humanity's biological and cultural heritage at the same time. In this chapter, we will analyze how cultural heritage legal instruments – such as the UNESCO Convention for the Protection of the World Cultural and Natural Heritage and the UNESCO (United Nations Educational, Scientific and Cultural Organization) Convention for the Safeguarding of the Intangible Cultural Heritage – can be used (and in some cases, have been used) to promote agrobiodiversity and food diversity in different and innovative ways.[1]

The UNESCO Convention for the Safeguarding of Intangible Cultural Heritage: interfaces with agrobiodiversity and food diversity

The Convention for the Safeguarding of the Intangible Cultural Heritage was adopted during the 2003 General Conference of UNESCO, and entered into force in 2006. It is the first binding multilateral instrument for the safeguarding of intangible cultural heritage, which is defined by the Convention (Article 2.1) as

> the practices, representations, expressions, knowledge and skills – as well as the instruments, objects, artifacts and cultural spaces associated therewith- that communities, groups and, in some cases, individuals recognize as part of their cultural heritage. This intangible cultural heritage, transmitted from generation to generation, is constantly recreated by communities and groups in response to their environment, their interaction with nature and their history, and provides them with a sense of identity and continuity.

According to the Convention, intangible cultural heritage is manifested *inter alia* in the following domains: oral traditions and expressions, including language as a vehicle of the intangible cultural heritage, performing arts, social practices, rituals and festive events, knowledge and practices concerning nature and the universe, and traditional craftsmanship (this is an inclusive, rather than exclusive, list, which means that it is not complete, and that member states may include other domains). "Safeguarding" means measures aimed at ensuring the viability of the intangible cultural heritage, including the identification, documentation, research, preservation, protection, promotion, enhancement, transmission, particularly through formal and nonformal education, as well as the revitalization of the various aspects of such heritage. The aims of the UNESCO Convention's are to safeguard the intangible cultural heritage; to ensure respect for the intangible cultural heritage of the communities, groups, and individuals concerned; to raise awareness at the local, national, and international levels of the importance of the intangible cultural heritage, and of ensuring mutual appreciation thereof; and to provide for international cooperation and assistance (Article 1).

The UNESCO Convention's main message is that cultural heritage does not end at monuments and collections of objects. It also includes traditions or living expressions inherited from our ancestors and passed on to our descendants, which are important in maintaining cultural diversity in the face of growing globalization and cultural homogenization. Intangible cultural heritage is traditional, contemporary, and living at the same time, and it does not represent inherited traditions from the past only, but also contemporary rural and urban practices in which diverse cultural groups take part. Just like culture in general, intangible heritage is constantly changing and evolving, and being enriched by each new generation. It provides a link from our past, through the present and into our future, and contributes to giving us a sense of identity and continuity. Intangible cultural heritage can only be recognized as such by the communities, groups, or individuals that create, maintain, and transmit it – without their recognition, nobody else can decide for them that a given expression or practice is their heritage.[2] According to the UNESCO Convention, member states must identify and define the various elements of intangible cultural heritage present in their territory with the participation of communities, groups, and NGOs, as well as involve them actively in the management of their heritage (Article 15).

For a country to safeguard its intangible cultural heritage, it must first identify those expressions and manifestations that can be considered intangible cultural heritage and make a record, or inventory, of them. These inventories may then serve as a basis for developing safeguarding measures for the manifestations and expressions of the intangible cultural heritage included, and described, in the inventory.

Some examples of intangible cultural heritage included in UNESCO's Representative List of the Intangible Cultural Heritage[3] are:

- the Chant of the Sybil on Majorca (Spain), which marks the annual Christmas vigil, and is sung by a boy or girl accompanied by two or more church altar boys or girls;
- Huaconada, a ritual dance performed in the village of Mito, in the province of Concepción in the central Peruvian Andes;
- Marimba music and traditional chants of Colombia's South Pacific region, which are the heritage of Afro-Colombian groups in Valle del Cauca, Cauca, and Nariño;
- the Indigenous festival El Día de los Muertos (Day of the Dead) in Mexico, which commemorates the transitory return to Earth of deceased relatives and loved ones;
- the oral and graphic expressions of the Wajapi Indigenous people (a Tupi-Guarani cultural–linguistic people who live in the Brazilian Amazon) Wajapi's graphic art is known as Kusiwa and its designs are applied with red vegetable dyes extracted from the roucou plant mixed with scented resins;
- oxherding and oxcart traditions in Costa Rica, and the Baltic Song and Dance Celebrations in Latvia, Estonia and Lithuania; and
- Kumiodori, a traditional musical theatre from the islands of Okinawa Prefecture in Japan.[4]

Safeguarding intangible heritage focuses on the processes involved in transmitting or communicating such heritage from generation to generation, rather than on the production of its concrete manifestations, such as a dance performance, a song, a musical instrument, or a craft.

Historically, UNESCO's Representative List of Intangible Cultural Heritage has focused primarily on performing arts and crafts. More recently, however, UNESCO seems to have accepted that culinary traditions are a cultural expression as fundamental to identity and worthy of recognition as dance, theater, or music.[5] As food heritage is directly associated with crop genetic diversity, it is worth mentioning that in 2010 three culinary systems were included in the list:

- Traditional Mexican cuisine was the first in the world to be considered intangible cultural heritage. It was considered to be an ancestral, ongoing community culture. According to UNESCO, traditional Mexican cuisine is a comprehensive cultural model, comprising farming, ritual practices, age-old skills, culinary techniques, and ancestral community customs. It is made possible by collective participation in the entire traditional food chain: from planting and harvesting to cooking and eating. The basis of the system is founded on corn, beans, and chili; unique farming methods such as *milpas* (rotating swidden fields of corn and other crops) and *chinampas* (man-made farming islets in lake areas); cooking processes such as nixtamalization (lime-hulling maize, which increases its nutritional value); and singular utensils including grinding stones and stone mortars. Native ingredients such as varieties of tomatoes, squashes, avocados, cocoa, and vanilla augment the basic staples. Mexican cuisine is elaborate and symbol laden, with everyday tortillas and tamales, both made of corn, forming an integral part of Day of the Dead offerings. Collectives of female cooks and other practitioners devoted to raising crops and traditional cuisine are found in the State of Michoacán and across Mexico. Their knowledge and techniques express community identity, reinforce social bonds, and build stronger local, regional, and national identities. Those efforts in Michoacán also underline the importance of traditional cuisine as a means of sustainable development.[6] Mexico's application to UNESCO stresses that, in addition to the foods themselves, the recipes to prepare them and food-related customs, Mexico is home to "a complex cultural system of agricultural practices, traditions and symbolisms imbued with religious meaning and steeped in ritual." Two very important traditional agricultural systems – *milpas* and *chinampas* – as well as diverse varieties of corn, chili, beans, tomatoes, squashes, avocados – were recognized as part of this cultural system. UNESCO's designation should contribute not only to foster national cultural identity and promote traditional cuisine, but also to the preservation of native plant species and agroecosystems.[7]

- The Mediterranean diet of Spain, Greece, Italy, and Morocco was also inscribed in the UNESCO list. It constitutes a set of skills, knowledge, practices, and traditions ranging from the landscape to the table, including the crops, harvesting, fishing, conservation, processing, preparation, and, particularly, consumption of food. The Mediterranean diet is characterized by a nutritional model that has remained constant over time and space, consisting mainly of olive oil, cereals, fresh or dried fruits and vegetables, a moderate amount of fish, dairy, and meat, and many condiments and spices, all accompanied by wine or infusions, always respecting beliefs of each community. It is a cultural system rooted in respect for the territory and biodiversity, and it ensures the conservation and development of traditional activities and crafts linked to fishing and farming in Mediterranean communities such as Soria in Spain, Koroni in Greece, Cilento in Italy, and Chefchaouen in Morocco. Women play a particularly vital role in the transmission of expertise, as well as knowledge of rituals, traditional gestures and celebrations, and the safeguarding of techniques.[8]

- The gastronomic meal of the French is another intangible cultural heritage inscribed in the UNESCO list. According to UNESCO, it is a customary social practice when celebrating important moments in the lives of individuals and groups, such as births, weddings, birthdays, anniversaries, achievements, and reunions. It is a festive meal bringing people together for an occasion to enjoy the art of good eating and drinking. The gastronomic meal emphasizes togetherness, the pleasure of taste, and the balance between human beings and the products of nature. Important elements include the careful selection of dishes from a constantly growing repertoire of recipes; the purchase of good, preferably local, products whose flavors go well together; the pairing of food with wine; the setting of a beautiful table; and specific actions during consumption, such as smelling and tasting items at the table. The gastronomic meal respects a fixed structure, commencing with an aperitif (drinks before the meal) and ending with liqueurs, and encompassing in between at least four successive courses, namely a starter, fish and/or meat with vegetables, cheese, and dessert.[9]

UNESCO is also partnering with Bioversity International and the Kenyan Ministry for National Heritage and Culture on a project called "Safeguarding Traditional Foodways of Two Communities in Kenya." The project recognizes that traditional foodways, both those related to everyday life as well as those associated with special occasions (such as rituals, social practices, and festive events), constitute an important part of the intangible heritage of communities, and that the diversity of foodways and related knowledge about nature in Kenya is at risk. The project aims to identify and inventory traditional foodways in two communities in Kenya (Eastern Pokot in the Rift Valley Province and Isukha in the Western Province), encourage these communities to appreciate traditional food practices, and raise awareness in Kenya about the endangered diversity of its traditional foodways and related knowledge about nature.[10] An additional purpose of the project is to promote plant genetic resources conservation and sustainable use.

In addition, in 2009, UNESCO's Intergovernmental Committee for the Safeguarding of the Intangible Cultural Heritage selected the project "Safeguarding Intangible Cultural Heritage of Aymara Communities in Bolivia, Chile and Peru" as one of the initiatives that best reflects the principles and objectives of the Convention. The project involves primary school teachers and students aged 5–11 years from the Aymara communities of Bolivia (La Paz, Oruro, and Potosí), Chile (Tarapacá, Arica, Parinacota, and Antofagasta), and Peru (Tacna, Puno, and Moquegua). Its activities will be carried out over a period of five years, with the aims of identifying and inventorying the traditional knowledge and oral traditions of Aymara communities, promoting and disseminating Aymara's oral and musical expressions, and reinforcing traditional knowledge related to the production of textile arts and traditional agricultural techniques. Once again, a project aimed at safeguarding cultural heritage promotes traditional agricultural techniques, and may also contribute to the conservation and sustainable use of agrobiodiversity and food security.

Registry of Intangible Cultural Heritage and Agrobiodiversity-Rich Systems in the Brazilian Amazon: a new perspective for the safeguarding of traditional agricultural systems

The Brazilian Constitution approved in 1988 expanded the concept of cultural heritage (Article 216) and explicitly recognized its dual nature – tangible and intangible. According to the Constitution, the Brazilian cultural heritage consists of goods, of material and immaterial nature, taken individually or as a whole, which bear reference to the identity and memory of the various groups that form the Brazilian society, including:

- forms of expression;
- ways of creating, making and living;
- scientific, artistic and technological creations;
- works, objects, documents, buildings, and other spaces intended for artistic and cultural expressions;
- urban complexes and sites of historical, natural, artistic, archaeological, paleontological, ecological, and scientific value.

Thus, cultural heritage includes not only material assets but also oral traditions and expressions, innovative ways of creating, making and living, social practices, and knowledge concerning nature, among others. The concept adopted by the Brazilian Constitution was that it is not possible to understand cultural heritage without taking into consideration the values invested in it and what it stands for – its immaterial or intangible dimension – and, likewise, the dynamics of intangible heritage cannot be understood without knowledge of the material culture which supports it. The constitutional definition comprehends cultural expressions that are not objects, but dynamic processes, and it values "living" heritage, rooted in the daily lives of communities. According to the Constitution, the Brazilian government must, in cooperation with society, promote and protect the Brazilian cultural heritage by means of inventories, registries, vigilance, protection decrees, expropriation, and other forms of precaution and safeguarding.[11]

On April 4, 2000, Presidential Decree No. 3551 created an intangible cultural heritage registry, which is divided into four different books:

1 Registry of Knowledge, in which practices and traditional craftsmanship, rooted in the daily lives of local communities, are registered. Examples of cultural heritage inscribed under this category are:
   a The traditional process of making ceramic pans in Goiabeiras (a county in the city of Vitoria, capital of the southeastern Brazilian state of Espírito Santo). These pans are used for making *moqueca capixaba* (a stew made with fish, onions, garlic, tomatoes, peppers, and cilantro, and which is a very traditional dish in this region of Brazil).
   b The traditional process of making *acarajé* by *baianas* (women from the state of Bahia, in northeastern Brazil). *Acarajé* is a dish made from peeled black-eyed peas formed into a ball and then deep-fried in *dendê* (palm oil), and it is served split in half and then stuffed with *vatapá* and *caruru* – spicy pastes made from shrimp, ground cashew nuts, palm oil, and other ingredients. *Acarajé* is the ritual food of the goddess Iansan, of the Afro-Brazilian religion known as *candomblé.*

   It is important to emphasize that in both examples described above, the Registry aims to safeguard the processes of making ceramic pans or *acarajé*, and not the objects themselves.
2 Registry of Cultural Expressions, in which literary, musical, plastic, scenic and ludic performances are registered. An example of cultural heritage inscribed under this category is *frevo* – a musical style and dance performed in the street carnival of Olinda, in the northeastern part of Brazil.
3 Registry of Celebrations, in which rituals and feasts which are marked by a collective coexistence, such as work, religion, entertainment, and others practices of social life, are registered. An example is the religious feast of Nossa Senhora de Nazaré, celebrated in Belém do Pará, in northern Brazil.
4 Registry of Cultural Spaces, in which marketplaces, fairs, sanctuaries, squares, and other locations in which collective cultural practices are registered. Examples include the Iauaretê Waterfall, which is a sacred site for Indigenous people of the Upper Negro River, located in the Iauaretê district of the municipality of São Gabriel da Cachoeira, in the Brazilian Amazon, and the Caruaru Fair, which is an agglomeration of many open-air markets in the city of Caruaru (in the state of Pernambuco, in northeastern Brazil), where just about anything is sold: fruits, vegetables, medicinal plants, livestock, handicrafts, embroidery, leather products, etc.[12]

The main objective of the Registry is to gather and systematize thoroughly all knowledge and documentation related to the cultural expression or practice for which recognition is sought, to enable its widespread diffusion and social recognition. The Registry has a declaratory nature (that is, it only declares or recognizes something that already exists as a cultural expression or practice) and it must always be supported by the social groups and local communities involved in its production and transmission. Cultural heritage safeguarded by the Registry does not necessarily generate goods and services with an economic value, despite their high cultural, symbolic, political, and social value.

When an intangible cultural heritage is registered, it receives the title of "cultural heritage of Brazil," and its registry creates the legal obligation for governmental agencies (at federal and state levels) to develop safeguarding actions and plans, aimed at supporting the continuity of its existence and transmission and at providing the necessary social and material conditions for its continuity. The Registry also takes into consideration the dynamic and evolving nature of intangible cultural heritage, and IPHAN[13] (the Brazilian federal agency in charge of safeguarding cultural heritage) must re-evaluate all registered heritage at least once every 10 years, in order to decide whether to revalidate the title of "cultural heritage of Brazil." Tradition is regarded as practices and cultural expressions which are transformed and updated through time, maintaining their essence and meanings for contemporary societies, that is, tradition does not refer necessarily to something old or belonging to the past.

Presidential Decree 3551/2000 also established the National Intangible Heritage Program, which adopts, as additional instruments to the Registry, the National Inventory of Cultural References (INRC) and safeguarding plans in which the most appropriate forms of safeguarding cultural heritage are defined, with options ranging from financial assistance for holders of knowledge to facilitating access to raw materials needed for crafts. The INRC is aimed at producing knowledge related to domains of social life to which meanings and values are attributed by a social group, and which constitute cultural cornerstones and identity references for that given social group. In addition to the categories of heritage already included in the registry, it includes buildings associated with certain uses, historical meanings, and urban images, regardless of their architectural or artistic value.

On November 8, 2010, IPHAN recognized, for the first time, an agrobiodiversity-rich traditional agricultural system as "intangible cultural heritage of Brazil," the traditional agricultural system of the Negro River region, in the northwest of the Brazilian Amazon. The request for recognition was filed by the Association of Indigenous Communities of the Middle Negro River (ACIMRN) in July, 2007, with the support of two interdisciplinary research programs on agrobiodiversity and associated traditional knowledge developed in the Brazilian Amazon.[14]

According to Laure Emperaire, Lúcia Van Velthem and Ana Gita de Oliveira (Emperaire *et al.*, 2008), in the context of the Negro River, an agricultural system may be understood as "a whole set of knowledge, myths, oral expressions, practices, products, techniques, artifacts and other associated manifestations and expressions which involve the management of spaces, the cultivation of plants, the transformation of agricultural products and local food systems." These authors explain that the notion of system links cultural heritage (to which recognition is sought) to a wider and more complex set of social relationships. Thus, it opens up the perspective of registering (by IPHAN) broader elements of Brazilian cultural heritage, such as the Registry of Traditional Agricultural Systems, among which is the Negro River, characterized by a set of interdependent elements rather than a single object or specific good. Emperaire *et al.* (2008) also add that:

> the traditional agricultural system of the Negro River is very rich in agrobiodiversity: studies carried out in two communities of the Middle Negro River – Tapereira and Espírito Santo – and in the city of Santa Isabel do Rio Negro, found 243 cultivated species and 73 cassava varieties. Each family cultivates between 17 and 97 different species and from 6 to 20 cassava varieties. In addition to the cassava genetic diversity, the studies also found a high diversity of peppers, pineapples, yams and bananas, which confirms the regional importance of the Negro River in terms of conservation of agricultural diversity.

To Emperaire *et al.* (2006), the Registry of the Negro River agricultural system as intangible cultural heritage is "a concrete example of how instruments and policies for safeguarding cultural heritage can be used in favor of agrobiodiversity, cultural diversity and local agricultural systems."

In Brazil, as in most developing countries, most native plant genetic resources are conserved on-farm, regardless of whether they are located within or outside the limits of protected areas, and these resources are not well represented in *ex situ* collections. Of the 250,000 accessions conserved in the germplasm banks of EMBRAPA (Brazil's principal agricultural research institution), for example, approximately 76 percent are of exotic species, and only the remaining 24 percent are of native species (Goedert, 2007, p. 33). On-farm conservation focuses its attention on agricultural crops of interest to farmers, and, as Clement *et al.* (2007, p. 515) explain, "so many people are involved with on-farm conservation because it is intrinsic to their social and economic organizations; knowing and conserving biological diversity over time and space is one of the main factors in their social reproduction." In the Brazilian Amazon, the Kayabi Indigenous people cultivate over 140 cultivars belonging to 30 different species (Silva, 2008); the Yanomami Indigenous people 40 (Milliken and Albert, 1999); and rubber tappers (*seringueiros*) in Upper Juruá (in the state of Acre, in the Brazilian Amazon) 17 manioc/cassava, 14 banana, and nine bean species (Pantoja *et al.*, 2002).

Another form of safeguarding products and processes associated with agrobiodiversity is through the development of inventories and the registry of local food systems as intangible cultural heritage, considering that food and culture are directly associated. IPHAN has also produced, between 2001 and 2006, the National Inventory of Cultural References of Manioc (Cassava) Flour (in the states of Pará, Bahia, Rio de Janeiro, Santa Catarina, and Paraná). Although cassava is cultivated in several parts of Brazil, and cassava flour is consumed throughout the country, each region of Brazil has its own methods of cultivation, preparation, and consumption, which are expressed in traditions and knowledge passed from one generation to another, and in the influence of different ethnic groups. There is a wide range of cassava species, which are used for different purposes and in different social and cultural contexts.

IPHAN is also working on an inventory of cultural references of the traditional method of making *cajuína*, a nonalcoholic drink made from blended cashew apple juice. The cashew apple (known in Brazil as *caju*) is the accessory fruit – also called false fruit or pseudocarp – of the native Brazilian cashew tree. *Cajuína* is crystal clear and it has a typical yellow–amber color that results from caramelization of the natural sugars present in cashew apple juice. The typical *cajuína* does not contain any chemical additive or preservative, as the drink is sterilized inside the bottle. *Cajuína* is very traditional in the northeast of Brazil, especially in the states of Ceará and Piauí.

In 2005, IPHAN also concluded the National Inventory of Cultural References of traditional practices of *tacacazeiras* (women who make *tacacá*, a traditional dish that is widely consumed in northern Brazil). *Tacacá* is a type of soup made with *jambu* (a native variety of paracress, whose leaves and flower heads contain an analgesic agent also used to numb toothaches), *tucupi* (a broth made with cassava), dried shrimps, and small yellow peppers. It is served extremely hot, in *cuias* (traditional handcrafted bowls), usually in the late afternoon. *Tacacazeiras* (*tacacá* makers) and their stands are part of the landscape in the streets of Belém (capital of the state of Pará, in the Brazilian Amazon), as pointed out by Maria Dina Nogueira (2006). Promoting these traditional foodways is a way of conserving agrobiodiversity, as many plant varieties used in the preparation of traditional dishes would not be cultivated otherwise.

Intangible cultural heritage includes not only songs, tales, and dances, but also agricultural knowledge, innovations, and practices held by traditional and local farmers, which range from different forms of cultivation (home gardens, swidden, agroforestry systems) to biological control of pests and diseases and genetic improvement of local plant varieties, through farmers' selection and breeding. Traditional and local knowledge associated with agrobiodiversity has been recognized as part of the Brazilian intangible cultural heritage and, therefore, must be safeguarded. However, both aspects of this cultural heritage – material/tangible (agroecosystems and cultivated plants) and immaterial/intangible (agricultural knowledge and practices) – are protected by the Brazilian Constitution (Article 216) and, therefore, safeguarding measures, such as actions and plans, must contemplate both of them, which are intrinsically linked and indissociable.

Recognition of traditional knowledge associated with maize diversity and of local foods as intangible cultural heritage in Peru

On December 23, 2009, Peru issued National Directorial Resolution (Resolución Directoral Nacional) No. 1986/INC,[15] declaring, as "national cultural heritage," the knowledge, practices and technologies associated with the traditional cultivation of maize in the Sacred Valley of Incas, in the Andes of Peru. This was the first time that an agricultural system was recognized as cultural heritage in Peru.

According to the above-mentioned Resolution, maize or *Sara* (in the Quechua language), despite being a cereal of Meso-American origin, has been cultivated in the Peruvian Andes for millennia, and 55 maize varieties can be found in this region. The remains of maize dating from 4,300 BC have been found in the region of Ayacucho (in the south-central Andes), and remains of maize dating from 2,500–1,800 BC have also been found in the coastal area of the country. Maize has always had great symbolic and cultural value for pre-Hispanic peoples. During the Inca Empire (called *tahuantinsuyo* by the Inca), maize was so important that holiday calendars, land divisions, and social relationships were all established around its sowing and harvesting cycles. The ubiquitous presence of maize in cultural expressions of Peruvian Indigenous peoples can be appreciated in designs (textiles, imagery, and ceramics) and the performing arts (dances, songs, and myths).

The Inca society developed sophisticated practices and technologies to produce maize: new varieties of maize, adapted to different geographic and climatic regions, were created, and complex irrigation and storage systems were developed. The Sacred Valley of Incas is located at an altitude between 2,600 and 3,050 meters above sea level, and its temperate climate is very favorable to the cultivation of maize. Currently, eight native maize varieties are cultivated, the best known of which is "giant white corn" (*maiz blanco gigante*). Maize is one of the most important elements of rituals and ceremonies occurring in the Sacred Valley. Each maize variety has its particular characteristics, and is used in a specific ceremony, for magical or medicinal purposes, or for the worship of deities. Maize harvesting is done in a collective manner, through reciprocal and communal work (called *ayni*), and ritual elements are also present in maize harvesting, such as the offering of a traditional drink (*challasqa)* to the first cut stems. Practices and technologies associated with the cultivation of maize are important because of both their symbolic and economic dimensions, according to the Resolution, which was enacted in accordance with Peruvian Law No. 28296, of 2004, and its regulation, approved through Supreme Decree No. 011-2006-ED. Peru ratified, in 2004, UNESCO's Convention for the Safeguarding of Intangible Cultural Heritage.

According to Peruvian Law No. 28296 (Ley General del Patrimonio Cultural de la Nación), the intangible/immaterial cultural heritage of the Peruvian nation is constituted of

> creations of a cultural community based on tradition, which are expressed by its members individually or in a collective manner, and which are recognized by the community as expressions of its cultural and social identity, including orally transmitted values, such as Indigenous languages and dialects and traditional knowledge, whether artistic, culinary, medicinal, technological, folklore or religious, and the collective knowledge of peoples and other cultural expressions or manifestations, which together constitute our cultural diversity.

This law also affirms that intangible cultural heritage belongs to the Peruvian nation, and that no natural or legal person can claim ownership over it, and that the communities that maintain and preserve intangible cultural heritage are their direct possessors. This law created the National Registry of Folklore and Popular Culture (Registro Nacional de Folclore y Culturas Populares), in which all tangible and intangible cultural heritage belonging to folklore and popular culture are registered. Safeguarding actions and policies are aimed at identifying, documenting, registering, inventorying, preserving, promoting, valorizing, transmitting, and revitalizing intangible cultural heritage. The official agency responsible for safeguarding intangible cultural heritage is Instituto Nacional de Cultura, and its Dirección de Registro y Estudio de la Cultura del Perú Contemporáneo (DREPC).

Maize, as a food, had already been included in the declaration of the Peruvian traditional food as cultural heritage.[16] Peruvian traditional food is based on several millennia of cultural development by diverse ethnic groups, and is one of the oldest food cultures in the world. Agricultural technologies, as well as the management of water, were highly developed by the Incas, which led to the domestication of a huge variety of plants and to the use of a large variety of fauna species. Peruvian food is a result of the country's rich biodiversity and cultural diversity. Other traditional foods are also recognized as part of the cultural heritage of Peru:

- *Pisco* is a grape brandy produced in Peru.[17] Grapes used for manufacture of *pisco* benefit from mild weather and from the tectonic formation of

the soil in the province of Pisco, extending to the valleys of Lima, Ica, Arequipa, Moquegua, and some valleys of Tacna with similar conditions.
- *Pisco sour*[18] is a cocktail containing *pisco*, lemon or lime juice, egg whites, syrup, and bitters.
- *Ceviche*, which is a dish made of cubed raw fish, lime or lemon juice, onion, and traditional Andean spices which include salt and chili, is a combination of Muslim and native Andean cuisines. *Ceviche* is consumed in all parts of Peru.[19]
- *Pachamanca* is a unique and ancient way of cooking certain foods (lamb, chicken, guinea pig, marinated in spices, as well as sweet potato, chili, beans, etc.), which are distributed within a hole in the ground and covered with rocks at high temperatures (the earthen oven is known as a *huatia*). *Pachamanca* is used mainly in the central Peruvian Andes, and has existed since the time of the Inca Empire.[20]

Peru has already recognized, as cultural heritage, traditional knowledge and use of the plant *ayahuasca*, for medicinal and religious purposes, especially in the Amazon region.[21]

The UNESCO Convention for the Protection of the World Cultural and Natural Heritage and the concept of "cultural landscapes"

The UNESCO Convention concerning the Protection of the World Cultural and Natural Heritage (known as the World Heritage Convention) was adopted by the General Conference of UNESCO at its 17th session, in 1972. The Convention's purpose is to ensure the identification, protection, conservation, presentation, and transmission to future generations of cultural and natural heritage of "outstanding universal value"[22] from the points of view of history, art, science, esthetics, ethnology, or anthropology. The World Heritage Convention has been ratified by 187 states. Among the 911 properties (sites) inscribed on the World Heritage List, 704 are cultural (of which 66 are cultural landscapes), 180 natural, and 27 mixed (natural/cultural) properties, located in 151 states (as of June 2010).[23] The World Heritage Committee is responsible for the implementation of the World Heritage Convention, defines the use of the World Heritage Fund,[24] and allocates financial assistance upon requests from states. It also has the final say on whether a property is inscribed on the World Heritage List.[25]

The World Heritage Convention was developed from the merging of two separate movements: the first focusing on the preservation of cultural sites and the other dealing with the conservation of nature. Reflecting an (almost) antagonistic perception of nature and culture, the convention originally divided World Heritage sites into two categories: natural and cultural heritage. According to Peter Fowler (2003), when the Convention was adopted, nature conservationists' thinking was along the lines that the less human interference there had been with an area, the "better" it was. Similarly, the cultural movement embraced mainly monuments, buildings, and ruins as isolated phenomena largely in the minds of architects, architectural historians, and those of an esthetic tendency, with little thought of context and the landscape itself. Most cultural heritage sites on the list were monuments and buildings, recognized as masterpieces of human creativity. On the other hand, natural heritage sites inscribed on the World Heritage List were mainly pristine and wilderness areas, with little human interference. Later, a "mixed" (cultural and natural) category of site was created, to enable the inscription of properties (sites) whose recognition could be justified under both natural and cultural criteria. However, this category also did not explore to any great extent the interfaces between biological and cultural diversity.

It was only in 1992 that the World Heritage Committee established a new and innovative category of site: the "cultural landscape," within the Operational Guidelines for the Implementation of the World Heritage Convention. This happened in the same year (1992) that the UNCED was held, in Rio de Janeiro, and the CBD was adopted.[26] The European Landscape Convention, also known as the Florence Convention, promotes the protection, management, and planning of European landscapes and organizes European cooperation on landscape issues. The convention was adopted on October 20, 2000, in Florence (Italy) and came into force on March 1, 2004. It is different from the UNESCO World Heritage Convention not only in its scope (which is regional, rather than international) but also because it covers all landscapes, even those which have no outstanding universal value. According to the European Landscape Convention, landscape means "an area, as perceived by people, whose character is the result of the action and interaction of natural and/or human factors."[27]

Cultural landscapes are justifiably included in the UNESCO World Heritage List when interactions between people and their environment are evaluated as being of "outstanding universal value." According to the Operational Guidelines for the Implementation of the World Heritage Convention:[28]

1 Cultural landscapes are cultural properties and they represent "the combined works of nature and of man." They are illustrative of the evolution of human society and settlement over time, under the influence of the physical constraints and/or opportunities presented by their natural environment and of successive social, economic and cultural forces, both external and internal.
2 Cultural landscapes should be selected on the basis both of their outstanding universal value and of their representativity in terms of a clearly defined geocultural region and also for their capacity to illustrate the essential and distinct cultural elements of such regions.

3 The term "cultural landscape" embraces a diversity of manifestations of the interaction between humankind and its natural environment.

4 Cultural landscapes often reflect specific techniques of sustainable land use, considering the characteristics and limits of the natural environment they are established in, and a specific spiritual relation to nature. Protection of cultural landscapes can contribute to modern techniques of sustainable land use and can maintain or enhance natural values in the landscape. The continued existence of traditional forms of land use supports biological diversity in many regions of the world. The protection of traditional cultural landscapes is therefore helpful in maintaining biological diversity.

5 Cultural landscapes fall into three main categories, namely:

i The most easily identifiable is the clearly defined landscape, designed and created intentionally by man. This embraces garden and parkland landscapes constructed for esthetic reasons which are often (but not always) associated with religious or other monumental buildings and ensembles. Some examples are Kew Gardens,[29] in the UK, the Garden Kingdom of Dessau-Wörlitz, in Germany, the Lednice–Valtice Cultural Landscape in the Czech Republic, and Kalwaria Zebrzydowska, the Mannerist Architectural Park and Landscape Complex, in Poland.

ii The second category is the organically evolved landscape. This results from an initial social, economic, administrative, and/or religious imperative and has developed its present form by association with and in response to its natural environment. Such landscapes reflect that process of evolution in their form and component features. They fall into two subcategories:

- A relict (or fossil) landscape is one in which an evolutionary process came to an end at some time in the past, either abruptly or over a period. Its significant distinguishing features are, however, still visible in material form. Examples are the Cultural Landscape and Archaeological Remains of the Bamiyan Valley, in Afghanistan, and the Blaenavon Industrial Landscape, in the UK.
- A continuing landscape is one that retains an active social role in contemporary society closely associated with the traditional way of life, and in which the evolutionary process is still in progress. At the same time it exhibits significant material evidence of its evolution over time. Examples are the Archaeological Landscape of the First Coffee Plantations in the Southeast of Cuba, the Tokaj Wine Region Cultural Landscape, in Hungary, the Puszta Pastoral Landscape of Hortobagy National Park, in Hungary, the Rice Terraces of the Philippine Cordilleras, and the Agricultural Landscape of Southern Öland, in Sweden.

iii The final category is the associative cultural landscape. The inclusion of such landscapes on the World Heritage List is justifiable by virtue of the powerful religious, artistic, or cultural associations of the natural element rather than material cultural evidence, which may be insignificant or even absent.[30] Examples are the Uluru-Kata Tijuta National Park, in Australia (also inscribed as a natural site), the Holy Valley (Ouadi Qadisha) and the Forest of the Cedars of God (Horsh Arz el-Rab) in Lebanon, the Royal Hill of Ambohimanga, in Madagascar, and the Tongariro National Park, in New Zealand (also inscribed as a natural site).

The World Heritage Committee's decision to create the "cultural landscape" category was an important development in linking conservation of natural and cultural heritage, as it embraced (UNESCO World Heritage Centre, 2003):

- recognition of the diversity of manifestations of the interaction between humankind and its natural environment;
- introduction of the term "sustainability" into the Operational Guidelines via "specific techniques of sustainable land-use";
- acceptance of the living heritage of Indigenous people and other local communities;
- introduction, into the Operational Guidelines, of traditional management practices and customary law as acceptable forms of protection for natural heritage;
- recognition of traditional forms of land use and of community stewardship;
- maintenance of biological diversity through cultural diversity; and
- consideration of spiritual relationships to nature.

According to Mechtild Rössler (2005, p. 37), cultural landscapes are at the interface between nature and culture, tangible and intangible heritage, biological and cultural diversity, and they represent "a tightly woven net of relationships that are the essence of culture and people's identity." He points out that the recognition of cultural landscapes gave value to land use systems that represent the continuity of people working the land over centuries and sometimes millennia to adapt the natural environment and retain or enhance biological diversity. According to Rössler (2005, p. 40), the key world crops were developed in the spectacular agricultural systems in the High Andes, terraced rice paddies in Asia,[31] or oasis systems in the Sahara, and the global importance of these systems and the genetic diversity of these cultural landscapes were recognized (as a result of their inscription as cultural landscapes).

"Cultural landscapes" and the safeguarding of traditional agricultural systems in the Philippines, Cuba, Hungary, Sweden, and Brazil

The "continuing" form of "organically evolved landscapes" represents the category of cultural landscape that is most adequate for safeguarding biodiversity-rich agricultural systems. Some examples of agricultural landscapes inscribed under the World Heritage List are:[32]

1 The Rice Terraces of the Philippine Cordilleras (date of inscription: 1995) were the first site to be included on the World Heritage Cultural Landscape list under the continuing organically evolved category. The rice terraces have existed on very steep slopes for an estimated 2,000 years. The maintenance of the living rice terraces reflects a cooperative approach of the whole community which is based on detailed knowledge of the rich diversity of biological resources existing in the Ifugao agro-ecosystem, a finely tuned annual system respecting lunar cycles, zoning and planning, extensive soil conservation, mastery of a most complex pest control regime based on the processing of a variety of herbs, and accompanied by religious rituals. In 2001, the terraces were included on the List of World Heritage in Danger, as uncontrolled tourism and the introduction of open market economy threatened both the natural heritage of the province and the traditional practices of its inhabitants.
2 The Archaeological Landscape of the First Coffee Plantations in the Southeast of Cuba were inscribed in 2000. The remains of the nineteenth and early twentieth-century coffee plantations in eastern Cuba are unique and eloquent testimony to a form of agricultural exploitation of virgin forest, the traces of which have disappeared elsewhere in the world. The production of coffee in eastern Cuba during the nineteenth and early twentieth centuries resulted in the creation of a unique cultural landscape, illustrating a significant stage in the development of this form of agriculture. The site consists of the remains of 171 historic coffee plantations on the steep and rugged slopes of mountain valleys at the base of Sierra Maestra.
3 The Agricultural Landscape of Southern Öland, in Sweden, was also inscribed in 2000. The southern part of the island of Öland in the Baltic Sea is dominated by a vast limestone plateau. Humans have lived here for some 5,000 years and adapted their way of life to the physical constraints of the island. The interaction between man and the natural environment in the south of Öland is of unique universal value. The continuity of land use goes back to the Stone Age, when people began farming this area. The use made of the land has not changed significantly since then, with arable farming and animal husbandry remaining the principal economic activity.

4 The Tokaj Wine Region Historic Cultural Landscape, in Hungary, was inscribed in 2002. The cultural landscape of Tokaj graphically demonstrates the long tradition of wine production in this region of low hills and river valleys. The intricate pattern of vineyards, farms, villages, and small towns, with their historic networks of deep wine cellars, illustrates every facet of the production of the famous Tokaj wines, the quality and management of which have been strictly regulated for nearly three centuries.
5 The Puszta Pastoral Landscape of Hortobagy National Park, in Hungary, was inscribed in 1999. The cultural landscape of the Hortobágy Puszta consists of a vast area of plains and wetlands in eastern Hungary. Traditional forms of land use, such as the grazing of domestic animals, have been practiced in this pastoral society for more than two millennia.

All of the above-mentioned examples show the possibility of using the cultural landscape category to promote and safeguard agrobiodiversity heritage sites (Singh and Varaprasad, 2008). The rich genetic heritage of crops and livestock associated with diverse agroecosystems, as well as agricultural techniques, practices, and innovations held by local communities, can be protected not only through conventional environmental law instruments but also via cultural heritage protection law, both internationally and nationally.

In addition to cultural landscapes "of outstanding universal value," recognized by UNESCO's International Convention on World Heritage, Brazil created a national legal instrument for the recognition of Brazilian cultural landscapes, known as *chancela* (the word *chancela* could be translated as a seal, or as an official stamp, or as a certificate granted to some cultural landscapes of particular value). Brazilian cultural landscapes are regulated by Administrative Order (Portaria) No. 127, of April 30, 2009, issued by the President of IPHAN (the Brazilian federal agency for cultural heritage protection). According to this Administrative Order, a Brazilian cultural landscape is a "peculiar portion of the national territory that is representative of the process of interaction of man with the natural environment, in which life and human science have made their marks or attributed values." The "Brazilian cultural landscape" is declared by a seal instituted by IPHAN, and any citizen is entitled to request the opening of an administrative process aimed at awarding a "Brazilian cultural landscape" seal to a specific site. The seal establishes a pact among government, civil society, and private companies, aimed at collectively managing the portions of the national territory that have been recognized as cultural landscapes. The seal must take into consideration the dynamism of culture and human activity and must be revalidated after, at most, 10 years. The objective of the seal is to contribute to the preservation of cultural heritage, complementing and integrating other existing safeguarding instruments.

On May 3, 2011, IPHAN (the Brazilian federal agency for cultural heritage protection), for the first time, granted the "Brazilian Cultural Landscape" seal to specific sites: the rural communities of Testo Alto (in the municipality of Pomerode) and Rio da Luz (in the municipality of Jaraguá do Sul), in the Itajaí Valley, in the state of Santa Catarina (southern Brazil). This is an important region for the conservation of agrobiodiversity and associated cultural diversity. In the Itajaí Valley, the small rural properties run by family small-scale farmers have been the key elements for development and sustainability of the rural properties since the colonial period. Even when immigrant colonies became economically developed and some of their members became wealthy, agricultural mini-properties remained as the main basis for the agricultural production system. All houses have vegetable gardens (containing tomatoes, cabbage, collard greens, lettuce, pumpkins, cucumbers, peanuts, peppers and herbal spices, all of which are abundantly used in meals) and orchards in the backyards, with avocados, persimmons, *carambolas* (starfruit), *jaboticabas* (a tough-skinned purple grapelike tropical fruit grown in Brazil), *pitangas* (a Brazilian spicy red fruit), oranges, lemons, guavas, and *araçás* (a yellow fruit that resembles a guava, but with a more acidic flavor). Banana trees and sugarcane are ubiquitous and palm trees' heart of palm is saved for special occasions. Generally, small lakes, with ducks and teals, are located nearby, and bamboo, because of its limitless uses, is indispensible. In the areas of Polish and Italian immigrants grapevines are nearly mandatory, enabling the production of traditional home-made wines (IPHAN, 2007). This is an example of the potential use of the category "cultural landscape" to protect traditional and local agricultural systems. Like Brazil, other countries could use the seal of cultural landscapes as an instrument to protect local and traditional agricultural systems which are rich in agrobiodiversity and representative of the country's cultural diversity.

Globally Important Ingenious Agricultural Heritage Systems (GIAHS): general overview of pilot agroecosystems in Peru, Chile, the Philippines, Magreb (Algeria, Morocco, and Tunisia), China, Kenya, and Tanzania

In 2002, FAO started an initiative for the conservation and adaptive management of Globally Important Agricultural Heritage Systems (GIAHS). The initiative aims to establish the basis for international recognition, dynamic conservation, and adaptive management of GIAHS and their agricultural biodiversity, knowledge systems, food and livelihood security, and cultures throughout the world.[33] GIAHS are defined as "remarkable land use systems and landscapes which are rich in globally significant biological diversity evolving from the co-adaptation of a community with its environment and its needs and aspirations for sustainable development."[34] The GIAHS program is supported by the Global Environment Facility (GEF, through the UNDP). Other partners include UNESCO, the International Center for the Study of the Preservation and Restoration of Cultural Property (ICCROM), the United Nations University (UNU), the International Fund for Agricultural Development (IFAD), Bioversity International, and IUCN (International Union for Conservation of Nature). Universities, private sector, and civil society organizations also participate in pilot projects.

The concept of GIAHS is distinct from, and more complex than, a conventional heritage site or a protected landscape. Even though the GIAHS project recognizes that the experience of UNESCO with the identification of World Heritage Sites and landscapes, particularly the category of continuing, organically evolved landscapes, has been useful to promote agrobiodiversity, it considers that such category would need to be complemented by a more agricultural focus and a land use system approach. The possibility of including agricultural systems under existing cultural landscape categories or eventually creating a new category of World Heritage is being explored by the GIAHS project, which aims to establish an internationally accepted system for the recognition of GIAHS.[35] The GIAHS project expects to facilitate mainstreaming of agrobiodiversity conservation in national biodiversity policies and plans to improve the capacity of countries (where pilot systems were established) to promote sustainable use of agrobiodiversity. It intends to benefit mainly local traditional family farming communities and Indigenous peoples. The project aims to create a GIAHS network and a long-term open-ended program. Approximately 200 agricultural systems have been identified, and pilot systems have been established in the following countries:[36]

1 The Andean agricultural system (Peru), in the valleys of Cuzco and Puno, in the proximity of the Inca city of Machu Picchu. Agricultural crops are divided in terraces, which reach 4,000 meters in altitude, and the region is the center of origin for potatoes (domesticated by the Aymara and Quechua Indigenous peoples), quinoa, cinchona, coca, amaranth, chili, and roots of great importance in regional diets, such as arracacha and yacón.[37]
2 The agricultural system in the Chiloé Archipelago, in southern Chile, is a center of origin and diversity for potatoes, and approximately 200 native potato varieties are still cultivated by the Huilliche Indigenous people, as well as a garlic variety (*ajo chilote*) that exists only in the volcanic soils of the Chiloé Archipelago.
3 The rice terraces in the province of Ifugao, in the Philippine mountains, which have also been recognized as a World Heritage Site by UNESCO under the "cultural landscape" category of the World Heritage Convention. It is an agroecosystem of high mountains, in which terraces

interact with a set of micro-basins, which function as irrigation and filtering systems. The rice terraces are located along the curves of the mountains. It is estimated that 565 rice varieties are conserved (Nozawa *et al.*, 2008).

4 Agricultural systems in the oases of the Magreb region (Algeria, Morocco, and Tunisia). The oases of the Magreb region are green islands flourishing in a constraining and harsh environment. They are home to a diversified and highly intensive and productive system, which has been developed over millennia. The Tamegroute oasis, in Morocco, also participates in the UNESCO program "Man and the Biosphere" and is part of the "biosphere reserve of southern Morocco oases." It is a highly diversified agricultural system, which is intensive and productive, developed over thousands of years. These areas produce dates, other fruits (pomegranates, figs, peaches, apples, grapes, etc.), legumes and vegetables, cereals, medicinal plants, etc.
5 Rice–fish agriculture in China. Fish are raised in flooded rice fields, in an ecologic symbiosis: fish provide fertilizer for rice, regulate micro-climatic conditions, soften the soil, disturb the water, and eat larvae and weeds in the flooded fields, and rice provides shade and food for fish. It is a very traditional agricultural system, which has existed since the Han Dynasty, 2000 years ago.
6 Hani rice terraces system, in China. These rice terraces are located in the Honghe Hani and Yi Autonomous Prefecture, in the southeast part of Yunnan Province. The typical ecological landscape of the Hani rice terraces is composed of the forest, village, terrace, and river. Hani minorities have lived in this landscape for over 1,300 years.
7 Wannian traditional rice culture system, in China. Wannian traditional rice is a remarkable old and prototypical variety. It is a variety unique to Heqiao village, and it can be grown only in the water, soil, and climatic conditions that prevail in Heqiao village. The traditional rice needs the perennial cold spring water for irrigation, and surrounding forests play a crucial role in soil and water conservation. The surrounding forests and paddy fields are part of the same biodiversity-rich agroforestry system;
8 Pastoral system and upland agroecosystem (Kenya, Tanzania). Over 75 percent of the African population lives in rural small-scale holdings and family farms. The aim of this project is to enhance the viability of smallholding and traditional agriculture and agropastoral systems and enhance food and nutritional security of Indigenous communities depending on these systems in Kenya and Tanzania. The project intends to address adaptive management and conservation of productive landscape of Masaai and Tapade communities.

During the identification stages of the GIAHS projects, many agricultural systems were identified, and met the criteria for GIAHS. Some of these systems have been listed as candidate system for GIAHS recognition, such as the traditional agroecosystems in the Carpathians region (Central and Eastern Europe), the Lemon Gardens (southern Italy), traditional agriculture in the Koraput region (Orissa, India), Soppina Bettas systems (Western Ghats, India), Qashqai nomadic pastoralism (Iran), Milpa-Solar systems (Mexico), Chinampa agricultural system (Mexico), Qanat irrigation systems and home gardens (Iran), and Wewe Irrigation system (Sri Lanka).

The GIAHS program aims to identify, define, and support forms of conservation and dynamic management of these agricultural systems, in order to allow local farmers to maintain biological diversity and at the same time conserve the natural resources necessary for their survival. It seeks to develop public policies and incentives for *in situ*/on-farm conservation of biodiversity and associated traditional knowledge. One of the characteristics of these traditional agricultural systems is precisely their wealth of agrobiodiversity: at least 177 distinct varieties of potatoes have been identified in the Peruvian Andes region, approximately 20 traditional rice varieties are found in fish–rice farming in China, and over 10 distinct varieties of dates have been identified in the Magreb oases. In addition to acknowledging the value and protecting traditional/local agricultural systems, the GIAHS program can provide inputs for discussions regarding creation of protected areas especially aimed at conservation of agrobiodiversity (also known as "agrobiodiversity reserves").

GIAHS, Amazonian dark earths, and agrobiodiversity

Agriculture heritage systems can be found all over the world, and the FAO program (mentioned above) intends to encompass 100–150 GIAHS worldwide. One of the proposals presented to FAO, for recognition as a GIAHS, was the Amazonian Dark Earths system. The proposal was presented by Brazilian research institutions (EMBRAPA, Instituto Nacional de Pesquisas da Amazonia, and Museu Emilio Göeldi) and a US university (Cornell University, Ithaca), but it was not selected as a pilot project. In the Americas, other proposals included the Agrarian System of the Wayana (French Guyana), the Little Colorado River Watershed (Arizona, USA), Conservation of the Milpa Traditional Agroecosystem and its related Cultural Values in Los Altos of Chiapas (Mexico), and Los Cuchumantanes of Huehuetenango (Guatemala).

Amazonian dark earths are dark-colored, highly fertile soils – they are a unique product of ingenious soil management by ancient Indigenous people. Their management, as practiced in the Amazon Basin, builds on a diverse and complex integration of organic soil amendments to maximize revenues and food quality, whilst minimizing resource degradation. Crop yields on Amazonian dark earths are several times higher than on adjacent soils and remain high for many years in a region that normally does not support more than one cropping cycle without massive fertilizer inputs. This resilience creates a remarkable livelihood security,[38] and many researchers believe that Amazonian dark earths maintain higher genetic and biological diversity than other types of soil and could potentially be the "reservoirs" of the rich Amazon agrobiodiversity (Clement *et al.*, 2003).

The formation of dark earths is related to social changes that took place in many parts of the Amazon and in the South American lowlands starting 10,000 years ago. Historically, archeological and paleobotanical studies aimed at the identification of plant domestication centers in the Americas focused their attention in Meso-America – Mexico, Guatemala, and Honduras – and in the central Andes. Contrasting with the plentiful data and research in these regions, Brazilian archaeological sites are still relatively unknown and little studied.[39] However, recent archaeological studies in the upper Xingu and central Brazilian Amazon, led by Eduardo Neves (2006), from the Museum of Archaeology and Ethnology of the University of São Paulo, and Michael Heckenberger (Heckenberger, 2005; Heckenberger *et al.*, 2008), of the University of Florida–Gainesville, show that before arrival of the Europeans the region was densely occupied by complex and highly productive societies which profoundly impacted the Amazon environment. These studies contradict the idea, widely disseminated, that the Amazon was, until European colonization, a pristine and untouched forest, sparsely populated by small tribes. In truth, the Amazon Basin was already occupied by different Indigenous peoples at the end of the fifteenth century, at the time of the beginning of European colonization of the Americas. This occupation, however, was not uniform; on the contrary, it varied through time and space. Precolonial populations adopted social and political patterns of organization which also varied greatly (Heckenberger *et al.*, 2003, 2007, 2008; Neves, 2006). In the central Amazon, where the Negro and Solimões rivers meet, archaeological traces of the Açutuba culture were found, which is thought to have inhabited the region next to Manaus for nearly 10 centuries, until 1,600 years ago. The Manacapuru culture is also believed to have lasted for approximately half a millennium in the same region. The upper Xingu region, in turn, sheltered Indigenous groups living in villages of up to 500,000 square meters and inhabited by up to 5,000 people (Viveiros de Castro, 2008). These populations built squares, roads, bridges, landfills, dams, etc., and caused profound impacts on the Amazon environment, changing forests into domesticated landscapes.

According to Charles Clement (1999a), of the 257 species that were cultivated in the Americas when Europeans arrived (in 1492), 138 were found in the Amazon. This rich genetic diversity, associated with the management and cultivation practices of pre-Columbus peoples, was fundamental in guaranteeing survival of Indigenous peoples in a region as ecologically complex as the Amazon. The brutal decline in the Amazon Indigenous population after the European invasion (between 90 percent and 95 percent of the population was eliminated by diseases brought by Europeans and slavery) provoked a drastic drop in Amazon genetic resources, as they were already in advanced stages of domestication and depended on humans for survival. The Amazon occupies half of South America but is often neglected when discussing American crop origins and diversity.

According to Clement *et al.* (2003), the Amazon is a rich and complex ecological, historical, and cultural mosaic, with high and patchily distributed agrobiodiversity, and dark earths of various types and qualities. Dark earths may generally maintain greater agrobiodiversity than other soil types, and they are more likely to harbor genetically distinct populations of plants that are no longer viable in the modern market. However, with much of native Amazonian agrobiodiversity already lost, and an unknown proportion that remains at risk, a priority should be placed on increasing our very limited general knowledge of this diversity.

Archaeologist Eduardo Neves (2006) explains that, in addition to plant domestication centers which were already known in the Americas (Meso-America and the Andes), the Amazon has gradually gained recognition as an independent South American domestication center. The list of Amazon domesticated plants is long and includes pineapples, *açai*, peanuts, papaya, manioc/cassava, and *pupunha*.[40] The upper Madeira Basin and its tributaries (in the present state of Rondonia) made up the domestication center for two of the most important plants grown in the Amazon: manioc/cassava and *pupunha*. Domestication of manioc/cassava led to the development of characteristics such as thicker and longer roots and sophisticated technologies, based on several instruments, such as the grater, the *tipiti* (cylindrical and elastic basket made of *arumã* fiber), and *cumatá* (large, tightly woven round baskets also made of *arumã*), which turn a poisonous plant into important food products such as *beiju*, flour, *tapioca*, and *caxiri* (a fermented manioc/cassava-based drink). Likewise, domestication of *pupunha* – a type of palm tree whose fruit is widely consumed throughout the Amazon – selected more robust fruit (Neves, 2006).

Plant genetic studies to identify local landraces, and ethnoecological studies of locally adapted agroecosystems where much of the remaining agrobiodiversity is preserved, are urgently needed in the Amazon (Clement *et al.*, 2003). The recognition of Amazonian dark earths as GIAHS would have helped to preserve this unique and invaluable agricultural system.

Notes

1 The Convention on the Protection and Promotion of the Diversity of Cultural Expressions (known as the Convention on Cultural Diversity) expressly recognizes the importance of traditional knowledge as a source of intangible and material wealth, and in particular the knowledge systems of Indigenous peoples, and "its positive contribution to sustainable development, as well as the

need for its adequate protection and promotion." It also recognizes the several forms of creation, production, and manifestation of cultural expressions and the need to adopt measures to protect and promote the diversity of cultural expressions. Therefore, it is also an instrument that can be used to protect traditional knowledge associated with agrobiodiversity.

2 Source: UNESCO, 'What is intangible cultural heritage?' (www.unesco.org/culture/ich/doc/src/01851-EN.pdf; accessed January 10, 2011).

3 The organs of the Convention are the General Assembly of States Parties and the Intergovernmental Committee for the Safeguarding of the Intangible Cultural Heritage. The Committee must establish, keep up to date, and publish a Representative List of the Intangible Cultural Heritage of Humanity, as well as a List of Intangible Cultural Heritage in Need of Urgent Safeguarding. In 2010, the following heritage were inscribed in the second list: Meshrep (a cultural event that includes performance arts, such as music, dance, drama, folk arts, acrobatics, oral literature, foodways and games, found among the Uygur people, who are concentrated largely in China's Xinjiang Uygur autonomous region); Ojkanje singing (a distinctive voice-shaking technique created by the throat, found in the Croatian regions of the Dalmatian hinterland); the watertight-bulkhead technology of Chinese junks (this technology permits the construction of ocean-going vessels with watertight compartments and it was developed in South China's Fujian Province); and the wooden movable-type printing of China (one of the world's oldest printing techniques, it is maintained in Zhejiang Province, where it is used in compiling and printing clan genealogies). For more information, see http://www.unesco.org/culture/ich/index.php?lg=en&pg=00011 (accessed January 10, 2011).

4 For a complete Representative List of the Intangible Cultural Heritage of Humanity, see http://www.unesco.org/culture/ich/index.php?lg=en&pg=00011 (accessed January 10, 2011).

5 In April 2009, the Section of Intangible Cultural Heritage of UNESCO held a special Expert Meeting on Culinary Practices in France.

6 Source: UNESCO, 'Traditional Mexican cuisine – ancestral, ongoing community culture, the Michoacán paradigm' (http://www.unesco.org/culture/ich/index.php?lg=en&pg=00011&RL=00400; accessed January 10, 2011).

7 Traditional Mexican cuisine was also included in the Inventory of the Intangible Cultural Heritage of Mexico maintained by the National Council for Culture and Arts.

8 Source: http://www.unesco.org/culture/ich/index.php?lg=en&pg=00011&RL=00394 (accessed January 10, 2011).

9 Source: http://www.unesco.org/culture/ich/index.php?lg=en&pg=00011&RL=00437 (accessed January 10, 2011).

10 Source: http://www.unesco.org/culture/ich/index.php?project_id=00176 (accessed January 10, 2011).

11 Source: http://www.planalto.gov.br/ccivil_03/Constituicao/Constituicao.htm; accessed January 10, 2011).

12 These are just a few examples. For a complete list of intangible cultural heritage inscribed in the four books of registry, see http://portal.iphan.gov.br/portal/montarPaginaSecao.do?id=12456&retorno=paginaIphan (accessed January 10, 2011).

13 For more information on IPHAN visit its website at www.iphan.gov.br.

14 The two programs are (1) "Traditional Management of Cassava in the Brazilian Amazon," carried out in the period 1998–2000, which was developed through a partnership between CNPQ (Brazilian National Council for Scientific and Technological Development), Instituto Socioambiental (a Brazilian NGO;

www.socioambiental.org), and Institut de Recherche pour le Développement (IRD, a French international development research institute), with financial support from Bureau des Ressources Génétiques (BRG), CNPQ, and IRD; and (2) "Local Populations, Agrobiodiversity and Associated Traditional Knowledge in the Brazilian Amazon" (PACTA), 2005–2009, which was developed through a partnership between CNPQ, Unicamp (State University of Campinas, in São Paulo State) and IRD, with the participation of the Association of Indigenous Communities of the Middle Negro River (ACIMRN), with financial support from IRD, CNPQ, Agence Nationale de la Recherche, Biodivalloc, and BRG.

15 To access the complete text of this Resolution, see http://intranet.inc.gob.pe/intranet/dpcn/anexos/74_1.pdf (accessed March 11, 2011).

16 Resolución Directoral Nacional No. 1362, of October 16, 2007: http://intranet.inc.gob.pe/intranet/dpcn/anexos/32_1.pdf.

17 http://intranet.inc.gob.pe/intranet/dpcn/anexos/2_1.pdf. *Pisco* is also produced in Chile, but there are differences in the way each country produces it. Both claim that *pisco* originated in their territory.

18 Resolución Directoral Nacional No. 1180, of September 7, 2007 (http://intranet.inc.gob.pe/intranet/dpcn/anexos/31_1.pdf).

19 Resolución Directoral Nacional No. 241, of March 23, 2004 (http://intranet.inc.gob.pe/intranet/dpcn/anexos/13_1.pdf).

20 Resolución Directoral Nacional No. 471, of July 8, 2003 (http://intranet.inc.gob.pe/intranet/dpcn/anexos/5_1.pdf).

21 Resolución Directoral Nacional No. 836, of June 24, 2008 (http://intranet.inc.gob.pe/intranet/dpcn/anexos/45_1.pdf).

22 According to the Operational Guidelines for the Implementation of the World Heritage Convention, "outstanding universal value" means cultural and/or natural significance that is so exceptional as to transcend national boundaries and to be of common importance for present and future generations of all humanity. As such, the permanent protection of this heritage is of the highest importance to the international community as a whole. Source: http://whc.unesco.org/archive/opguide08-en.pdf (accessed January 20, 2011).

23 Source: http://whc.unesco.org/en/list (accessed January 20, 2011).

24 A key benefit of ratification of the World Heritage Convention, particularly for developing countries, is access to the World Heritage Fund. Annually, about US$4 million is made available to help states to identify, preserve, and promote World Heritage Sites. Emergency assistance may also be made available for urgent action to repair damage caused by human-made or natural disasters. The List of World Heritage in Danger comprises particularly threatened sites and the attention and the funds of both the national and the international community are focused on the conservation needs of these sites. Source: http://whc.unesco.org/en/list (accessed January 20, 2011).

25 The World Heritage Committee meets once a year, and consists of representatives from 21 of the states that are party to the Convention. The 21 states that make up the current World Heritage Committee are Australia, Bahrain, Barbados, Brazil, Cambodia, China, Egypt, Estonia, Ethiopia, France, Iraq, Jordan, Mali, Mexico, Nigeria, Russian Federation, South Africa, Sweden, Switzerland, Thailand, and the United Arab Emirates. Source: http://whc.unesco.org/en/committee (accessed January 20, 2011).

26 Two years later, in 1994, IUCN (the World Conservation Union, which is the advisory body of the World Heritage Committee for natural sites) recognized "protected landscapes/seascapes" on an equal footing with other categories of protected areas. These areas are designated "Category V" in IUCN's system for categorizing protected areas, and they are defined as "areas of land, with coast and sea as appropriate, where the interaction of people and nature over time has

produced an area of distinct character with significant aesthetic, ecological and/or cultural value, and often with high biological diversity'. There are similarities between UNESCO's World Heritage "cultural landscape" category and IUCN Category V of protected areas (protected landscapes/seascapes), especially in terms of the emphasis placed on human–nature interaction. However, in the latter, the emphasis is on biodiversity conservation and ecosystem integrity, whereas in UNESCO's cultural landscapes the emphasis is on human history, continuity of cultural traditions and social values and aspirations. In addition, the fundamental criterion for recognition of a World Heritage Cultural Landscape is that of "outstanding universal value," whereas there is less stress placed on outstanding qualities in the case of IUCN Category V protected landscapes/seascapes. Source: Philips (2005).

27 Source: http://conventions.coe.int/Treaty/EN/Treaties/Html/176.htm (accessed January 20, 2011).

28 Source: Operational Guidelines for the Implementation of the World Heritage Convention (http://whc.unesco.org/archive/opguide08-en.pdf; accessed January 20, 2011).

29 Though inscribed as a "clearly defined landscape, designed and created intentionally by people," Kew Gardens is considered, by some world heritage experts, as an associative landscape as well, because of its contributions to the study of plant diversity and economic botany, since its creation in 1759.

30 A complete list of cultural landscapes inscribed in the World Heritage List can be found at http://whc.unesco.org/en/culturallandscape (accessed January 20, 2011).

31 The Japanese Agency for Cultural Affairs established, in October 2000, a Committee on the Preservation, Development and Utilization of Cultural Landscapes Associated with Agriculture, Forestry and Fisheries. According to a report released by this Committee on June 12, 2003, terraced rice fields, known domestically as *semmaida* (literally, one thousand rice fields), were designated "places of scenic beauty" under the name of Obasute (Tagoto no Tsuki), in Koshoku City, Nagano Prefecture. Another rice terrace in Wajima City, Ishikawa Prefecture, was designated as a "place of scenic beauty" under the name of Shiroyone no Semmaida. Source: http://www.bunka.go.jp/english/pdf/nourinsuisan.pdf (accessed January 20, 2011).

32 Source: http://whc.unesco.org/en/culturallandscape (accessed January 20, 2011).

33 Source: http://www.fao.org/nr/giahs (accessed January 20, 2011).

34 Source: http://www.fao.org/nr/giahs/whataregiahs/definition/en/ (accessed January 20, 2011).

35 Source: http://www.fao.org/nr/giahs/giahsproject/summary-features/en/ (accessed January 20, 2011).

36 Source: www.fao.org/nr/giahs (accessed January 20, 2011).

37 For more information on Andean agricultural systems, see Tapia (2009).

38 For further information about the proposal "Terra Preta Soils (Amazonian Dark Earths) as a Candidate System for the GIAHS Programme," see ftp://ftp.fao.org/sd/SDA/GIAHS/brazil_terrapreta_submission.pdf (accessed January 10, 2010).

39 To learn more about paleobotanical studies in Brazil, and about evidence of plant domestication found in the country, particularly in the interior of Minas Gerais and the coast of Rio de Janeiro, see Dias (2001).

40 Pupunha is the fruit of a palm tree typical of the northern region of Brazil, whose stem is used as a source of heart of palm. The palm hearts are often harvested from basal offshoots which do not kill the trees.

Bibliography

Altieri, M.A. and Koohafkan, P. (2002) 'Globally Important Ingenious Agricultural Heritage Systems (GIAHS): extent, significance and implications for development' (ftp://ftp.fao.org/sd/SDA/GIAHS/backgroundpaper_altieri.pdf; accessed January 20, 2011).

Amoêda, R., Lira, S. and Pinheiro, C. (2010) *Heritage 2010 – World Heritage and Sustainable Development*, Green Lines Institute for Sustainable Development, Barcelos, Portugal.

Clement, C. (1999a) '1492 and the loss of Amazonian crop genetic resources. I. The relation between domestication and human population decline', *Economic Botany*, vol. 53, pp. 188–202.

Clement, C. (1999b) '1492 and the loss of Amazonian crop genetic resources. II. Crop biogeography at contact', *Economic Botany*, vol. 53, pp. 203–216.

Clement, C., Mccan, J. and Smith, N. (2003) 'Agrobiodiversity in Amazonia and its relationship with dark earths', in Lehmnan, J., Kern, D., Glaser, B. and Woods, W. (eds.) *Amazonian Dark Earths: Origin, Properties, Management*, Kluver Academic Publishers, Dordrecht, pp. 159–178.

Clement, C., Rocha, S., Cole, D. and Vivan, J. (2007) 'Conservação on farm', in Nass, L. (ed.) *Recursos Genéticos Vegetais*, EMBRAPA Recursos Genéticos e Biotecnologia, Brasília, pp. 513–544.

Denevan, W.M. (2001) *Cultivated landscapes of Native Amazonia and the Andes*, Oxford University Press, Oxford.

Dias, O. (2001) 'A produção de vegetais no Brasil antes de Cabral', *Boletim do Instituto de Arqueologia Brasileira*, no. 11.

Dillehay, T.D., Rossen, J., Anres, T.C. and Williams, D.E. (2007) 'Preceramic adoption of peanut, squash and cotton in Northern Peru', *Science*, vol. 316, pp. 1890–1893.

Emperaire, L. (2008) 'O manejo da agrobiodiversidade: o exemplo da mandioca na Amazônia', in Bensusan, N. (org.) *Seria Melhor Mandar Ladrilhar? Biodiversidade: Como, Para Que e Por Quê*, UnB, Brasília, and IEB, Peirópolis, pp. 337–352.

Emperaire, L. and Peroni, N. (2007) 'Traditional management of agrobiodiversity in Brazil: a case study of manioc', *Human Ecology*, vol. 35, no. 6, pp. 761–768.

Emperaire, L., Velthem, L.H.V. and Oliveira, A.G. (2008) 'Patrimônio cultural imaterial e sistema agrícola: o manejo da diversidade agrícola no médio Rio Negro (AM)', study presented at the 26th Meeting of the Brazilian Anthropology Association, held between June 1 and 4, 2008, in Porto Seguro, Bahia, Brazil.

Fowler, P. J. (2003) *World Heritage Cultural Landscapes: 1992–2002* (World Heritage Series, no. 6), UNESCO, Paris.

Glaser, B. and Woods, W. (eds.) (2004) *Amazonian Dark Earths: Explorations in Space and Time*, Springer, Berlin.

Goedert, C. (2007) 'Histórico e avanços em recursos genéticos no Brasil', in Nass, L. (ed.) *Recursos Genéticos Vegetais*, EMBRAPA Recursos Genéticos e Biotecnologia, Brasília, pp. 25–60.

Harrop, S. (2005) 'Globally Important Ingenious Agricultural Heritage Systems: an examination of their context in existing multilateral instruments' (ftp://ftp.fao.org/sd/SDA/GIAHS/Harrop_GIAHS_summary.pdf; accessed January 20, 2011).

Heckenberger, M. (2005) *The Ecology of Power: Culture, Place and Personhood in the Southern Amazon, A.D. 1000–2000*, Routledge, New York.

Heckenberger, M., Kuikuro, A., Kuikuro, U.T., Russell, J.C., Schmidt, M., Fausto, C. and Franchetto, B. (2003) 'Amazonia 1492: pristine forest or cultural parkland?' *Science*, vol. 301, pp. 1710–1714.

Heckenberger, M., Russell, J.C., Toney, J.R. and Schmidt., M. (2007) 'The legacy of cultural landscapes in the Brazilian Amazon: implications for biodiversity'. *Philosophical Transactions of the Royal Society*, vol. 362, pp. 197–207.

Heckenberger, M., Russell, J.C., Fausto, C., Toney, J.R., Schmidt, M., Pereira, E. and Kuikuro, A. (2008) 'Pre-Columbian urbanism, anthropogenic landscapes, and the future of the Amazon', *Science*, vol. 321, pp. 1214–1217.

Instituto do Patrimônio Histórico e Artístico Nacional (IPHAN) and Fundação Nacional de Arte (Funarte) (2003) *O Registro do Patrimônio Imaterial: Dossiê Final das Atividades da Comissão e do Grupo de Trabalho Patrimônio Imaterial*, Brasília.

Instituto do Patrimônio Histórico e Artístico Nacional (IPHAN) (2007) Superintendência Regional de Santa Catarina, *Roteiros Nacionais de Imigração*, sob a supervisão e coordenação de Dalmo Vieira Filho e Maria Regina Weissheimer, Florianópolis, Santa Catarina, Brazil (unpublished).

Japan Agency for Cultural Affairs, Committee on the Preservation, Development and Utilization of Cultural Landscapes Associated with Agriculture, Forestry and Fisheries (2003) 'Report of the Study on the Protection of Cultural Landscapes Associated with Agriculture, Forestry and Fisheries' (http://www.bunka.go.jp/english/pdf/nourinsuisan.pdf; accessed January 20, 2011).

Katz, E. (2008) 'Alimentação indígena na América Latina: Comida invisível, comida de pobres ou patrimônio culinário?', study presented at the 26th Meeting of the Brazilian Anthropology Association, held between June 1 and 4, 2008, in Porto Seguro, Bahia, Brazil.

Lira, S. and Amoêda, R. (2010) *Constructing Intangible Heritage*, Green Lines Institute for Sustainable Development, Barcelos, Portugal.

Londres Fonseca, M.C. (1996) 'Da modernização à participação: a política federal de preservação dos anos 70 e 80', *Revista do Patrimônio Histórico e Artístico Nacional*, vol. 24, pp. 153–160.

Londres Fonseca, M.C. (2003) 'Para além da pedra e cal: por uma concepção ampla de patrimônio cultural', in Abreu, R. and Chagas, M. (orgs.) *Memória e Patrimônio: Ensaios Contemporâneos*, DP & A, Rio de Janeiro, pp. 56–75.

Milliken, W. and Albert, B. (1999) *Yanomami, a Forest People*, Royal Botanical Gardens, Kew, London.

Neves, E. (2006) *Arqueologia da Amazonia*, Jorge Zahar, Rio de Janeiro.

Nogueira, M.D. (2006) 'Mandioca e farinha: identidade cultural e patrimônio nacional', in *Agrobiodiversidade e Diversidade Cultural*, Ministério do Meio Ambiente, Brasília, pp. 25–27.

Nozawa, C., Malingan, M., Plantilla, A. and Ong, J. (2008) 'Evolving culture, evolving landscapes: the Philippine rice terraces', in Amend, T., Brown, J., Kothari, A., Philipps, A. and Solton, S. (eds.) *Protected Landscapes and Agrobiodiversity Values* (Values of Protected Landscapes and Seascapes, a series published by the Protected Landscapes Task Force of IUCN's World Commission on Protected Areas), IUCN, Gland, Switzerland.

Pantoja, M., Almeida, M.B., Conceição, M.G., Lima, E.C., Aquino, TV., Iglesis, M.P. and Mendes, M.K. (2002) 'Botar roçados', in Carneiro da Cunha, M. and Almeida, M. (orgs.) *Enciclopédia da Floresta. O Alto Juruá: Práticas e Conhecimentos das Populações*, Companhia das Letras, São Paulo, pp. 249–283.

Phillips, A. (2005) 'Landscape as a meeting ground: category V protected landscapes/ seascapes and world heritage cultural landscapes', in Brown, J., Mitchell, N. and Beresford, M. (eds.) *The Protected Landscape Approach: Linking Nature, Culture and Community*, IUCN, Gland, Switzerland, and Cambridge, UK, pp. 19–35.

Ramakrishnan, P.S. (2004) 'Globally Important Ingenious Agricultural Heritage Systems (GIAHS): an eco-cultural landscape perspective (ftp://ftp.fao.org/sd/SDA/ GIAHS/backgroundpapers_ramakrishnan.pdf; accessed January 20, 2011).

Ribeiro, R. (2007) *Paisagem Cultural e Patrimônio*, IPHAN, Rio de Janeiro.

Röessler, M. (2005) 'World heritage cultural landscapes: a global perspective', in Brown, J., Mitchell, N. and Beresford, M. (eds.) *The Protected Landscape Approach: Linking Nature, Culture and Community*, IUCN, Gland, Switzerland, and Cambridge, UK, pp. 37–46.

Sauer, C. (1986) 'As plantas cultivadas na América do Sul tropical', in Ribeiro, B. (org.) *Suma Etnológica Brasileira: Etnobiologia*, 3rd edn, Vozes, FINEP, Petrópolis, Brazil, pp. 59–90.

Singh, A. and Varaprasad, K.S. (2008) 'Criteria for identification and assessment of agro-biodivesity heritage sites: evolving sustainable agriculture', *Current Science*, vol. 94, no. 9, pp. 1131–1138 (http://www.ias.ac.in/currsci/may102008/1131. pdf; accessed January 20, 2011).

Silva, G.M. (2008) 'Uso e conservação da agrobiodiversidade pelos índios Kayabi do Xingu', in Bensusan, N. (org.) *Seria Melhor Mandar Ladrilhar? Biodiversidade: Como, Para que e Por Quê*, Universidade de Brasília and IEB Brasília, pp. 317–336.

Tapia, M. (1999) *Agrobiodiversidad en los Andes*, Fundacion Friedrich Ebert, Lima.

UNESCO World Heritage Centre (2003) *Cultural Landscapes: The Challenges of Conservation* (World Heritage Paper no. 7) (http://unesdoc.unesco.org/ images/0013/001329/132988e.pdf; accessed January 20, 2011).

UNESCO and Centro Regional para la Salvaguardia del Patrimonio Cultural Inmaterial de America Latina (2008) *Estado del Arte del Patrimonio Cultural Inmaterial*, Cusco, Peru.

Viveiros de Castro, E. (2008) 'Amazônia antropizada', in *Almanaque Brasil Socioambiental: Uma Nova Perspectiva Para Entender o País e Melhorar Nossa Qualidade De Vida*, Instituto Socioambiental, São Paulo, pp. 102–103.

Von Droste, B., Plachter, H. and Rössler, M. (eds.) (1995) *Cultural Landscapes of Universal Value: Components of a Global Strategy*, Gustav Fischer Verlag, Stuttgart, in cooperation with UNESCO.

12

AGROBIODIVERSITY AND PROTECTED AREAS

Over the last decade, promoting and supporting the on-farm management of genetic resources, in local farming systems, home gardens, orchards, or other cultivated areas of high diversity, has become firmly established as a key component of crop conservation strategies (FAO, 2010, p. 4). The importance of *in situ* conservation of crop wild relatives for food security and climate change has also been extensively recognized, with surveys and inventories being carried out worldwide. However, the role of protected areas in promoting the conservation and sustainable use of agrobiodiversity has been greatly underestimated. At the international level, there is still no specific category of protected area for agrobiodiversity conservation, whether for *in situ* conservation of crop wild relatives[1] or for on-farm management of agrobiodiversity-rich farming systems (though other categories of protected areas have occasionally been used for agrobiodiversity conservation).

The number of protected areas in the world has grown from approximately 56,000 in 1996 to about 70,000 in 2007, and the total area covered has expanded in the same period from 13 to 17.5 million km^2, but areas with the richest agrobiodiversity (such as centers of origin and/or diversity) have received significantly less protection than the global average, according to *the Second Report on the State of the World's Plant Genetic Resources for Food and Agriculture* (FAO, 2010, p. 34). This report adds that a significant number of PGRFA, including crop wild relatives and useful plants collected from the wild, occur outside protected areas (in cultivated fields, orchards, grasslands, etc.) and consequently do not receive any form of legal protection.

From a legal point of view, it is very clear that the concept of biodiversity (the variability among living organisms from all sources) includes agrobiodiversity,[2] and that protected areas can be established to promote *in situ* and on-farm conservation of agricultural biodiversity. According to the CBD, *in situ* conservation means the "conservation of ecosystems and natural habitats and the maintenance and recovery of viable populations of species in their natural surroundings and, *in the case of domesticated or cultivated species, in the surroundings where they have developed their distinctive properties*" (emphasis added). The ITPGRFA is more incisive in this regard. According to its Article 5(c,d), member countries must "promote and support farmers and local communities' efforts to manage and conserve on-farm their plant genetic resources for food and agriculture," as well as "promote *in situ* conservation of wild crop relatives and wild plants for food production, including in protected areas, by supporting, *inter alia*, the efforts of Indigenous and local communities." All countries that have ratified the Treaty are, therefore, obliged to promote *in situ* and on-farm conservation of agrobiodiversity within and outside the limits of protected areas.

As several authors have already pointed out (Phillips, 2005; Rössler, 2005), there has been a shift in the conservation paradigm, from designating exceptional natural sites without people (pristine or near-pristine areas) to recognizing the value of natural heritage sites in a landscape context that includes people. That is, more emphasis has been placed on human–nature interaction, and the importance of protected areas that focus on lived-in landscapes has been increasingly recognized. At the international level, some examples of this paradigm shift are the inclusion, in 1992, of the "cultural landscape" category in UNESCO's World Heritage List and the recognition, in 1994, by IUCN (the World Conservation Union), of "protected landscapes/seascapes" (where there is an interaction between people and nature) on an equal footing with other categories of protected area. Globally Important Agricultural Heritage Systems (GIAHS) are also an important development in this direction (see Chapter 11).

Following this trend, an agrobiodiversity reserve (or heritage site) or a protected "agrobiodiverse landscape" category could be created, focusing specifically on the management of agricultural biodiversity and on the recognition of traditional agricultural techniques, practices, and knowledge. Most initiatives to conserve agrobiodiversity inside protected areas have taken place in parks, biological reserves, etc., which were not conceived for the conservation of crops and cultivated areas. According to the *Second Report on the State of the World's Plant Genetic Resources for Food and Agriculture* (FAO, 2010), very little survey or inventory work has been carried out on PGRFA in protected areas, compared with other components of biodiversity in these areas. *In situ* conservation of wild species of agricultural importance occurs mainly as an unplanned result of efforts to protect particular habitats or species, and the conservation of crop wild relatives has been relatively neglected, according to the *Second Report on the State of the World's Plant Genetic Resources for Food and Agriculture* (FAO, 2010). However, the report does give some examples of conservation of wild species, such as the Erebuni Reserve in Armenia, which was established to conserve populations of cereal wild relatives; the Flusslandschaft Elbe Biosphere Reserve in Germany, which is important for the *in situ* conservation of wild fruit crop genetic resources and perennial ryegrass; and sanctuaries that have been established in the Garo Hills of Meghalaya in India to conserve the native diversity of wild *Citrus* and *Musa* species (FAO, 2010).

The creation of an internationally recognized protected area category especially aimed at conserving agrobiodiversity would help to promote public awareness about the importance of agricultural diversity and its implications for food security, nutrition, health, social equity, and environmental sustainability. Historically, most conservation efforts and policies have focused on wild biodiversity, as if conserving wild plants and animals were more important to humanity than conserving the diversity of agricultural crops, such as rice, beans, corn, potatoes, which are part of our day-to-day diets. In addition, the lack of an integrated approach to agricultural and environmental policies leads, in many cases, to separate and unarticulated efforts to conserve threatened wild species and ecosystems, by environmental agencies, and *ex situ* conservation of domesticated crops, by agricultural research institutions, without much interaction between them.

The creation of agrobiodiversity reserves would also draw more attention from policy-makers to the relevance of identifying "agrobiodiversity hotspots," surveying and inventorying areas with high agricultural diversity (especially crop wild relatives, traditional/local landraces, and ingenious farming systems), and defining criteria and methods for the identification and adaptive management of such areas. Through these initiatives, scientific understanding of on-farm management of agrobiodiversity would increase, as would the recognition of the value of local seed systems and social networks in maintaining plant genetic diversity.

Many protected areas (such as national parks and wilderness areas) may contain wild relatives of cultivated plants and landraces, even if they were not especially created for this purpose, and agrobiodiversity may, eventually, benefit from the creation of such protected areas. However, a category of protected area especially designed to promote on-farm management of agrobiodiversity should allow sustainable use of plant genetic resources and focus on the interaction between humans and nature and on adaptive management, which does not occur with conventional protected areas categories, such as national parks and wilderness areas (which are more focused on ecosystems conservation for scientific research and/or environmental monitoring). The special legal status of "agrobiodiversity reserves" would impose restrictions on activities (such as overgrazing, excessive use of herbicides, widening of roads and other infrastructure works, removal of home gardens and orchards) that have negative impacts on agricultural biodiversity, as well as establish stronger protection for water and soil resources and more severe biosecurity regulations, in order to avoid possible contamination by genetically modified crops. The Cartagena Biosafety

Protocol acknowledges the "crucial importance" of centers of origin and genetic diversity for plants, and, consequently, the need to adopt special measures to avoid the impacts of GMOs on these centers. Thus, an agrobiodiversity reserve, especially aimed at *in situ*/on-farm conservation of agrobiodiversity, would have to impose stricter biosafety rules.[3]

As it is traditional and local agricultural systems that conserve and manage most of agricultural diversity, an agrobiodiversity reserve could be created only with the support and involvement of farmers who live in the designated area. The involvement of farmers would also be essential in the development and implementation of management plans and actions, which should promote an integrated (and complementary) approach to scientific knowledge and traditional knowledge systems. Agrobiodiversity reserves must meet broader objectives of sustainable local development and social inclusion, rather than just environmental conservation, in order to be socially and politically sustainable, especially in developing countries.

Agrobiodiversity reserves need not necessarily be established in publicly owned lands, and it does not make sense, evidently, to expropriate the lands of farmers within their limits. Agrobiodiversity reserves could be established through agreements with local farmers who live in the designated areas and manage agrobiodiversity-rich agroecosystems. They could be compensated for the conservation of crop diversity through the payment of environmental services, for instance.[4] When farmers do not hold property titles over their areas (which is very common in Brazil and in other Latin American countries), but they are in possession of their cultivated lands, the creation of agrobiodiversity reserves would also contribute to land tenure reform and social equity. In Brazil, for example, there are special categories of protected areas (extractive reserves and sustainable development reserves) that are aimed at promoting the conservation and sustainable use of biodiversity, as well as local development. They were included in the Brazilian National System of Protected Areas (called SNUC – Sistema Nacional de Unidades de Conservação da Natureza, regulated by Federal Law 9.985/2000[5]). Extractive reserves are publicly owned lands, and the federal state signs contracts with associations that are representatives of traditional populations (rubber tappers, Brazil nut collectors, fishermen, etc.) who live in the protected area. Such contracts do not grant property titles to these populations, but they grant the right to sustainably use natural resources of extractive reserves, under certain restrictions (mining and hunting are forbidden, and timber exploration can take place only in exceptional situations; only subsistence agriculture and livestock raising can occur).

According to the *Second National Report on the State of Brazil's Plant Genetic Resources for Food and Agriculture*, published by the Brazilian Ministry of Agriculture in 2009, there are 56 federal extractive reserves, predominantly in the Amazon, where species such as *Bertholletia excelsa*

(Brazil nuts), *Hevea brasiliensis* (rubber tree), *Euterpe oleracea* and *Euterpe precatoria* (açaí), *Theobroma grandiflorum* (cupuaçu), and *Attalea speciosa* (babaçu), can be found. However, in the case of Cerrado (Brazilian savanna) species that are important for regional diets and extractive activities, such as pequi (*Caryocar brasiliensis*), cagaita (*Eugenia dysenterica*), baru (*Dipteryx alata*), buriti (*Mauritia flexuosa*), macaúba (*Acrocomia aculeata*), and mangaba (*Hancornia speciosa*), there are no protected areas specifically aimed at their *in situ* conservation. The report focuses mainly on wild species, and considers that extractive and sustainable development reserves, as well as national forests, can serve as "genetic reserves" and promote *in situ* conservation of plant genetic resources. However, extractive reserves were not conceived for agrobiodiversity conservation and sustainable use, and some of the rules governing such reserves may be incompatible with traditional farming systems and agricultural practices and methods.

One of the main arguments against the creation of "agrobiodiversity reserves" has been that traditional and local agricultural systems are widely scattered worldwide and that it would be too complex to define which sites should be turned into "agrobiodiversity reserves." Protected areas will always be insufficient to conserve the planet's biodiversity (whether wild or domesticated), because the processes that generate and maintain diversity take place in a scale that transcends the usual dimensions of protected areas. In addition, agrobiodiversity reserves alone will not be enough to avoid the impacts of monocultures and industrialized agricultural models, especially if they are mere "islands" surrounded by unsustainable agricultural practices.

However, albeit insufficient, protected areas have played an important role in conservation of wild biodiversity *in situ* and could also be relevant for sustainable management of agrobiodiversity. The domesticated component of biodiversity has been historically neglected by public policies, and the creation of instruments specifically aimed at its conservation would encourage the generation of deeper knowledge about the biological and sociocultural processes that sustain and enrich agrobiodiversity, and could also generate other sources of income for traditional, agroecological, and local farmers, through agrotourism, enhancing the value of agrobiodiversity products, etc.

Regardless, however, of the creation of a special protected area category for agrobiodiversity, it is important to make better use of existing protected areas to promote conservation of agricultural biodiversity. The first step would be to make an inventory of wild relatives of cultivated plants and local varieties found in protected areas, in order to evaluate risks and threats to their conservation and develop *in situ* and on-farm conservation and management actions and policies. Management plans of protected areas rarely include actions specifically dedicated to agricultural biodiversity, and this situation needs to be changed. Additionally, sustainable use of agrobiodiversity resources should be promoted, as conservation is directly associated with dynamic use and management of plant genetic resources.

The first protected area established principally for the preservation of a crop wild relative (a species of teosinte,[6] a maize wild relative, called *Zea perennis*), as well as traditional agricultural systems and landraces was the Biosphere Reserve of Sierra de Manantlán, in the states of Jalisco and Colima, in Mexico.[7] The reserve was created by means of presidential decree in 1988, and in the same year the UNESCO international program "Man and Biosphere"[8] also approved its designation as a biosphere reserve. In 1984 the state of Jalisco had already bought land in Las Joyas, where there is a large teosinte population, and the Natural Laboratory Las Joyas de Sierra de Manantlán was installed, managed by the University of Guadalajara. The rest of the reserve is made up of land belonging to Indigenous communities (20 percent), private land (40 percent), and communal, collective use lands called *ejidos*.[9] The University of Guadalajara's Manantlán Institute of Ecology and Biodiversity Conservation (IMECBio) and the International Maize and Wheat Improvement Center (CIMMYT) develop programs for conservation of teosinte populations in the reserve, which is one of the most important protected areas in Mexico (Maxted *et al*., 2010). Currently, teosinte and traditional maize varieties in Mexico are faced with the serious problem of contamination from genetically modified corn, similar to the situation in Brazil.

The history behind the creation of the Sierra de Manantlán Biosphere Reserve is curious:[10] In 1976, Professor of Botany, Dr Hugh Iltis, at the University of Wisconsin–Madison in the United States, sent to botanists around the world a New Year's card in the form of a poster of *Zea perennis* (maize species) on which he had written "extinct in the wild." Wild populations of *Z. perennis* had last been seen in 1921 in western Mexico, by two US Department of Agriculture botanists, who introduced the species to university greenhouses in the United States. Since then several other botanists had tried, and failed, to relocate the wild population. One poster was placed on a bulletin board at the University of Guadalajara by a local taxonomist, who urged her students: "Go and find this teosinte, and prove that gringo Iltis wrong." One undergraduate student took up the challenge and went back to the plant's last known location in western Mexico and found the long-lost *Zea perennis*.

An even more important discovery was made. On being told that *Z. perennis* was growing in another location the student, Rafael Guzman, collected more seed. However, this teosinte (known locally as *milpilla*) turned out to be a new species – *Z. diploperennis*. Unlike *Z. perennis*, this species freely interbreeds with maize, which raised the possibility that the crop could be grown for several years from one rootstock and, perhaps more importantly, it appeared to be tolerant of seven maize viruses and the only member of the genus *Zea* that is immune to three of them. The discovery of

Z. diploperennis and the subsequent declaration of a biosphere reserve has led to intensive research into the biodiversity, and specifically the flora of the reserve. The existence of *Z. diploperennis* and other crop wild relatives is likely to be due to the traditional agricultural practices of slash-and-burn cultivation (*coamil*) and cattle ranching. The management practices and objectives of the reserve thus stress the need to conserve traditional agricultural systems. The *coamil* system is planned to continue in areas within the reserve so that the *Z. diploperennis* populations can survive. Over 2,700 plant species have been recorded in the area, of which 40 percent are endemic to Mexico, but virtually all teosinte populations are either threatened or endangered (Maxted *et al.*, 2010).

Another interesting experience is that of the "Parque de la Papa" (Potato Park), in the region of Cuzco, in southern Peru. The Parque de la Papa was created by six communities of the Quéchua Indigenous people (Amaru, Chawaytire, Cuyo Grande, Pampallaqta, Paru-Paru, and Sacaca), as a community-based, agrobiodiversity-focused conservation area (Argumedo, 2008). It includes 8,661 hectares of community-managed land, which ranges between 3,200 and 5,000 meters above sea level. The Parque de la Papa is a living, dynamic "library" of genetic diversity of potatoes and traditional agricultural systems. Nearly 1,200 traditional potato varieties are known, named, and managed by local people in the region where the park is located. A typical farm plot may contain 250–300 varieties, and approximately 900 varieties of native potato are grown in the park area (Argumedo, 2008).[11] Potatoes are so important in Peru that in 2005 the Peruvian government (through Decreto Supremo 009-2005-AG) declared May 30 as "National Potato Day," when several celebrations are held.[12] The region is also a center of diversity for agricultural crops such as quinoa[13] and amaranth,[14] and traditional farming includes raising domesticated animals such as guinea pigs, llamas, and alpacas.

In 2004, the Association of the Parque de la Papa Communities signed an agreement with the International Potato Center (IPC), one of the international agricultural research centers, for repatriation of 410 native potato species that originate from that region and which were conserved in IPC germplasm banks. The objective of the park is to protect not only biological/genetic resources, but also social and cultural systems and processes which produced (and continue to produce) agricultural diversity. Some income alternatives have been developed, and others considered, such as agroecotourism, environmental education, creation of a visitor center with different potato varieties on display and a restaurant with typical dishes, sales of colorful potato varieties in local markets, etc. (Andersen and Winge, 2008).

Agrobiodiversity reserves or "zones" are provided for in Peruvian legislation: Supreme Decree 068-2001-PCM, which regulated Law 26839/1997[15] (on conservation and sustainable use of biological diversity), created the possibility that "agrobiodiversity zones" be established in Indigenous lands (Articles 38 and 39 of Supreme Decree 068-2001-PCM). However, agrobiodiversity zones are not part of the national system of protected areas in Peru, established by Law 26834/1997, which does not have a special category aimed at conserving agrobiodiversity (Ruiz Muller, 2009). Manuel Ruiz Muller also defends the creation of an internationally recognized category of protected area aimed at promoting agrobiodiversity hotspots (Ruiz Muller, 2009).

Peruvian agrobiodiversity zones are based on the concept of "Indigenous biocultural heritage", developed by ANDES (Association for Nature and Sustainable Development). A Peruvian Law (28477/2005) also established a list of native crops that are considered "natural heritage of the nation," and a National Registry of Native Potato Varieties was created in 2008. The Cuzco regional government enacted a local norm (Ordenanza Regional 010-2007) that forbids sales, cultivation, use, and transportation of genetically modified potato varieties in the territory of Cuzco.

A study carried out by the Worldwide Fund for Nature (WWF), called "Food Stores: Protected Areas Conserving Crop Wild Relatives and Securing Future Food Stocks" (Maxted *et al.*, 2010), analyzed important ecoregions for conservation of agricultural genetic diversity. The study pointed out the importance and strategic role of protected areas for *in situ* conservation of wild relatives of cultivated plants and local varieties. It concluded, however, that the current global system of protected areas has been inefficient in protecting agricultural biodiversity, both from a standpoint of the location of protected areas and the form through which they are managed. The main finding of the study was that most centers of diversity of the main agricultural crops are poorly protected. Of the 34 ecoregions that are important for conservation of agricultural genetic diversity which were analyzed by the study, 29 ecoregions (or 82 percent of them) have less than 10 percent of their extensions covered by protected areas, and six ecoregions (18 percent) have 1 percent or less. Many protected areas contain populations that are essential for conservation of the diversity of plant genetic resources, but the management plan and actions for conservation of these areas do not include them specifically.

The study focuses specifically on the role of protected areas for *in situ* conservation of crop genetic diversity, and presents a list of protected areas worldwide that are located in important ecoregions for conservation of agricultural biodiversity. It also contains a list of wild relatives of cultivated plants and local varieties in existing protected areas, and case studies in Turkey, Mexico, Africa, Vietnam, Peru, and India.

The study highlights the fact that Turkey has adopted several measures to promote *in situ* conservation of agricultural biodiversity, as it is located in a region which is one of the birthplaces of agriculture (known as the Fertile Crescent)[16] and has an extraordinarily rich and diversified flora. Turkey is located in two Vavilovian centers of diversity for agricultural crops: the

Mediterranean and Near East. A project with the goal of *in situ* conservation of plant diversity in the country, supported by the Global Environment Facility (GEF), and with participation by the Turkish Ministries of Agriculture, Forests and Environment and local communities, initially identified three sites for the establishment of genetic reserves (Ceylanpinar, the Kazdag National Park, and the Mountains of Amanus, Bolkar, and Aladag) as well as priority species for conservation [wild relatives of wheat, rye, lentils, chick peas, wild plum species, a nut species (*Castanea sativa*) and three tree species, *Abies equi-trojani*, *Pinus brutia*, and *P. nigra*]. Later, 22 genetic reserves were created, and management plans were developed for each, with the objective of maintaining the greatest possible genetic diversity in each reserve. Local communities participated in development of the management plans, in order to guarantee that they would continue to have access to genetic reserves and traditional practices and activities. Some traditional practices were recognized as essential for the conservation of agrobiodiversity, such as animal pastures, important for scattering of seeds and their germination in later years.

In Vietnam, an *in situ* and on-farm agrobiodiversity conservation project was also developed,[17] with the participation of local communities. Vietnam is also located in a Vavilovian center of diversity for agricultural crops, and in 1995 an Action Plan for Biodiversity was approved, with agrobiodiversity as one of its priorities. The plan emphasizes on-farm management of varieties that are adapted to local environmental conditions and the participation of farmers in the definition and implementation of conservation actions. Species that were given priority are rice, taro, tea (*Camellia* and *Ilex*), mug beans, citrus fruits and two native fruits: lychee and longan. Eight genetic reserves were selected, two of which are in protected areas (Ba Vi National Park and Huu Lien Natural Reserve) and the other six are made up of cultivated ecosystems.

Another study conducted by IUCN (the World Conservation Union), called "Protected Landscapes and Agrobiodiversity Values" (Amend *et al.*, 2008), analyzed case studies from different parts of the world that illustrate the role played by protected areas in conservation of agrobiodiversity and associated agricultural knowledge and practices. The study focused mainly on protected areas category V: protected landscapes. Protected landscapes are defined by IUCN as "areas in which interactions between mankind and nature throughout time have produced characteristics which set them apart due to aesthetic, ecological and/or cultural value, and, usually, high biological diversity." This interaction between man and nature is recognized as vital for protection, maintenance, and evolution of protected landscapes. IUCN established six categories of protected areas, and protected landscapes are just one of these categories, although they are pointed out as being the best suited for *in situ* conservation of agrobiodiversity hotspots, with participation of local communities. Different categories and types of protected areas used for *in situ* conservation of agrobiodiversity are presented in case studies in Ecuador (one case study was carried out on farmscapes of the Quijos River in the tropical Andes and another on the Chimborazo Faunal Production Reserve), Spain (on the Garrotxa Volcanic Zone Natural Park), Peru (on the Potato Park), Germany (on the UNESCO Biosphere Reserve Rhön), the Philippines (on rice terraces), Serbia (on Stara Planina Mountain Nature Park), Canada (on the Gaspé Peninsula of Québec), Ethiopia (on the Borana conserved landscape), England (on national parks and areas of outstanding natural beauty), Nepal (on the districts of Bardiya, Kailali, and Kanchanpur), and the United States (Canyon de Chelly National Monument and Navajo farming traditions).

Notes

1 The global project "*In situ* conservation of crop wild relatives through enhanced information management and field application," supported by UNEP and coordinated by Bioversity International, is also developing actions and plans aimed at promoting *in situ* conservation of crop wild relatives in protected areas, according to FAO's *Second Report on the State of the World's Plant Genetic Resources for Food and Agriculture* (FAO, 2010). The project works in Armenia, Bolivia, Madagascar, Sri Lanka, and Uzbekistan, and its report is available at http://www.unep.org/dgef/LinkClick.aspx?fileticket=pOTyo5g97ns%3D&tabid=1710&mid=7564 (accessed December 10, 2010).

FAO's Commission on Genetic Resources for Food and Agriculture (CGRFA) commissioned a study on the establishment of a global network for *in situ* conservation of crop wild relatives, which was carried out by Nigel Maxted and Shelagh Kell, and is one of the thematic background studies that is part of FAO's *Second Report on the State of the World's Plant Genetic Resources for Food and Agriculture* (FAO, 2010). The study proposed conservation priorities and specific locations in which to conserve the most important wild relatives of 14 of the world's major food crops (finger millet, pearl millet, garden pea, cowpea, barley, sweet potato, cassava, potato, maize, rice, banana/plantain, sorghum, wheat, and faba bean). The report points out that about 9 percent of the crop wild relatives of these 14 crops require urgent conservation.

In Brazil, the Ministry of Environment is developing a project aimed at mapping creole/local varieties and wild relatives of the following crops: cotton, peanuts, rice, pumpkins, manioc/cassava, maize, and *pupunha* (a palm tree typical of the northern region of Brazil). See Coradin (2006). Between 2005 and 2007, the Ministry of Environment coordinated the project "Identification of Species of the Brazilian Flora of Current and Potential Economic Value, Used Locally and Regionally: Plants for the Future," but it did not focus specifically in protected areas.

2 The CBD defines domesticated or cultivated species as "species in which the evolutionary process has been influenced by humans to meet their needs."

3 In this book, we will not focus on biosafety regulations and on the impacts of GMOs, as these issues have been extensively explored by other researchers. For more information about the Cartagena Protocol, see Mackenzie *et al.* (2003). For more information on biosafety rules in Latin American countries (Peru, Ecuador, Bolivia, Colombia, Venezuela, and Mexico), see Lapeña (2007).

In Brazil, Federal Law 11105/2005 establishes biosafety rules and creates the National Biosafety Council. It was regulated by Decree 5591/2005. Law 11460/2007 forbids the cultivation of GMOs in Indigenous territories and protected areas (with the exception of a protected area category called "Área de Proteção Ambiental"). It also establishes rules for GMOs cultivated in buffer zones of protected areas.

The Administrative Order No. 21, issued on January 13, 2005, by the Brazilian Ministry of Agriculture, recognizes some locations, municipalities, and states (listed in its annex) where genetically modified (or contaminated with traces of genetic modification) cotton (*Gossypium hirsutum*) cannot be cultivated.

4 For more information, see FAO, 'The State of Food and Agriculture 2007: paying farmers for environmental services' (http://www.fao.org/publications/sofa/sofa2007/en/; accessed January 20, 2011).

5 For more information on extractive reserves (in Portuguese, *reservas extrativistas*) see Santilli (2005, 2010).

6 Teosinte is a group of large grasses of the genus *Zea* (maize), found in Central and South America. There are five recognized species of teosinte: *Zea diploperennis*, *Z. luxurians*, *Z. mays*, *Z. nicaraguensis*, and *Z. perennis*. The maize gene pool consists mainly of cultivated *Z. mays* and several related wild species (*Z. diploperennis* and *Tripsacum* spp.). *Z. diploperennis* and *Z. perennis* are perennial, whereas all other genera are annual. Source: FAO (1997) *The State of the World's Plant Genetic Resources for Food and Agriculture*, FAO, Rome.

7 Source: 'Mexico: Sierra de Manantlán "from a one-species project to an ecosystem project" ', in Maxted *et al.* (2010).

8 Biosphere reserves are sites established by countries and recognized under UNESCO's Man and the Biosphere program to promote sustainable development based on local community efforts and sound science. After their designation, biosphere reserves remain under national sovereign jurisdiction, yet they share their experience and ideas nationally, regionally and internationally within the World Network of Biosphere Reserves. There are currently 564 reserves in 109 countries. Source: http://www.unesco.org/new/en/natural-sciences/environment/ecological-sciences/biosphere-reserves/ (accessed January 10, 2011). For more information on Sierra de Manantlán and other Mexican protected areas, see http://www.conanp.gob.mx (accessed January 10, 2011).

9 *Ejidos* are communal, collective use lands of great importance in the agricultural life of Mexico. The *ejidos* system was widely used during the time of the Aztec Empire to promote collective use of lands. Spanish conquerors tried to put an end to this system, which was re-introduced by President Lázaro Cárdenas in 1934, as an important component of land tenure reform. The review of the Mexican Constitution, which took place in 1991, eliminated peasants' rights to request access to land by means of the *ejidos* system, in which government expropriated private land or set apart public land for collective use by peasants.

10 Source: 'Mexico: Sierra de Manantlán, from a one-species project to an ecosystem project', in Maxted *et al.* (2010).

11 According to the *Catalog of Native Potato Varieties of Huancavelica, Peru*, published by the International Potato Center (CIP) and the Federación Departamental de Comunidades Campesinas (FEDECH), there were, in 2006, 144 native varieties of potatoes in the Huancavelica Department, in center-south Peruvian Andes. See also Graves (2001).

12 The United Nations declared 2008 the "International Year of the Potato."

13 Quinoa is a highly nutritious grain originated from Colombia, Peru, and Chile, widely consumed in the Andes high plateau.

14 Amaranth (also known as *kiwicha* in Quéchuan language) is also a highly nutritive grain, rich in protein, calcium, and zinc. Prior to Spanish colonization, it was nearly as widespread in the Americas as maize. See also *Lost Crops of the Incas* (National Research Council, 1989), which contains information about

many plants domesticated by the Incas (such as *oca*, *maca*, *arracacha*, *tarwi*, *nuñas*, and *lucuma*), which were of great importance for local diets and were replaced by agricultural crops imposed by the Spanish, such as wheat and rye. Bermejo and León (1992) describe agricultural crops of the Americas which were marginalized in the past 500 years, as well as processes that led to loss of genetic diversity.

15 Law 26839/1997 and Supreme Decree 068-2001-PCM can be seen at http://www.farmersrights.org/database/peru.html (accessed January 18, 2009).

16 The Fertile Crescent covers southeastern Turkey, Palestine, Israel, western parts of Syria, eastern parts of Iraq and western parts of Iran, and it is a center of origin and diversity for many important agricultural crops, such as wheat and rye. It is also believed that many important animals – such as cows, goats, sheep, and pigs – were first domesticated in this region.

17 The project "*In situ* Conservation of Native Landraces and their Wild Relatives in Vietnam" was supported by the Global Environment Facility (GEF) and took place between 2002 and 2005. To learn more, see www.un.org.vn/undp/projects/vie01g35/index.htm (accessed February 2, 2009). See also Thi Hoa *et al.* (2005).

Bibliography

Amend, T., Brown, J., Kothari, A., Phillips, A. and Stolton, S. (eds.) (2008) *Protected Landscapes and Agrobiodiversity Vales*, IUCN, Gland, Switzerland, and GTZ, Kasparek Verlag, Heidelberg.

Andersen, R. and Winge, T. (2008) 'Success stories from the realization of farmers' rights related to plant genetic resources for food and agriculture', The Farmers' Right Project, Fridtjof Nansen Institute Report 4/2008, Lysaker, Norway (http://www.farmersrights.org/resources/global_works.html; accessed February 2, 2009).

Argumedo, A. (2008) 'The Potato Park, Peru: conserving agrobiodiversity in an Andean Indigenous Biocultural Heritage Area', in Amend, T., Brown, J., Kothari, A., Phillips, A. and Stolton, S. (eds.) *Protected Landscapes and Agrobiodiversity Vales*, IUCN, Gland, Switzerland, and GTZ, Kasparek Verlag, Heidelberg.

Bermejo, H. and León, J. (eds.) (1992) *Cultivos Marginados: Otra Perspectiva de 1492* (Colección Producción y Protección Vegetal no. 26), FAO, Rome.

Centro Internacional de la Papa (CIP) and Federación Departamental de Comunidades Campesinas (FEDECH) (2006) *Catálogo de Variedades de Papa Nativa de Huancavelica*, Lima, Peru.

Coradin, L. (2006) *Parentes Silvestres das Espécies de Plantas Cultivadas*, Ministério do Meio Ambiente, Brasília.

EMBRAPA (Brazilian Agricultural Research Corporation), Ministry of Agriculture, Livestock and Food Supply and FAO (2009) *The State of Brazil's Plant Genetic Resources* (http://labexkorea.wordpress.com/2010/02/05/plant-genetic-resources-for-food-and-agriculture-in-brazil/; accessed December 10, 2010).

FAO (2010) *Second Report on the State of the World's Plant Genetic Resources for Food and Agriculture*, Commission on Genetic Resources for Food and Agriculture, FAO, Rome (http://www.fao.org/agriculture/crops/core-themes/theme/seeds-pgr/sow/en/; accessed December 10, 2010).

Harrop, S. (2007) 'Traditional agricultural landscapes as protected areas in international law and policy', *Agriculture, Ecosystems & Environment*, vol. 121, no. 3, pp. 296–307.

Hunter, D. and Heywood (eds.) (2011) *Crop Wild Relatives: A Manual of In Situ Conservation*, Bioversity International and Earthscan, London.

Iriondo, J.M., Maxted, N. and Dulloo, M.E. (eds.) (2008) *Conserving Plant Diversity in Protected Areas*, CAB International, Wallingford, UK.

Lapeña, I. (2007) *Semillas Transgénicas en Centros de origen y Diversidad*, Sociedad Peruana de Derecho Ambiental, Lima.

Mackenzie, R., Burhenne-Guilmin, F., La Viña, A.G.M. and Werksman, J. (2003) *An Explanatory Guide to the Cartagena Protocol on Biosafety* (IUCN Environmental Policy and Law Paper no. 46), IUCN Environmental Law Centre, Gland, Switzerland.

Maxted, N., Kell, S., Ford-Lloyd, B. and Stolton, S. (2010) 'Food stores: protected areas conserving crop wild relatives and securing future food stocks', in Stolton, S. and Dudley, N. (eds.) *Arguments for Protected Areas: Multiple Benefits for Conservation and Use*, Earthscan, London.

National Research Council (1989) *Lost Crops of the Incas*, National Academy Press, Washington, DC.

Phillips, A. (2005) 'Landscape as a meeting ground: Category V protected landscapes/seascapes and world heritage cultural landscapes', in Brown, J., Mitchell, N. and Beresford, M. (eds.) *The Protected Landscape Approach: Linking Nature, Culture and Community*, IUCN, Gland, Switzerland, and Cambridge, UK, pp. 19–35.

Raves, C. (ed.) (2001) *The Potato, Treasure of the Andes: from Agriculture to Culture*, International Potato Center, Lima.

Rössler, M. (2005) 'World heritage cultural landscapes: a global perspective', in Brown, J., Mitchell, N. and Beresford, M. (eds.) *The Protected Landscape Approach: Linking Nature, Culture and Community*, IUCN, Gland, Switzerland, and Cambridge, UK, pp. 37–46.

Ruiz Muller, M. (2009) *Las Zonas de Agrobiodiversidad y el Registro de Cultivos Nativos en el Perú: Aprendiendo de nosotros mismos*, Sociedad Peruana de Derecho Ambiental, Lima, and Bioversity International, Rome.

Santilli, J. (2005) *Socioambientalismo e Novos Direitos: Proteção Jurídica à Diversidade Biológica e Cultural*, Instituto Socioambiental (ISA), Peirópolis, and Instituto Internacional de Educação para o Brasil (IEB), São Paulo.

Santilli, J. (2010) 'Human-inhabited protected areas (HIPAs) and the law: integration of local communities and protected areas in Brazilian Law', *Journal of Sustainable Forestry*, vol. 29, no. 2–4, pp. 390–402.

Scurrah, M., Andersen, R. and Winge, T. (2008) 'Farmers' rights in Peru: farmers' perspectives', The Farmers' Right Project, Fridtjof Nansen Institute Report 16/2008, Lysaker, Norway (http://www.farmersrights.org/resources/regional_works.html; accessed February 2, 2009).

Thi Hoa, L.T., Dinh, N., Thi Ngoc Hue, N.,Van Ly, N. and Ngoc Hai Ninh, D. (2005) '*In situ* conservation of native lychee and their wild relatives and participatory market analysis and development: the case of Vietnam', *ISHS Acta Horticulturae*, no. 665, pp. 125–140 (www.actahort.org/books/665/665_15.htm; accessed February 2, 2009).

13

GEOGRAPHICAL INDICATIONS FOR AGROBIODIVERSITY PRODUCTS?

Case studies in France, Mexico, and Brazil

Geographical indications can also be a tool to enhance the value of agrobiodiversity products and promote sustainable rural development. This legal and economic instrument is aimed at identifying and adding value to products and services associated with certain territories, conceived in their natural and cultural dimensions. Geographical indications establish a connection between products (or services) and their territorial identity, and the quality, characteristics, and reputation of origin-based products are essentially associated with their geographical origins.

Geographical indications create special markets for differentiated products that are linked to traditions and cultural identities and seek their own form of competition in markets dominated by globalized and standardized products. Standardized products are "soulless," with no cultural roots, and can be found in stores and supermarkets anywhere in the world. Origin-based products (protected by geographical indications) form part of the cultural and biological heritage of the countries/regions from which they originate. Some of the many classical examples of products with geographical indications and which are named after the places where they originated include tequila, Parmigiano-Reggiano cheese, Darjeeling tea,[1] champagne, Roquefort cheese, Parma ham, feta cheese, and port wine.

France was a pioneer country in granting legal protection to geographical names, passing its first law on geographical indications – known in France as *appellations d'origine contrôlées* (AOC) – as far back as 1919,[2] initially for wine and then for all agrifood products (in 1990). Many other European countries, especially in southern Europe (Italy, Spain, Portugal, and Greece),[3] also use geographical indications to promote rural development, and in 1992 the EU regulated geographical indications and designations of origin for agricultural products and foodstuffs (with the exception of wines and spirits), through Council Regulation (EC) No. 2081/92, which was replaced by Council Regulation (EC) No. 510/2006.[4] In

1994, the obligation to recognize and protect geographical indications as IP rights was included in the WTO's TRIPS Agreement. As all WTO member countries have to sign and implement the TRIPS Agreement, geographical indications became recognized and legally protected in many countries after this agreement came into effect, in 1996 (according to Article 65.1 of the TRIPS Agreement).[5]

Geographical indications are defined by this agreement as "indications which identify a good as originating in the territory of a Member, or a region or locality in that territory, where a given quality, reputation or other characteristic of the good is essentially attributed to its geographical origin" (Article 22). Although the TRIPS Agreement establishes a definition for geographical indications, countries have some flexibility to adopt their own definitions in national laws, as long as they grant effective protection against any use of geographical names that misleads the public as to the true place of origin of the good and/or constitutes an act of unfair competition.[6] The legal protection granted to geographical indications not only prevents misuse (and imitations) of geographical names on national and international markets, but it also enables the codification of the whole production process and promotes a link between producers and consumers. Geographical indications recognition also tends to reinforce local identity and pride in the product.

However, the TRIPS Agreement establishes higher levels of protection for wines and spirits than for other goods. The TRIPS Agreement forbids the use of geographical indications identifying wines and spirits not originating in the place indicated by the geographical indication even when the true origin of the wine or spirit is indicated or the geographical indication is used in translation or accompanied by expressions such as "kind," "type," "style," "imitation," or the like (Article 23.1[7]). For example, the name "cognac" cannot be used to describe products that are not produced in the French region of Cognac (as cognac is a protected geographical indication), even if identified by, for example, the expression "cognac produced in New Jersey." In the case of other goods, provided the true origin is indicated and this eliminates the possibility of misleading the consumer, the false geographical name may be used, for example "Espelette-type peppers, produced in New Jersey" (Espelette is a French protected geographical indication). Two main issues are being debated in the TRIPS Council: the establishment of a multilateral system of notification and registration of geographical indications for wines and spirits and the extension of the higher level of protection provided for wines and spirits to other products.

The TRIPS Agreement also sets forth that there is no obligation to protect geographical indications that are not or cease to be protected in their country of origin, or which have fallen into disuse in that country (Article 24.9). This means that the protection of geographical indications must first be established at the national level before it can be obtained at the international level. There is also no obligation to protect geographical indications that have become "generic" or common names,[8] and many disputes concerning what is the "generic" nature of a geographical name have already arisen. Camembert cheese is an example: as Camembert has been used for over a century to define a type of cheese that is produced in several countries, the producers in the region of Camembert (Normandy, France) have not been able to reserve the exclusive right to use the name. The protection has been granted only to "Camembert de Normandie," as a protected designation of origin, but not to Camembert alone, since it became a generic or common name.[9] In the case of feta cheese, producers from other countries (especially Denmark) also tried to have the name "feta" considered a generic one. However, EU courts ruled in favor of recognition of this denomination of origin to Greece, considering that feta cheese is associated with a type of cheese traditionally made in some areas of Greece (mainland and the island of Lesbos), from lamb milk or a mixture of lamb and goat milk, through a natural and traditional method of press-free drying. Thus, the designation "feta" can be used only by producers established in Greece.

Geographical indications must not be confused with trademarks, which are also IP rights protected under the TRIPS Agreement. Trademarks are distinctive signs which are used by companies to identify themselves and their products or services to consumers. Trademarks distinguish products or services from other, identical, similar or related ones which are made by different companies or persons. Trademarks are not associated with specific territories, and characteristics of the products or services are not linked to their regions of origin, unlike geographical indications. They differentiate products according to the companies which make them, not to their geographical origins. Producers established in a region delimited by a protected geographical indication may not move to a different region and continue to use the same geographical name, or even negotiate the use of the name to other producers who are not established in that region (geographical indications are collective and nontransferable rights[10]), or whose products do not have qualities associated exclusively or essentially with the delimited territory. Trademarks can be negotiated by their owners, and products with registered trademarks may be manufactured in any geographical region.

Conflicts can occur between the owner of a prior registered trademark[11] and local producers who want to protect their geographical indications. According to the TRIPS Agreement (Article 22.3), member states must refuse or invalidate the registration of a trademark that contains a geographical indication with respect to goods not originating in the territory indicated (if use of the indication in the trademark for such goods misleads the public as to the true place of origin). However, when a trademark has been applied for or registered in good faith before the geographical indication is protected in its country of origin, the registration of the trademark (which contains a geographical name) cannot be invalidated by member states (Article 24.5). This happened in the case of rooibos herbal tea (also known as red bush tea), which is specific to the southwest of South Africa: rooibos was registered as a trademark in the United states by a rooibos exporter in 2001, although there is no specific geographical indication protection system (only collective trademarks) in South Africa. This situation made it difficult for South Africans to export rooibos to the United States, and led to litigation that resulted in an out-of-court settlement at a cost to the industry of about US$1 million.[12]

Now that the main characteristics and purposes of geographical indications have been presented, we will discuss its potential use to add value to agrobiodiversity products and to integrate *in situ*/on-farm strategies for conservation and sustainable use of agricultural biodiversity. It should not be assumed that geographical indications will always be an adequate instrument to enhance the value of agrobiodiversity products, and some experiences may be useful to highlight the socioenvironmental, cultural, and economic conditions needed for such purpose. There are both positive and negative examples, and we will start presenting the experiences of Laurence Bérard (anthropologist) and Philippe Marchenay (ethnobiologist), researchers from the Centre National de la Recherche Scientifique (CNRS),[13] involving French geographical indications (*appellations d'origine contrôlées – AOC*). Then we will show the negative impacts of the geographical indication of tequila, in Mexico.

In France, many products are designated by their places of origin, that is, the geographical name of the place where they were made, and some examples are Beaufort and Comté cheeses (made in the French regions with the same name) and Roquefort cheese (which has become so well known that some people assume it is just a type of cheese, not realizing that it is the name of a French town and region where it is produced). Although geographical indications were not developed for this purpose, Laurence Bérard and Philippe Marchenay defend their use for promoting conservation of biological and cultural diversity, by means of adding value to products that hold intrinsic relations with their territory (*terroir*), in both its environmental and cultural dimensions, and which are associated with traditional practices and knowledge (*savoir-faire*) and a collective memory. They give some French examples (Bérard and Marchenay, 2004, 2008; Bérard *et al.*, 2005):

- *Ardèche chestnuts*. For centuries nut production was one of the main activities in the Ardèche department (in the center–south of France), and local communities learned to identify, select, and graft a large variety of chestnuts, with sizes, shapes, and organoleptic qualities that vary from place to place, according to local practices and customs. Some chestnut varieties from northern Ardèche are traditionally eaten boiled with every meal, replacing bread, and in the south other varieties have become a staple food. Social, cultural, and economic life in the Ardèche always revolved around chestnut production, a local product that is intimately connected with the territory (*terroir*). When chestnut production declined, introduction of hybrid varieties was considered, to meet the requirements of technical and commercial criteria. However, this would have been a complete distortion of the entire traditional system, which would have changed from agroforestry to intensive cultivation, and many producers protested. In order to create greater appreciation for their traditional form of production, farmers requested recognition of a denomination of origin (granted in 2006), covering 19 varieties of exclusively local chestnuts. The use of hybrids and chemical fertilizers is prohibited, and only traditional agroforestry is permitted. The denomination of origin protects, as a unit, the chestnut groves, local varieties, and methods of tree husbandry as well as the landscape.

- *Cider, calvados, and poiré from Normandy.*[14] Cider is a fermented drink made from apple juice, *poiré* is made from pears, and *calvados* from apples which are fermented into cider and then distilled. Therefore, it is not possible to talk about calvados and not talk about cider and a cultivation system known as *pré-verger*, in which orchards are associated with animal pastures. The same space provides fruits (apples and pears) for production of beverages and grass for cattle (used in production of cheese and other milk derivatives, as well as beef). Local farmers hold a wide variety of knowledge associated with interactions among animals, pasture, trees, and fruit. Apples can be classified as sweet, bittersweet, bitter, acid, acidulated, etc. A total of 177 varieties have been identified. Some of the products associated with cider protected by AOC are *calvados*, *calvados* Pays d'Auge, *calvados* Domfrontais, and Normandy *pommeau*. Among milk products, Normandy Camembert and butter and fresh milk from Isigny stand out. The AOC for the Domfront *poiré*, recognized in 2002, establishes *plant de blanc* as the main pear variety, which is traditionally used in that region, although other local varieties may also be used. The AOC defines how plant resources (pear trees) and associated agroecosystems (orchards) should be managed, providing norms regarding pruning, plantation density (which should be under 150 trees per hectare), associations between orchards and pastures, etc. In this case, products protected by AOC aim to ensure continuity of local productive systems. Previous AOCs, such as Crau hay (1999) and Comté cheese (1958), also resulted from a systemic view which transcends the product, focusing on the agroecosystem as a whole.
- *Espelette peppers, Puy green lentils, Cévennes sweet onions, and Paimpol beans.* All of these food products are protected by AOC which allow farmers to reproduce their own seeds, an important exception to the general French rule, which is extremely restrictive regarding use and production of seeds by farmers themselves. In all four cases, farmers can

> multiply and use seeds produced in their properties, as long as they are not sold as such (as seeds). This makes it possible for farmers to select and improve seeds of local varieties for sexual reproduction, making use of traditional knowledge and practices. This permission recognizes the role of farmers as plant improvers, and they do not depend on intermediaries, which, in commercial systems, are responsible for seed multiplication. By selecting seeds in accordance with differentiated criteria (plant size, shape, volume, taste, fruit color, etc.), farmers maintain plant diversity. AOCs were recognized in the following years: 2000 (Espelette peppers), 1996 (Puy green lentils), 2003 (Cévennes sweet onions), and 1998 (Paimpol beans).

If French researchers can give positive examples of use of geographical indications to enhance the value of agrobiodiversity products, the same is not true in the assessments made by Mexican researchers Ana Valenzuela-Zapata and Jorge Larson, regarding social and environmental impacts of denominations of origin for tequila and mezcal in Mexico. Both tequila and mezcal[15] are distilled beverages made from agave, a plant with juicy stems and leaves that is rich in carbohydrates and which resembles a gigantic pineapple. Mexico is a center of origin and diversity for agave, a plant adapted to volcanic soils and arid and semiarid regions, which constitute 45.5 percent of the Mexican territory, and the large diversity of both cultivated and wild agaves make up the natural and cultural landscape of the country.[16]

The denomination of origin for tequila was one of the first in Latin America, recognized by a 1974 Mexican law,[17] and the denomination of origin for mezcal was recognized later, in 1994. Tequila is the name of a town located in the state of Jalisco, but the denomination of origin for tequila includes not only the Jalisco territory, but also Guanajuato, Michoacán, Nayarit, and Tamaulipas. According to the code of practices of the tequila denomination of origin, only the *Agave tequilana* species (also known as blue agave) can be used for production of this distilled drink, and the denomination covers both tequila (in which sugarcane sugars can substitute up to 49 percent of blue agave) and 100 percent tequila (made exclusively from blue agave).

According to Bowen and Valenzuela-Zapata (2008), prohibition of the use of agave species other than blue agave in tequila production resulted in the virtual disappearance of other agave, which led to severe loss of diversity of species and varieties of this plant. Genetic homogeneity of agave crops made them more vulnerable to pests and diseases, and large-scale industrial tequila production, mainly for export, contributed to the disaggregation of traditional agricultural systems and to the replacement of diversified agricultural systems by large monocultures of blue agave. Furthermore, there was a sharp increase in the use of chemical pesticides instead of traditional agricultural practices, and growing farm mechanization eliminated many jobs (Bowen and Valenzuela-Zapata, 2008). The region delimited by the denomination of origin is home to 105 companies, mostly foreign, and approximately 30 million blue agave plants (60 million tons) are consumed annually in the production of nearly 205 million liters of tequila (Valenzuela-Zapata *et al.*, 2006; Valenzuela-Zapata, 2007). According to Jorge Larson (2007a,b), diversity of agave species is considered a drawback in the industrial production chain of tequila, and farmers became mere suppliers of labor. Criteria for quality, imposed vertically to ensure sanitary security and homogeneity, turned tequila into an industrial chemical product. Nevertheless, traditional tequilas still exist, although their production is isolated and clandestine.

The fact that the denomination of origin for mezcal, approved in 1994, follows the same path as tequila's geographical indication, regarding monoculture and loss of diversity of species and varieties of agave, raises concerns. Only five species of agave are allowed to be used for production of mezcal, although the Mexico Committee on Biodiversity has already identified 12 species of agave that can be used in production of mezcal and tequila, in addition to others which have not yet been studied. The region covered by the denomination of origin for production of mezcal includes the states of Durango, Guerrero, Oaxaca, San Luis Potosí, and Zacatecas, but mezcal is also produced in other regions, such as southern Puebla state and Morelos (not included in the denomination of origin). Mezcal cannot be sold under this name (mezcal) when produced outside the region of the denomination of origin, which excludes other producers. According to Larson (2007b), the denomination of origin favors "*dé-localisation*" ("delocalization" or loss of local identity) of mezcal, standardizes its production, and disregards not only the rich biological diversity, but also the countless techniques of preparation, cooking, fermentation, and distillation held by local communities. Valenzuela-Zapata also points out that there is a vast range of clandestine mezcals, in which the diversity of species and varieties of agaves used and conserved in production of this distilled drink is much larger than in mezcals produced in accordance with official regulations. According to Valenzuela-Zapata (2007), varietal pureness is promoted as increasing the quality of tequila and mezcal, and the rules for both denominations of origin disregard the diversity of varieties and species of agave and the different practices, knowledge, and cultural processes associated with them.

In Brazil, there have been very few studies into the use of geographical indications to promote agrobiodiversity conservation, and sustainable use. However, Delphine Vitrolles, Luiz Mafra, and Claire Cerdan (Vitrolles *et al.*, 2006) were responsible for two case studies, both in Minas Gerais, involving the following products: (1) coffee from the Minas Gerais Cerrado (Brazilian savanna), which was formally recognized as a geographical indication by INPI (the Brazilian institute for industrial property, responsible for registering geographical indications) in 2005; and (2) Serra da Canastra cheese, which still has not been officially awarded a geographical indication, but the possibility is being discussed by its producers, with the support of the state's technical assistance and rural extensions institutions (EMATER and EPAMIG), universities, local government, and French cooperation.

Minas Gerais is a Brazilian state (in the southeast) that is nationally renowned for its high production of milk and handcrafted cheeses. The reputation of "Minas cheese" is widespread in Brazil. Generally, Brazilians associate "Minas cheese" with a type of white cheese, whether or not it is produced in the state of Minas Gerais. However, some regions in the state of Minas Gerais are traditional producers of Minas cheese, which is made in a small-scale, nonindustrial manner, using raw milk, curds, and a natural yeast (called *pingo*). These regions were characterized and identified by IPHAN (Brazilian federal cultural heritage agency) as being Serro, Serra da Canastra, and Salitre (or Upper Paranaíba), when IPHAN carried out the necessary studies for the registration of the "non-industrial process of making Minas cheese" as Brazilian immaterial cultural heritage.[18] According to IPHAN, the "non-industrial process of making Minas cheese" is a form of "traditional knowledge and an outstanding feature of cultural identity of these Minas Gerais regions."

The study carried out by Delphine Vitrolles, Luiz Mafra, and Claire Cerdan (Vitrolles *et al.*, 2006) focused on Serra da Canastra cheese, and these researchers reported the following findings. Production and consumption of cheese in the Canastra mountain range are part of the history of this region, which was occupied mainly in the eighteenth century, during a historical period known as the "mining cycle" (when colonization focused on mining activities in the southeast of Brazil). Portuguese immigrants who came to this territory produced their cheese according to traditional practices and techniques that they had brought from a region known as the Estrela mountain range, in Portugal. Serra da Canastra cheese has been made in accordance with a traditional and empirical method for over 200 years, and the region known as the Canastra mountain range (Serra da Canastra) is made up of seven municipalities – Bambuí, Delfinópolis, Medeiros, Piumhi, São Roque de Minas, Tapiraí, and Vargem Bonita – which have common environmental, sociocultural, and economic characteristics. According to the IBGE (Brazilian Statistics and Geographical Studies Institute, 2002), 69.9 percent of the properties in this region are less than 100 hectares, which shows the predominance of small-scale family farming, and in many towns traditional cheese production is the main source of employment income for farmers. It is thus an activity that makes it possible for many families to continue making a living, without having to migrate to big cities. Serra da Canastra cheese is cylindrical, with a diameter of approximately 15 centimeters, whitish when fresh, with a slim yellow rind after a few days of maturation, and it is produced using raw milk. The characteristics of Serra da Canastra cheese result from a combination of traditional techniques and methods of production and geographical, climatic, and soil conditions which are associated, to some extent, with the fact that the region is the origin of the São Francisco River (Vitrolles *et al.*, 2006).

Nevertheless, families that produce Serra da Canastra cheese face several difficulties and unfair competition, because the geographical name "Canastra" is a registered trademark of industrial producers[19] who make cheese using pasteurized milk. This often prevents consumers from being able to differentiate handcrafted Serra da Canastra cheese (made with raw milk and traditional techniques) from industrial cheese, as industrial producers also use the name "Canastra," taking advantage of its reputation among consumers. Furthermore, federal sanitary regulations establish requirements (such as pasteurization of the milk and maturation time) which are incompatible with the traditional manner of producing cheese. In order to overcome this problem, the State of Minas Gerais passed Law 14185, of 2002, which allows the use of raw milk in traditional Minas cheese. These state regulations were created as a result of the recognition that pasteurized milk completely changes the properties of handcrafted Minas cheese, as well as the relationship between the end product and the natural environment in which the animals are raised. However, the Minas Gerais state law with specific regulations for handcrafted Minas cheese only partly solved the problem, as Serra da Canastra cheese is sold not only in Minas Gerais, but also in São Paulo and Rio de Janeiro, and interstate and international cheese trade is regulated by federal laws, which do not include such an exception for a traditional cheese-making process.

The second case studied by Vitrolles *et al.* (2006) focused on the geographical indication recognized for "coffee from the Minas Gerais Cerrado," which is produced in social, cultural, and economic contexts entirely different from traditional Minas cheese. These researchers point out that coffee from the Minas Gerais Cerrado (savanna) is not a traditional product, and that this region only became occupied by coffee growers in the 1970s, when coffee growers from other states, especially Paraná (south of Brazil), sought land with favorable climate conditions for coffee growing. As lands in the Cerrado were considered to be of low fertility, coffee growers managed to buy large areas where coffee could be cultivated in an intensive and mechanized manner. The specific characteristics of Minas Gerais Cerrado coffee are mainly associated with environmental conditions in this region, which is located in a continental zone of the Brazilian territory that is somewhat shielded from the effects of oceanic fluctuations. This region has a peculiar distribution of sunlight, which ensures uniform growth of the coffee. Thus, Minas Gerais Cerrado coffee has an intense aroma, delicate acidity, and a sweet but full flavor, characteristics that add economic value. Minas Gerais Cerrado coffee is sold as raw material to large roasting industries, and its production is aimed mainly at the international market (United States, EU, and Japan). Regulations for coffee production are very strict, in order to meet international standards. The registration of the geographical indication by the Cerrado Council of Coffee Grower Associations (Conselho das Associações dos Cafeicultores do Cerrado – CACCER) is part of a strategy for competitive insertion in the international market, with value added to a differentiated product. Norms for coffee production also include several social and environmental demands, and the geographical area covered by the geographical indication includes the regions of Triângulo Mineiro, Upper Paranaíba and part of upper and northwest São Francisco.

The different social, cultural and economic contexts and different motivations of social stakeholders involved in the production of Serra da Canastra cheese and Minas Gerais Cerrado coffee are pointed out by Vitrolles *et al.* (2006): Serra da Canastra cheese is a traditional product produced by a method that has been passed from one generation to the next, and is associated with a strong identity and cultural heritage. Geographical indication is an instrument to recognize the value of handcrafted cheese, which is threatened by legal regulations that restrict the use of raw milk and increasing competition from industrial cheese. Furthermore, the geographical indication claim aims to make economically feasible an activity that represents the main source of income for the vast majority of family farmers inhabiting the Minas Gerais cheese-making regions. For coffee growers, the geographical indication aims to highlight the close bond between the coffee and the environmental characteristics of the territory in which it is produced, to give it a competitive edge in an international market that increasingly seeks "special" or gourmet coffees with differentiated aroma and taste.

Brazilian agriculture (like in many Latin American countries) is essentially marked by a sharp duality of agricultural models – small-scale family farming and industrial agribusiness – and the coexistence of these two models creates diverging political, social, and economic interests, which have important impacts on the definition of public policies in support of geographical indications. There are two ministries in charge of policies for agricultural and rural development in Brazil:[20] the Ministry of Agriculture, Livestock and Food Supply (Ministério da Agricultura, Pecuária e Abastecimento – MAPA), whose mission is to "encourage growth of agricultural production and development of agribusiness," and the Ministry of Agrarian Development (Ministério do Desenvolvimento Agrário – MDA), which is responsible for agrarian reform and family farming policies, and these two ministries often develop contradicting public policies. MAPA created the Coordination of Geographical Indications for Agricultural Products, which carried out studies on potential geographical indications in Brazil and identified some products, such as Serrano cheese, wines, and grapes in Rio Grande do Sul; Serrano cheese and Goethe grapes in Santa Catarina; coffee in Paraná; handcrafted Minas cheese, cachaça,[21] and yams in Minas Gerais; curd cheese, *cajuína*,[22] handcrafted cachaça, buriti[23] sweets, and cashew nuts in Ceará; and cacao, flour, ornamental fish, *guarana*,[24] *açaí*,[25] Brazil nuts, and *cupuaçu*[26] in Amazonia; among others.[27]

In Acre (Brazilian Amazon), among other products, manioc (cassava) flour was identified as a potential product for geographical indication recognition. However, Mauro de Almeida *et al.* (2009) point out, in a study for the Biodivalloc program,[28] that public policies aimed at supporting geographical indications in Brazil have not integrated the environmental and cultural dimensions of local production processes, and have focused primarily on sanitary regulations, homogeneity of the product, and visibility of the package. This perspective does not favor conservation of biological diversity and the associated cultural heritage, which includes traditional knowledge and practices. Delphine Vitrolles (2009) also analyzed the use of geographical indications[29] to enhance the value of a local speciality, Serrano cheese from Rio Grande do Sul (in southern Brazil), and concluded that it tends to promote the exclusion of most producers and the loss of cultural identity and diversity. The standardization of production norms tends to lead to the homogenization of the product, undermining its typicality (Vitrolles, 2009).

Public policies involving geographical indications must take into consideration the different agricultural models that exist in most developing countries, so that these instruments can benefit not only large producers, but also family, traditional, and agroecological farmers. It is not enough to make local products more lucrative if there are no mechanisms to ensure that the added value will benefit all stakeholders involved in the local production system, and not only intermediaries. There must be fair distribution along the entire production and supply chains, as well as at all market stages. Producers' organizations must represent the various social categories and interests of the geographical indication system, and act with transparent and balanced rules, allowing the participation of all stakeholders in political decisions, which includes the producers of raw material and ingredients used in the geographical indication system. A balanced representation of stakeholders in the geographical indication system should ensure that smaller producers have an equal voice, and that benefits are equitably distributed.

Additionally, public policies aimed at promoting geographical indications must consider not only quality standards, but also the social, cultural, and environmental sustainability of products. When the rules in the "codes of practices" are established, they must guarantee the sustainability of the geographical indication system and prevent overexploitation of natural and local resources if the geographical indication becomes a commercial success. The specificity of certain geographical indication products relies on the use of native plant varieties and breeds, frequently threatened with extinction. Traditional production techniques often participate in preserving traditional landscape features, as well as avoiding land and soil degradation, and the "codes of practices" must allow enough legal space for such techniques. Codes of practices must contemplate rules specifically aimed at promoting diversity, in order to avoid excessive homogeneity and/or industrialization, which destroys local identity and typicality. Phytosanitary regulations should also incorporate diversity as a value, and seek a balance between human health and food security standards and the recognition of the value of local and traditional practices which are relevant for agricultural diversity.

Geographical indications will be a tool in favor of agrobiodiversity conservation and sustainable use only if this is a specific goal of public policies and of all stakeholders involved in the geographical indication system. In order to avoid common negative effects of geographical indications, such as the exclusion of certain agrifood products because protection is granted only to a few – to the detriment of others – it is important to think of new geographical indication models that are focused not only on specific agrifood products, but on agricultural systems as a whole, comprehending all of its elements and interrelations and its entire environmental and cultural diversity. A systemic approach will promote diversity and heterogeneity of genetic resources, whereas, if only one resource is promoted, there is a high risk of specialization and loss of genetic diversity.

Geographical indications for origin-based products must be seen as part of a comprehensive and integrated rural development strategy. Geographical indications can be a starting point for the development and promotion of an entire geographical/cultural heritage, as well as a whole basket of goods and services associated with it. When a differentiated basket of local products and services is promoted, instead of just one specific product, the focus is placed on the territorial identity of a group of products that are closely associated with local history and culture (Pecqueur, 2003; Fonte and Acampora, 2007). This means that local stakeholders can use geographical indication products and the specific local resources linked to them (local gastronomy, traditions, landscapes, etc.) and their reputation as a tool to increase the competitiveness of the entire local and economic system. The geographical indication products will be elements of territorial identity for all local actors and assume the role of catalysts in an integrated rural development strategy.[30] In this way, highly valued public interests – such as agrobiodiversity conservation and sustainable use, cultural heritage, including food heritage and diversity, consumers' protection, local development, etc. – will be promoted, and not just the interests of private IP holders.

However, it is important to consider that not all agrobiodiversity products have great commercial and economic potential, despite their high environmental, social, and cultural value, and that policies aimed at differentiated insertion of agrobiodiversity products in the market will always have limited impact, and should be only part of broader public policies for agrobiodiversity conservation and use.

Notes

1 Darjeeling tea is cultivated, processed, and manufactured in the hilly areas of Darjeeling district, in the state of West Bengal in India.

2 In 1905, a French law gave the government the task of identifying areas in which farming production could be eligible for *appellation d'origine contrôlée* (AOC), and the first decree delimitating the champagne production area was approved in 1908. In 1919, a French law established geographical indications as collective IP, but did not establish a legal definition of geographical indications. In 1935, a new decree law established a national committee for wine-growing geographical indications, which became in 1947 the Institut National des Appellations d'Origine (INAO). Sources: www.inao.gouv.fr (accessed January 15, 2010), Bérard and Marchenay (2008), and Roncin and Boulineau (2005).

Since 2006, INAO has been responsible not only for AOCs, but also for all designations of quality and origin for agrifood products in France. These include *Spécialités Traditionelles Garanties* (guaranteed traditional specialities, which are not associated with the area of origin, but with a traditional composition of a product, or which are produced according to a traditional production method, and are regulated, at the European level, by Regulation 509/2006); since 1960, *Label Rouge* (red label, which guarantees that a product is of superior quality to similar products, and exists in France); and *Agriculture Biologique* (agroecological agriculture, which respects the environment and farmers' autonomy, and is regulated, at the European level, by Regulation 834/2007). INAO changed its name to Institut National de l'Origine et de la Qualité, but it kept the same abbreviation (INAO).

3 More than 10,000 legally protected geographical indications exist worldwide. Developing countries together account for about 10 percent of the total, and the other 90 percent of geographical indications come from the 30 OECD countries. The countries/areas with the largest number of protected geographical indications are: the EU, 6,021 (5,200 wines and spirits, 821 foods); the United States, 910 (730 wines, 100 spirits, 80 foods); Switzerland, 682 (660 wines and spirits, 22 foods); New Zealand, 600 (550 wines, 50 foods); Australia, 427 (all wines); and China, 403 (mostly foods but 23 wines and spirits and other products). Chile has 82 (80 wines and spirits, two foods); India 45 (foods and other products, no wines or spirits); and Mexico 10 (foods, wines and spirits, and other products). In the EU, the main origins of protected geographical indications are: (excluding wines and spirits) Italy, 174; France, 162; Spain, 119; Portugal, 114; and Greece, 86. Source: International Trade Center (2009) 'Guide to Geographical Indications: linking products and their origins' (http://www.intracen.org/publications/Free-publications/Geographical_Indications.pdf; accessed November 15, 2010). This guide includes case studies on Antigua coffee (Guatemala), Blue Mountain coffee (Jamaica), Darjeeling tea (India), Gobi Desert camel wool (Mongolia), Kona coffee (Hawaii), mezcal (Mexico), Nariño coffee (Colombia), and Veracruz coffee (Mexico).

4 According to European Council Regulation No. 510/2006, "designation of origin" means the name of a region, a specific place, or, in exceptional cases, a country used to describe an agricultural product or a foodstuff originating in that region, specific place, or country, the quality or characteristics of which are essentially or exclusively due to a particular geographical environment with its

inherent natural and human factors and the production, processing and preparation of which take place in the defined geographical area. According to the same Regulation, "geographical indication" means the name of a region, a specific place, or, in exceptional cases, a country used to describe an agricultural product or a foodstuff originating in that region, specific place, or country and which possesses a specific quality, reputation, or other characteristics attributable to that geographical origin, and the production and/or processing and/or preparation of which take place in the defined geographical area.

That is, for designation of origin protection, all phases of the production process must be localized inside the production area and the quality of the product must be strictly related to a particular geographical environment with its inherent natural and human elements. The geographical indication covers agricultural products and foodstuffs closely linked to a geographical area, where at least one of the stages of production, processing, or preparation takes place within the given area. Source: http://eur-lex.europa.eu/LexUriServ/LexUriServ.do?uri=OJ:L:2006:093:0012:0025:en:PDF (accessed January 15, 2010).

5 The WIPO (World Intellectual Property Organization, the UN agency specialized in IP) is responsible for the administration of a number of international treaties that deal partly or entirely with the protection of geographical indications, notably the Paris Convention for the Protection of Industrial Property (1883, with several later revisions) and the Lisbon Agreement for the Protection of Appellations of Origin and their International Registration (1958). For more information, see http://www.wipo.int/geo_indications/en/ (accessed January 15, 2010).

6 According to the TRIPS Agreement (Article 22.4), the geographical indication protection is also applicable against a geographical indication which, although literally true as to the territory, region, or locality in which the goods originate, falsely represents to the public that the goods originate in another territory. For example, there is a city named Paris (Parisi, actually) in the state of São Paulo (Brazil), but if a product label says "wine produced in Paris," it will be misleading the public as to the true origin of the wine, even if it is literally true as to the name of the city where it was produced.

7 Article 23.3 of the TRIPS Agreement requires that each member state determine the practical conditions under which the homonymous indications will be differentiated from each other in order to avoid misleading consumers.

8 Source: *Linking People, Places and Products: A Guide for Promoting Quality Linked to Geographical Origin and Sustainable Geographical Indications*, FAO and SINER-GI (Strengthening International Research on Geographical Indications) (http://www.foodquality-origin.org/guide/guide.pdf; accessed January 12, 2010).

9 Another example of a conflict involving a "generic" name is the Cotija cheese, from Mexico. The genuine Cotija cheese reputation has been under threat by producers using the designation "Cotija type" for cheeses that may have been produced outside the original production area. Consequently, the name "Cotija" is often used in a generic way. "Cotija type" cheeses are often made by industrial processes (through intensive production, without maturation and with fillings, etc.) and as a result they tend to be cheaper, although the taste can still be distinguished from authentic Cotija cheese. In order to preserve the Jalmich mountain farmers' distinctive way of life and to ensure a sustainable income for their products without having to relocate from the region, the producers of Cotija cheese have been engaged since 1999 in a process of qualification. They have sought to obtain legal protection for the reputation of authentic Cotija cheese through the use of a denomination of origin. The Mexican Intellectual

Property Office rejected the denomination of origin request in 1994, as it considered the denomination to be generic and registered the name "Cotija region of origin" under a collective trademark. Thus, other producers can still use the name "Cotija" for cheese even if it may have been produced elsewhere. There are increasing concerns over the potential for a shift away from local production, the transfer of IP rights away from the local community, as well as misuse of the name by other producers outside the area who do not comply with local standards and code of practice. Source: 'Case Study 1: Generic name or not? A GI product with a collective trademark, Cotija Cheese (Mexico)', in *Linking People, Places and Products: A Guide for Promoting Quality Linked to Geographical Origin and Sustainable Geographical Indications*, FAO and SINER-GI (Strengthening International Research on Geographical Indications) (http://www.foodquality-origin.org/guide/guide.pdf; accessed January 12, 2010).

10 In principle, geographical indications are protected as of the date of registration until the conditions of registration cease to exist, and there is no need to renew the registration.

11 Nearly seven years ago, Ethiopia's coffee sector launched a plan to take better advantage of its coffee diversity (Ethiopia is a center of origin and diversity for coffee). Ethiopia applied for trademark registrations of its world-renowned speciality coffee brands (Sidamo, Harar, and Yirgacheffe) in the United States, Canada, and other countries. At the same time, Ethiopia began negotiating with coffee roasters to sign agreements acknowledging the right of Ethiopians to control these brands.

Ethiopia succeeded in registering its coffee brands in the EU, Japan, and Canada. In the United States, Ethiopia registered Yirgacheffe, but Starbucks Coffee, a large multinational company, opposed the registration of the Sidamo and Harar brands, which are sold at high prices, as gourmet coffees, in its stores all over the world. After a lengthy campaign, with support from Oxfam (a development agency), Starbucks Coffee agreed to sign an agreement with the government of Ethiopia, acknowledging the rights this country has over its coffee brands and establishing conditions for Starbucks Coffee to distribute and sell these Ethiopian coffee brands. Currently, Ethiopia has successfully registered trademarks in Canada, the EU, the United States, and Japan. Source: www.oxfam.org.au/media/article.php?id=285 (accessed January 12, 2010).

12 Source: 'When a GI is registered outside of the territory, the example of Roobois' (Box 4), in *Linking People, Places and Products: A Guide for Promoting Quality Linked to Geographical Origin and Sustainable Geographical Indications*, FAO and SINER-GI (Strengthening International Research on Geographical Indications) (http://www.foodquality-origin.org/guide/guide.pdf; accessed January 12, 2010).

13 Bérard and Marchenay are members of the Eco-Anthropology and Ethnobiology research unit of CNRS (Centre National de la Recherche Scientifique). They are joint heads of a research team titled "Ressources des Terroirs – Cultures, Usages, Sociétés," based in Bourg-en-Bresse, in France. For further information, see www.ethno-terroirs.cnrs.fr.

14 Normandy is a region in northwestern France, currently divided into two administrative regions: Lower Normandy, which includes the departments of Calvados, Manche and Orne, and Upper Normandy, which includes the departments of Eure and Seine-Maritime.

15 The word mezcal originates from *metl* (agave) + *calli* (cooked) in *Náhuatl*. *Náhuatl* is a group of languages and dialects in the Aztec linguistic family. Currently, it is spoken mainly in central Mexico. According to the *Ley General*

de Derechos Linguísticos de los Pueblos Indígenas, passed in Mexico in 2003, *Náhuatl*, as well as other Indigenous languages, are recognized as national languages.

A large family of distilled drinks made from agave are known as mezcal, to which sugar can also be added. Initially, tequila was considered a type of mezcal (called "mezcal tequila wine"), but it ended up acquiring its own identity and being called tequila only.

Prior to the Spanish invasion, mezcal (cooked agave) was widely used as sugar, before the introduction of sugarcane.

16 In 2006, the agave landscape and ancient industrial facilities of tequila were recognized by UNESCO as world heritage, under the category "cultural landscape." For further information, see http://whc.unesco.org/en/list/1209 (accessed January 12, 2010).

17 Other Mexican products protected by denominations of origin include *bacanora*, *sotol*, and *charanda* (all three are distilled spirits), the Ataulfo mango, Veracruz and Chiapas coffee, and Talavera pottery. In South America, other examples of agrifood products protected by geographical indications are *quinua real del Altiplano*, from Bolivia; *blanco gigante del Cuzco*, a local maize variety from Peru; and pisco, a type of distilled liquor made from several varieties of grapes from southern Peru and northern Chile. Pisco was registered as a geographical indication by both countries (Peru and Chile).

18 The "non-industrial process of making Minas cheese" was recognized by IPHAN (a federal agency) as immaterial cultural heritage on June 13, 2008. According to Decree 3551, of 2000, knowledge and manufacturing techniques embedded in the daily lives of communities can be registered in the Registry of Knowledge (which is one of the cultural heritage safeguarding tools). Handcrafted Serro cheese was also registered as immaterial cultural heritage of the State of Minas Gerais on August 7, 2002, based on Decree 42505, of 2002, which provides for registration of immaterial cultural heritage of the State of Minas Gerais.

19 According to the National Institute of Industrial Property (INPI), the "Canastra" trademark is registered on behalf of Piumhiense Cooperative Society of Milk Products Ltd. According to Article 124 of Law 9279/1996 (which regulates trademarks, geographical indications, and other IP rights in Brazil), INPI must refuse or invalidate trademarks that contain false geographical indications and mislead the public as to the true place of origin of the product. However, INPI still has not invalidated this trademark (as of February 10, 2011).

20 In Brazil, there are two types of geographical indications: indications of origin (*indicações de procedência*) and denominations of origin (*denominações de origem*), according to Law 9279 of 1996, Article 176 (http://www.planalto.gov.br/ccivil_03/Leis/L9279.htm; accessed January 20, 2010). Indications of origin correspond to the geographical name of the country, town, region, or location in a given territory which became known as a center of extraction, production, or manufacturing of a product or service. Therefore, an indication of origin requires only a connection between the product and a given geographical location, regardless of its intrinsic characteristics and qualities. In contrast, a denomination of origin requires not only that there is a connection between the product and its geographical location, but also that the product has unique characteristics that are related to the territory, including natural aspects (climate, soil, vegetation, etc.) and cultural aspects (traditional knowledge, practices, processes and techniques, etc.) The right to use geographical indications is collective, inalienable, and extends to all producers established in the delimited territory. The registration of a geographical indication can be

requested only by associations, cooperatives, or other collective entities that represent the producers of a particular geographical area.

In Brazil, the National Institute of Industrial Property (Instituto Nacional de Propriedade Industrial – INPI) is responsible for registration of geographical indications, which are recognized as collective IP rights. INPI has already registered the following Brazilian geographical indications: (1) wine from the Vinhedos Valley, a wine-producing region in the Rio Grande do Sul mountains (in southern Brazil), in 2002; (2) coffee from the Minas Gerais Cerrado, in 2005; (3) beef and its derivatives, from the Pampa Gaucho da Campanha Meridional (an ecosystem of Rio Grande do Sul), in 2006; (4) handcrafted cachaça from Paraty, in 2007; (5) grapes and mango from Vale do Submédio São Francisco, in 2009; (6) leather products from Vale do Sinos (state of Rio Grande do Sul), in 2009; (7) white, red, and sparkling wines from Pinto Bandeira (state of Rio Grande do Sul), in 2010; and (8) rice from the Litoral Norte Gaúcho (north coastal area of Rio Grande do Sul), in 2010. Source: http://www.inpi.gov.br/menu-esquerdo/indicacao/igs-registradas/nacionais/ (accessed February 10, 2011).

21 *Cachaça* is a liquor made from fermented sugarcane. It is the most popular distilled alcoholic beverage in Brazil.

22 *Cajuína* is a nonalcoholic drink prepared from cashew apple juice in a traditional manner, especially in the Brazilian states of Piauí and Ceará (northeast).

23 The *Buriti* tree is one of the biggest palm trees in the Amazon, with nutritious fruit and nuts.

24 *Guaraná* is a berry that grows in Venezuela and the northern parts of Brazil, and it is very rich in caffeine. Sateré-Mawé is the name of a Brazilian Indigenous people. In Brazil, native Sateré-Mawé *guaraná* is one of the local agrifood products that the Slow Food movement has been working with. Slow Food is an international membership organization that defends biodiversity in food supply, promotes food and taste education, and connects sustainable co-producers through events and building networks. Slow Food's activities are aimed at protecting traditional and sustainable quality foods and defending the biodiversity of cultivated and wild varieties as well as cultivation and processing methods. Slow Food's approach to agriculture, food production, and gastronomy is based on a concept of food quality defined by three interconnected principles: good – fresh and flavorsome seasonal diet that satisfies the senses and is part of our local culture; clean – food production and consumption that does not harm the environment, animal welfare, or our health; and fair accessible prices for consumers and fair conditions and pay for small-scale producers. For further information, see www.slowfood.com (accessed January 10, 2010). In Brazil, Slow Food initiatives have involved several local agrifoods, such as *baru* nuts, *canapu* beans, red rice, *cagaita*, *umbu*, *pequi*, and *mangaba* (the last four are native fruits of the Brazilian Cerrado/savanna), among others (www.slowfoodbrasil.com; accessed January 10, 2010).

25 *Açaí* is a species of palm tree found in South America, and its fruits and hearts of palm are widely used for food. Ripe fruits are a deep-purple color.

26 *Cupuaçu* is a type of fruit tree native of the Amazon Rainforest. It is commonly used to make juices, jams, and ice-cream.

27 These are only a few examples. For a complete list of products, see http://www.agricultura.gov.br/portal/page/portal/Internet-MAPA/pagina-inicial/desenvolvimento-sustentavel/indicacao-geografica (accessed January 10, 2010).

28 The Biodivalloc program "*Des productions localisées aux indications géographiques: quels instruments pour valoriser la biodiversité dans les pays du sud*" (From localized products to geographical indications: which instruments

to value biodiversity in southern countries?) aims to analyze local biodiversity management processes and evaluate how legal instruments (such as geographical indications and other distinctive signs) can be adapted to promote conservation of biological and cultural diversity and meet the needs and expectations of local communities. The program adopts an interdisciplinary perspective, and it analyzes concrete case studies: in Ethiopia, impacts of geographical indications on the diversity of "forest coffees" and home gardens are analyzed; in Senegal, ecocertifications for sea and mangrove products are analyzed; in Niger, interfaces between localized products and trademarks associated to national and regional parks are analyzed; and in South Africa, interfaces between geographical indications and specific ecosystems are assessed. In the Brazilian Amazon, the analysis is centered on geographical indications for local products, such as manioc/cassava flour from Cruzeiro do Sul (Acre), and on the use of cultural heritage instruments to safeguard traditional agricultural systems of the Rio Negro region. In India, the use of geographical indications to value agroforestry systems is evaluated. The Biodivalloc program is developed by the joint research unit (UMR 208) for Local Heritage of the Institut de Recherche pour le Développement (IRD)/Muséum National d'Histoire Naturelle (MNHN), with support from the Agence Nationale de la Recherche/ Institut Français de la Biodiversité (2006–2009).

29 The geographical indication for Serrano cheese has not yet been officially recognized (as of February 10, 2011).

30 Sources: "Taking into account environmental and social issues in the code of practice' and "Extended territorial strategies for increasing rural development', in *Linking People, Places and Products: A Guide for Promoting Quality Linked to Geographical Origin and Sustainable Geographical Indications*, FAO and SINER-GI (Strengthening International Research on Geographical Indications) (http://www.foodquality-origin.org/guide/guide.pdf; accessed January 12, 2010). This study mentions several cases of geographical indication products based on biodiversity resources, such as *chivito* Criollo del Norte Neuquino (goat, Argentina), cacao Arriba (Ecuador), cherry of Lari (Italy), and Jinhua ham (China).

Bibliography

Almeida, M. de, Carneiro da Cunha, M., Eloy, L., Emperaire, L., Katz, E., Simoni, J., Rizzi, R., Santilli, J. and Velthem, L.H.V. (2009) 'L'indication géographique, un instrument de pérennisation des productions localisées en Amazonie? Le cas de la farine de manioc de Cruzeiro do Sul (Acre, Brésil)', paper presented at the international symposium Localizing Products: A Sustainable Approach for Natural and Cultural Diversity in the South?, June 9–11, 2009, Paris, UNESCO, CIRAD, IRD, and MNHN (http://www.mnhn.fr/colloque/localiserlesproduits/index_va.php; accessed January 12, 2010).

Autrepart. Revue de Sciences Sociales au Sud, no. 50, 2009/2. *Les Produits de Terroir au Service de la Diversité Biologique et Culturelle*? (www.cairn.info/revue-autrepart-2009.2.htm).

Bérard, L. and Marchenay, P. (2004) *Les Produits de Terroir: Entre Cultures et Règlements*, Centre National de la Recherche Scientifique (CNRS), Paris.

Bérard, L. and Marchenay, P. (2008) *From Localized Products to Geographical Indications: Awareness and Action*, Centre National de la Recherche Scientifique (CNRS), Ressources des Terroirs, Cultures, Usages, Sociétés, UMR

Eco-Anthropologie et Ethnobiologie, Bourg-en-Bresse, France (www.ethno.terroirs.cnrs.fr; accessed January 12, 2010).

Bérard, L., Cegarra, M., Djama,M., Louafi, S., Marchenay, P., Roussel, B. and Verdeaux, F. (eds.) (2005) *Biodiversity and Local Ecological Knowledge in France*, INRA, CIRAD, Institut du Développement Durable et des Relations Internationales, and Institut Français de la Biodiversité, Paris.

Boivert, V. (2006) 'From the conservation of genetic diversity to the promotion of quality foodstuff: can the French model of "appellation d'origine contrôlée" be exported?' (Collective Action and Property Rights Working Pape no. 49), IFPRI, Washington, DC (www.capri.cgiar.org; accessed September 30, 2008).

Bowen, S. and Valenzuela-Zapata, A. (2008) 'Les appellations d'origine et les durabilités sócio-economique et ecologique: le cas de la tequila au Mexique', *Cahiers Agricultures*, vol. 17, no. 6, pp. 552–560.

Cerdan, C., Vitrolles, D (2008) 'Valorisation des produits d'origine: contribution pour penser le développement durable dans la Pampa Gaucha au Brésil', *Géocarrefour*, vol. 83–3, pp. 191–200.

Cerdan, C., Vitrolles, D., Delfosse, C., Velloso, C.Q., Nabinger, C. and Silva, A.L. (2009) 'La diversité biologique et culturelle dans les demarches de qualité et de valorization de l'origine au sud Brésil', *Autrepart*, vol. 50, pp. 153–166.

FAO and SINER-GI (Strengthening International Research on Geographical Indications) *Linking People, Places and Products: A Guide for Promoting Quality Linked to Geographical Origin and Sustainable Geographical Indications*, FAO and SINER-GI (Strengthening International Research on Geographical Indications) (http://www.foodquality-origin.org/guide/guide.pdf; accessed January 12, 2010).

Fonte, M. (2010) 'The construction of origin certification: knowledge and local food', in Fonte, M. and Papadopoulos, A.G. (eds.) *Naming Food after Places: Food Relocalisation and Knowledge Dynamics in Rural Development*, Ashgate, Farnham, UK, pp. 149–171.

Fonte, M. and Acampora, T. (2007) 'Productos típicos, estratégias de desarrollo rural y conocimiento local', *Revista Opera (Observatorio de Políticas, Ejecución y Resultados de la Administración Pública)*, no. 7, edición especial *Territorios con Identidad Cultural. Perspectivas desde América Latina y la Unión Europea*, RIMISP (Centro Latinoamericano para el Desarrollo Rural), Universidad Externado de Colombia, Università di Napoli Federico II, pp. 191–212.

Hermitte, M.A. (2001) 'Les appellations d'origine dans la genèse des droits de la propriété intellectuelle', in Moity-Maïzi, P., Sainte Marie, C., Geslin, P., Muchnik, J. and Sautier, D. (eds.). *Systèmes Agroalimentaires Localisés: Terroirs, Savoir-Faire, Innovations* (Etudes et Recherches sur les Systèmes Agraires et le Développement, vol. 32), INRA and CIRAD, Versailles, pp. 195–206

Institut National de l'Origine et de la Qualité (INAO) (2006) *Local Flavour: A Tour of French Origin-linked Products*, Inao and Hachette, Paris (www.inao.gouv.fr; accessed January 12, 2010).

Kop, P.V., Sautier, D. and Gerz, A. (eds.) (2006) *Origin-based Products: Lessons for Pro-poor Market Development*, Royal Tropical Institute, Amsterdam, and CIRAD, Montpellier, France.

Lages, V., Braga, C. and Morelli, G. (orgs.) (2004) *Territórios em Movimento: Cultura e Identidade como Estratégia de Inserção competitiva*, Relume Dumará, Rio de Janeiro, and Sebrae, Brasília.

Larson, J. (2007a) 'Relevance of geographical indications and designations of origin for the sustainable use of genetic resources' (www.underutilized-species.org/Documents/PUBLICATIONS/gi_larson_lr.pdf; accessed January 12, 2010).

Larson, J. (2007b) '(Dé) localisation du mezcal', *Courrier de la Planète*, no. 83, pp. 54–57.

Marie-Vivien, D. (2010) 'Le droit des indications geographiques en Inde: un pays de l'ancien monde face aux droits français, communautaire et international', PhD thesis, Ecole des Hautes Etudes en Sciences Sociales, Centre d'études des normes juridiques – Yan Thomas, Paris, France.

Olszak, N. (2001) *Droit des Appellations d'Origine et Indications de Provenance*, Éditions Tec & Doc, Paris.

Pecqueur, B. (2003) 'Qualidade e desenvolvimento territorial: a hipótese da cesta de bens e de serviços territorializados', *Eisforia*, edição especial *Desenvolvimento Territorial Sustentável: conceitos, experiências e desafios teórico-metodológicos*, Universidade Federal de Santa Catarina, Centro de Ciências Agrárias, Programa de Pós-Graduação em Agroecossistemas, vol. 1, no. 1, pp. 135–153.

Ranaboldo, C. and Schejtman, A. (2008) (eds.) *El Valor del Patrimonio Cultural: Territorios Rurales, Experiencias y Proyecciones Latinoamericanas*, Instituto de Estudios Peruanos, Lima, Peru; Rimisp (Centro Latinoamericano para el Desarrollo Rural), Proyecto de Desarrollo Territorial Rural com Identidad Cultural, Santiago.

Roncin, F. and Boulineau, F. (2005) 'Valorisation des "légumes de terroir" para les signes officiels d´identification de la qualité et de l´origine (AOP-IGP): Comment est-ce possible? Et est-ce réalisable?' in Actes du colloque "Les Légumes: un Patrimoine à Transmettre et à Valoriser", Association Française pour la Conservation des Espèces Végétales, Angers, France, September 7–9, 2005.

Santilli, J. (2005) 'As indicações geográficas e as territorialidades específicas das populações tradicionais, povos indígenas e quilombolas', in Lages, V., Lagares, L, and Braga, C. (orgs.) *Valorização de Produtos com Diferencial de Qualidade e Identidade: Indicações Geográficas e Certificações para Competitividade nos Negócios*, Sebrae, Brasília.

Thévenod-Mottet, E. (2010) 'Geographical indications and biodiversity', in Lockie, S. and Carpenter, D. (eds.) *Agriculture, Biodiversity and Markets*, Earthscan, London, pp. 201–212.

Valenzuela-Zapata, A. (2007) 'Las denominaciones de origen tequila y mezcal y la biodiversidad en el genero Agave sp', paper presented at the seminar 'Biodiversity and Denominations of Origin: A Global Issue', April 18, 2007, Coimbra, Portugal (www.esac.pt/cernas/comunicacoes_BioDO/3.%20Ana%20Valenzuela_PDF.pdf.; accessed January 12, 2010).

Valenzuela-Zapata, A., Marchenay, P., Bérard, L. and Foroughbakhch, R. (2006) 'Conservación de la diversidad de cultivos en las regiones con indicaciones geográficas: los ejemplos del tequila, mezcal y calvados', in Álvarez, A.M. (coord.) *Agroindustria Rural y Território. Los Desafíos de los Sistemas Agroalimentarios Localizados, Tomo I*, Universidad Autónoma del Estado de México, Toluca, México, pp. 127–154.

Vitrolles, D. (2009) 'Comment concilier préservation d'un patrimoine culturel et gastronomique et normalization d'une production traditionnelle: l'exemple du fromage Serrano au sud du Brésil', paper presented at the international symposium Localizing Products: A Sustainable Approach for Natural and Cultural Diversity in the South?, June 9–11, 2009, Paris, UNESCO, CIRAD, IRD and MNHN (http://www.mnhn.fr/colloque/localiserlesproduits/index_va.php; accessed January 12, 2010).

Vitrolles, D., Mafra, L. and Cerdan, C. (2006) 'Enjeux et perspectives de développement des indications géographiques au Brésil, une analyse à partir des deux produits de l'Etat du Minas Gerais', paper presented at the 3rd International Colloquium of Groupement d'Intérêt Scientifique Syal (Systèmes Agroalimentaires Localisés), Baeza, Spain, October 18–21, 2006.

Websites

www.ethno-terroirs.cnrs.fr
www.terroirsetcultures.org
www.enitac.fr/cerpat
www.gis-syal.agropolis.fr
www.sitesremarquablesdugout.com
www.wipo.int/geo_indications/en/
www.origin-food.org
www.origin-gi.com
www.prodottitipici.com
www.arsia.toscana.it
www.rimisp.org/territorioeidentidad2

CONCLUSIONS

The world's agrarian and agricultural universe is extremely complex owing to the existence of not only a huge diversity of landscapes and ecosystems (with different environmental features), but also of different types of farmers, who have different livelihoods and develop different survival and production strategies. The reference to a "duality of agricultural models" (industrial agribusiness vs. small-scale farming) is commonplace, but, in reality, our agricultural biological and cultural diversity comprises not a duality, but rather a multiplicity of farming systems (Indigenous, traditional, family, peasant, organic, as well as different types of monocultures).

The coexistence of agricultural models with conflicting political, social, and economic interests has been, nonetheless, overlooked by most agricultural laws and policies, which tend to treat rural spaces as if they were all homogeneous and uniform. Such an approach has left very little space for the social, cultural, and economic survival of local and traditional agricultural systems, which are the very ones that conserve and sustainably use agrobiodiversity. Various laws are promoting, more and more, a single productivist, industrial agriculture paradigm, based on narrow genetic diversity, and they tend to undermine social and cultural processes that generate and enrich agricultural biodiversity.

Seed laws that meet mainly the needs of commercial large-scale production systems and rigid protection of IP rights over plant varieties contribute to reducing agricultural biodiversity and associated cultural diversity, aggravating the impacts of monocultures. In many developing countries, public policies and regulations (inspired by those in northern countries) have promoted an agricultural and rural development paradigm based on the production of commodities for export, underestimating the importance of local and traditional agricultural systems for food and nutritional security, sustainable rural development, and social equity. This approach is reflected in a set of agricultural laws approved in recent years and in the very limited legal space they leave for family, traditional, and local farming systems.

How can this situation be changed? The legal framework for agrobiodiversity is still under construction, internationally and in most countries, which gives all social stakeholders a valuable opportunity to reflect and discuss how policies, legislation, and regulations can contribute more effectively to the *in situ* and on-farm management and sustainable use of agrobiodiversity, as well as to food security, cultural diversity, and social equity. Many legal instruments discussed in this book are still in the process of being elaborated, formulated and/or revised or changed, both nationally and internationally. Other instruments still need to be constructed and conceived, and others were developed at the international level but must be implemented at the national level. This book presents a set of conclusions and propositions to be considered by all social stakeholders who are involved in the development of rural and agricultural policies and regulations, which are summarized below:

1 Seed laws must adopt one of the following options: they must either regulate only formal seed systems, leaving local and farmers' seed systems out of their scope, or they must regulate formal and local seed systems differently, considering their specificities and distinct characteristics. If only seeds of commercial uniform varieties are allowed to be produced, the use of locally adapted crops will be greatly discouraged and may even become illegal in some circumstances. Seeds determine, in large part, the agricultural model to be adopted, and if only seeds of uniform high-yielding varieties are made available, this will be the only farming system imposed on all farmers, with great losses for agrobiodiversity, farmers, and consumers. In addition, the choice of seeds is determined not only by agronomical, but also by social, cultural, and environmental factors. For farmers to be able to choose their seeds, public policies must promote widespread diversification of seeds and leave more legal space for local seed systems, rather than impose, artificially, a single commercial system, especially in developing countries.

   Seed laws were conceived for long supply chains, in which production, distribution, and trading of seeds involve many intermediaries between producers and consumers, and in which seed consumers (farmers) have no direct relations with producers (large seed producers and traders). The distance between seed producers and consumers is enormous, and no trust, collaboration, or reciprocity exists between them. Strict regulations established for long supply chains are not justified, however, when seeds are produced and sold at the local level and farmers have direct access to seed producers. These rules are even less justifiable when farmers produce their own seeds or receive them from other local farmers, through their social networks or through short supply chains, such as direct sales. If, originally, the intent of seed laws was to avoid the dissemination of poor-quality seeds, such original purposes have been overstepped. New priorities, such as agrobiodiversity conservation and sustainable use, must be taken into consideration and determine the revision and adaptation of seed laws.

2 The exchange and sale of seeds through local social networks, as well as the use of farm-saved seeds, are essential components of local small-scale farming systems in many developing countries, and restrictions to these traditional practices undermine the social and cultural processes that maintain and enrich agrobiodiversity. Therefore, in developing countries, any revisions or changes in national plant variety protection laws (to adapt them to the 1991 Act of the UPOV Convention) should be considered only after the implementation of the ITPGRFA (in countries that have ratified it), which recognizes farmers' rights, so that inconsistencies between these two instruments are avoided.
3 The principle of sovereignty of countries of origin over their own genetic resources, set forth by the CBD, was an attempt to balance out historical inequalities in north–south relations related to the appropriation of these resources by northern countries through IP rights. The CBD established a bilateral system of ABS, based on the assumption that there are "providers" and "recipients" (or users) of genetic resources, and that they must establish, contractually, the conditions for ABS. But how can farmers, used to sharing and promoting the exchange of genetic materials, knowledge and agricultural experiences via social networks define to whom these resources belong? The CBD did not provide a solution to the negative impacts of IP rights on biodiversity, and, at the same time, legitimized them (indirectly). Ironically, access to genetic resources and associated knowledge became, in many circumstances, even more restricted. The special nature of PGRFA was not considered by the CBD, which led to the adoption of the ITPGRFA, the first legally binding international instrument to deal exclusively with PGRFA, and to recognize farmers' rights (leaving their implementation to national laws). The multilateral ABS system established by the Treaty was an important accomplishment, but it has limited scope and there are still many challenges for its effective implementation at international and national levels.
4 The Nagoya Protocol is an instrument for the implementation of the ABS provisions of the CBD, and it applies only to genetic resources and associated traditional knowledge that are covered by the CBD. It does not apply to such resources when they are covered by other specialized international ABS instruments. So far, the ITPGRFA is the only international agreement to establish specialized (multilateral) ABS provisions. The Nagoya Protocol does *not* apply to PGRFA that are subject to the multilateral system of ABS that is established by the ITPGRFA. However, plant genetic resources that are not included in Annex I (of ITPGRFA) or which are used for any chemical, pharmaceutical, and/or other non-food/feed purposes fall under the scope of the CBD and of the Nagoya Protocol, as they are not subject to the multilateral system of the ITPGRFA. Only genetic materials that are in the public domain and under the management and control of contracting parties are excluded

from the scope of the CBD and the Nagoya Protocol, which means that genetic materials held by private companies, NGOs, and farmers (on their fields and collections) are not part of the multilateral system (unless they decide to put their genetic materials on the multilateral system, but this is voluntary and not mandatory) and, therefore, such genetic materials are included in the scope of the Nagoya Protocol. While they remain outside the multilateral system, they are subject to the rules of the CBD and of the Nagoya Protocol. However, if contracting parties to the Treaty (ITPGRFA) decide to widen the scope of the multilateral system, other crops and forages that are included in the multilateral system will also be left out of the CBD and of the Nagoya Protocol bilateral system of ABS.

The Nagoya Protocol also sets out that an internationally recognized certificate of compliance will serve as evidence that the genetic resource which it covers has been accessed in accordance with prior informed consent and that mutually agreed terms have been established. Contracting parties to the Nagoya Protocol must establish checkpoints and decide on what type of information should be requested at such checkpoints. The SMTA used for genetic material transfers made under the multilateral system established by the ITPGRFA should be considered to be an internationally recognized certificate of compliance to be presented by resource users at all relevant checkpoints.

5 Most countries that have national ABS laws, inspired by the CBD, do not establish a special regime for PGRFA. National ABS laws must clarify to which ABS regime PGRFA (listed in Annex I) that are found *in situ* are subject. They must also specify if any Annex I crops existing in *in situ* conditions will be placed in the multilateral system. Whether through administrative or legislative acts, countries have to make decisions on the following:

   a Will they apply the same norms and conditions that regulate ABS for wild genetic resources to PGRFA or will they create a differentiated ABS regime for PGRFA (whether included or not in Annex I and in the multilateral system, and whether found *in situ* or *ex situ*)?

   b Will they establish national benefit-sharing funds covering all PGRFA, whether found *in situ* or *ex situ* (that are not included in the international multilateral system)? If so, the governing bodies of such national benefit-sharing funds should include representatives not only of government, but also of small-scale and traditional farmers, who must be the main beneficiaries of such benefit-sharing funds. A certain percentage of the value of all seeds that are sold in the country could be allocated to these funds. Such percentage should be paid directly by the (national and international) private seed sectors (and not by national governments of developing countries), as a special tax/contribution imposed upon users of plant genetic resources, and the funds should be used for agrobiodiversity conservation and sustainable use, similarly to the international fund established by the ITPGRFA.

c Will they establish collective benefit-sharing mechanisms, such as the above-mentioned benefit-sharing funds, not only at the national level, but also at the regional level, involving countries of the same geographical region, that share common resources (such a possibility is already set out in the Nagoya Protocol)?

6 Countries could also establish, in their national ABS laws for PGRFA, some important reciprocity rules that the international multilateral system was not able to establish because of lack of consensus. For instance, countries could establish that access to *ex situ* collections held by public institutions would be granted only to private institutions that also make their own *ex situ* and *in situ* collections available to public institutions. Furthermore, national ABS laws could establish that private institutions can access plant genetic resources found in *in situ*, on public domain lands, only if they commit themselves (through a legally binding instrument) to make such resources available to public institutions.

7 Many countries have ratified Convention No. 169 of the ILO, concerning Indigenous and Tribal Peoples in Independent Countries, others have signed the UN Declaration on the Rights of Indigenous Peoples and/or adopted national legislation recognizing special rights to their ethnic minorities and local communities. In these countries, access to genetic materials is subject to the prior informed consent of these groups. Such materials cannot be considered to be in the public domain and under the management and control of contracting parties to the Treaty. Genetic resources held by these groups can only be accessed or placed in the multilateral system after their prior informed consent has been granted.

8 Achieving a balance between international and national action in relation to farmers' rights is still a challenge, but some progress has been made in this regard. The ITPGRFA did not establish international parameters that have to be necessarily adopted by contracting parties, which reflects the lack of consensus regarding how to implement farmers' rights. The Treaty could have kept some flexibility, to allow countries to adapt farmers' rights to local contexts, but it should have set minimum international standards. For agrobiodiversity conservation and sustainable use, it is crucial to ensure sufficient legal space within seed laws, IP legislation and ABS laws for farmers to continue saving, using, exchanging, and selling farm-saved seeds and propagating material. It is also important to consider that farm-saved seed includes both local traditional varieties and improved varieties that have been further adapted and developed by farmers. Article 6 of the ITPGRFA, which deals with sustainable use of plant genetic resources, is strongly linked to farmers' rights (established in Article 9). Article 6 is very clear that contracting parties

must "develop and maintain appropriate policy and legal measures that promote sustainable use of agrobiodiversity," and this includes, among other measures, "reviewing and adjusting breeding strategies and regulations concerning variety release and seed distribution" (Article 6.2g). It is important to ensure a broad genetic basis of crops, and both seed laws and plant breeders' rights must contain exceptions allowing small-scale/local farmers to save, exchange, use, and sell farm-saved seeds (of protected or public domain varieties) to other small-scale (local, traditional and agroecological) farmers, as long as it takes place in local markets and among local farmers. As plant breeders, farmers must also benefit from the "breeders' exemption," which allows the use of plant varieties (including those developed by the formal sector and protected by plant breeders' rights) as sources of diversity to develop new varieties.

9 Collective benefit-sharing mechanisms must be developed, and farmers' rights must be recognized as essentially collective, and not individual, rights. The most important benefit for farmers is probably the creation of legal space for farmers to save, use, exchange, produce, and sell their seeds, free of legal obstacles and restrictions that do not consider the specificities of local/traditional systems. Other collective benefit-sharing mechanisms must be promoted, such as public policies aimed at valuing and strengthening local and traditional agricultural systems, and at adding environmental and cultural value to local/regional products, through geographical indications and certifications; participation in all decision-making processes (that impact agrobiodiversity) at the international, national, regional, and local levels; strengthening of local/traditional farmers' capacity to participate in local and national agricultural markets; participatory plant breeding, seed fairs and local seedbanks, managed by farming communities; seed exchange networks; recognition of agrobiodiverse traditional/local agricultural systems as cultural landscapes, with the adoption of measures to safeguard such systems, at the national and international levels; establishing of national benefit-sharing funds to support on-farm conservation and sustainable use of agrobiodiversity (which would complement the international benefit-sharing fund created by the Treaty); collective payment for environmental services; creation of a special category of protected area aimed at conserving agrobiodiversity, among others. Links between agrobiodiversity conservation and sustainable local development must be promoted and strengthened.

10 Farmers' rights to participate in decision-making must be interpreted in a broad and inclusive manner, and must include participation in any political decisions with impacts on local agricultural systems and PGRFA. This includes all instances of power (international, federal, state, provincial, and municipal) and all policy areas impacting farmers' rights (land reform, rural credits, subsidies and insurance policies, food

and nutritional security, agricultural technology and research priorities, local/rural development policies, environmental policies, water management and use policies, etc.). National ABS laws must also include the participation of farmers' representatives in decisions regarding their implementation. These representatives must participate not only in decisions related to access authorizations to plant genetic resources and traditional knowledge, held by farming communities, but also in the definition of national policies on ABS for plant genetic resources.

11 Small-scale farmers and pastoralists are the custodians of much of the world's animal genetic diversity, and ensuring that they can continue to perform this role is essential. The specific nature of animal genetic resources in international and national legal frameworks must be taken into consideration, and legislation and public policies specifically aimed at promoting animal genetic diversity and at protecting livestock keepers' rights must be developed and enforced. It is also important that a common global pool of animal genetic resources be established, through networks of public *ex situ* collections (currently, there are very few, but they could be created) and public support for on-farm conservation of local breeds by livestock keepers.

12 The open source software and the "commons" movements have a lot in common with traditional and local farmers and their seeds' systems: both defend sharing, rather than exclusion. Like the computer programmers who started the open software movement, many farmers are unsatisfied with laws that conflict with local rules of access and sharing of resources and knowledge. They want not only protection against misappropriation of their resources and knowledge, but also to ensure enough legal space to maintain biological, social, and cultural processes that are essential for their food security, such as seed saving and exchange through social networks. An open source license should be considered to create a "protected commons" space in which sharing of plant genetic resources among farmers and plant breeders is allowed but protected from misappropriation.

13 Crops and agroecosystems must be regarded as cultural artifacts, molded by man, and legal instruments conceived to safeguard cultural heritage can be useful to recognize and promote agrobiodiverse farming systems and all their components, both tangible (such as landscapes and cultivated plants) and intangible (agricultural techniques, practices and knowledge). Intangible cultural heritage registries, cultural landscapes recognition and protection, as well as the safeguarding of globally important ingenious agricultural heritage systems, can all be important instruments to enhance agrobiodiversity conservation and sustainable use.

14 The role of protected areas in promoting the conservation and sustainable use of agrobiodiversity has been greatly underestimated. At the

international level, there is still not a specific category of protected area for agrobiodiversity conservation, whether for *in situ* conservation of crop wild relatives or for on-farm management of agrobiodiversity-rich farming systems (though other categories of protected areas have been eventually used for agrobiodiversity conservation). Agrobiodiversity reserves alone will not be enough to avoid the impacts of monocultures and industrialized agricultural models, especially if they are mere "islands" surrounded by unsustainable agricultural practices. Albeit insufficient, protected areas have played an important role in the conservation of wild biodiversity *in situ* and could also be relevant for agrobiodiversity conservation. A specific category of protected area, aimed at promoting *in situ* and on-farm conservation of agricultural biodiversity, should be created, both at international and national levels. Regardless of the creation of a specific protected area category for agrobiodiversity, it is important to make better use of existing protected areas to promote conservation of agricultural biodiversity.

15 Geographical indications could also be an interesting tool to enhance the value of agrobiodiversity products and promote sustainable rural development. However, geographical indications will only be a favorable means toward agrobiodiversity conservation and sustainable use if this is a specific goal of public policies and of all stakeholders involved in the geographical indications system. In order to avoid common negative effects of geographical indications, such as the exclusion of certain agrifood products because protection is granted only to a few, to the detriment of others, it is important to think of new geographical indications models, which are focused not only on specific agrifood products, but on agricultural systems as a whole, comprehending all of its elements and interrelations and its entire environmental and cultural diversity. A systemic approach will promote diversity and heterogeneity of genetic resources, whereas if only one resource is promoted there is a high risk of specialization and loss of genetic diversity. Geographical indications for origin-based products must be seen as part of a comprehensive and integrated rural development strategy, and can be a starting point for the development and promotion of an entire geographical/cultural heritage, as well as a whole basket of goods and services associated with it. In this way, highly valued public interests – such as agrobiodiversity conservation and sustainable use, cultural heritage, including food heritage and diversity, consumers' protection, local development, etc. – will be promoted, and not just the interests of private IP rights holders.

INDEX

For Product Safety Concerns and Information please contact our EU representative GPSR@taylorandfrancis.com
Taylor & Francis Verlag GmbH, Kaufingerstraße 24, 80331 München, Germany

www.ingramcontent.com/pod-product-compliance
Ingram Content Group UK Ltd.
Pitfield, Milton Keynes, MK11 3LW, UK
UKHW022008180926
475701UK00003B/257

* 9 7 8 1 1 3 8 6 8 0 3 3 3 *